The Conduct of War

The Conduct of War

An Introduction to Modern Warfare

Samuel B. Payne Jr.

Basil Blackwell

Copyright © Samuel B. Payne, Jr., 1989

First published 1988
First published in USA 1989

Basil Blackwell Ltd
108 Cowley Road, Oxford, OX4 1JF, UK

Basil Blackwell Inc.
432 Park Avenue South, Suite 1503
New York, NY 10016, USA

British Library Cataloguing in Publication Data
Payne, Samuel B.
  The conduct of war : an introduction
  to modern warfare
  1. Military operations. Strategy
  I. Title
  355.4'3

  ISBN 0–631–15531–7
  ISBN 0–631–15532–5 Pbk

Library of Congress Cataloging in Publication Data
Payne, Samuel B.
  The conduct of war : an introduction to modern warfare / Samuel
Burton Payne, Jr.
    p.   cm.
  Bibliography: p.
  Includes index.
  ISBN 0–631–15531–7 : $55.00 (est.).   ISBN 0–631–15532–5 (pbk.) :
$18.00 (est.)
  1. War. 2. Nuclear warfare. 3. United States—National security.
4. Europe—National security. 5. Soviet Union—National security.
6. Sea-power. 7. Guerrilla warfare.   I. Title.
U21.2.P335 1988
355'.02—dc19   88–16652

Typeset in 10 on 11 pt Ehrhardt
by Photo-graphics, Honiton, Devon
Printed in Great Britain by Billing & Sons (Worcester) Ltd

To my father
Samuel B. Payne
with love and admiration
and to Dee G. Appley
my favorite pacifist

Contents

List of Figures and Maps

Figures

Maps

Acknowledgements

It is my very pleasant duty to thank those who have helped me write this book.

Kenneth Hooker, Jr. read the entire manuscript and made detailed editorial notes. Much of whatever grace and clarity the book exhibits is due to him, and its remaining infelicities are – no doubt – due to my pig-headed failure to follow his advice. Professor George W. Rathgens, Professor James Turner Johnson, Dr. Colin S. Gray, and Dr. Judith E. Soukup each read one or more chapters; Caroline Kotonias-Payne, Steve Birkhold, John Bushnell, Jim Coombs, David Kirsten, and William Pierson also read portions of the manuscript.

Staff members at the Library of Congress, the New York Public Library, and the libraries of the University of Massachusetts, the University of Virginia, and Ferrum College helped me to do research at their institutions. I am particularly grateful to Lyn Wolz of Ferrum College for her indefatigable pursuit of inter-library loan books for my use. Mel Hoskin, Michael Hoskin, Karl Wozniak, and Earl G. Skeens, Jr. gave, loaned, or helped me find research materials. My colleagues Richard Smith, Matthew Campbell, Kip Lornell, and Martha Brandt gave me the benefit of their expert advice.

The Trustees and Administration of Ferrum College gave me a year's sabbatical to write this book and allowed me to teach half-time for another year. Alice Hooker gave me many a night's lodging at her home while I was doing research in Washington, DC. Suzanne Halwas typed the manuscript, met every deadline, and cheerfully endured a multitude of revisions. Kathy Spradlin and Lisa Norman at Ferrum College xeroxed many copies of the manuscript and assisted me in other ways. Finally, my editors at Basil Blackwell – René Olivieri, Helen Pilgrim, Ruth Bowden, and Elaine Leek – have been most helpful and cooperative.

Ferrum College,
Ferrum, Virginia
August 22, 1988

Introduction
The Conduct of War

*It is inherent in the very concept of war that everything that occurs must
originally derive from combat. . . .*

    *The whole of military activity must therefore relate directly or indirectly
to the engagement. The end for which a soldier is recruited, clothed, armed,
and trained, the whole object of his sleeping, eating, drinking, and marching
is* simply that he should fight at the right place and the right time.

                                  Carl von Clausewitz, *On War*

This is a book about how to wage war. There are two sides to national
security policy. One is what the great Prussian military theorist Carl von
Clausewitz termed "preparation for war." It encompasses the recruitment and
training of soldiers, the selection and education of officers, the procurement
of weapons and other military equipment, the administration of military
bureaucracies, and whatever else is needed to build and sustain a military
force. Von Clausewitz compares all this to the work a swordsmith does in
forging a sword. The other aspect of national security policy – and the subject
of this book – is the "conduct of war." This is like the use a fencer makes
of his sword in fighting a duel. The fencer does not need to know how to
make a sword, but he must know the sword's capabilities and characteristics,
the strokes and counterstrokes he can deal with it, and the capabilities,
characteristics, and intentions of his adversary. Similarly, a general fighting a
battle must know the capabilities and characteristics of his own forces, the
tactics he can employ, and the adversary's forces and intentions. Above all,
he must know why he is fighting the battle, what his objective is. All these
are aspects of the conduct of war.

    Armies exist to fight. The purpose of preparation for war is to make possible
the successful conduct of war. Even a strategic nuclear force whose sole
purpose is to prevent war serves that purpose only if it could conceivably be
used to wage a nuclear war. *"Nuclear weapons can prevent aggression only if there
is a possibility that they will be used."*[1]

    To recognize this principle is much easier than consistently to observe it
in practice. Military officers and civilian defense professionals are constantly
tempted to devote most of their energy to developing and sustaining their
armed forces and thus to neglect consideration of how those forces might be
used to wage a war. Armed forces are always preparing for war but they wage
war only from time to time. Some have not fought for more than a century.
An officer can enjoy a long, successful career training soldiers, procuring

weapons, and otherwise preparing for war without ever hearing a shot fired in anger. He may come to think of war as a tiresome interruption of the true business of his life which is to train soldiers and procure weapons. One of the most important objectives of military education must be to combat this very natural tendency and focus attention on the conduct of war.

Nobody should be light-hearted or emotionless about war. Even a very small war kills people; a nuclear war might be not only the greatest disaster ever suffered by the human race, but also the last. Peace is precious. Sometimes, though, it may have to be sacrificed to preserve that which is even more precious:

> There may come a time when there is no higher expression of the values by which we live than the fight for them. The hatred of all things military is finally a sign that you do not believe in what you are, that you do not believe that you have something to lose.[2]

This book is written by an American, primarily as a guide to United States national security policy. Its author cannot separate himself from the nation to which he belongs, nor would he wish to do so. On the other hand, "war is an act of human intercourse."[3] A government does not wage war in isolation from the rest of the international community; it wages war against an enemy, and often in partnership with an ally. The conduct of any war requires knowledge of the opponent, and waging a coalition war also requires knowledge of one's allies. Therefore, this book devotes considerable attention to the United States's most formidable adversary, the Soviet Union, and to the United States's allies, particularly in Western Europe. It also includes a short chapter on a military strategy, territorial defense, that appeals primarily to countries which are neutrals in the Cold War. Their ability to defend themselves also has an impact on United States national security policy.

Notes

1 This is part of the "usability paradox" as the authors of *Living With Nuclear Weapons* formulate it. The other part is *"but we do not want to make them so usable that anyone is tempted to use one."* Harvard Nuclear Study Group, *Living With Nuclear Weapons* (New York: Bantam Books, 1983), p. 34 (emphasis in original).

2 Leon Wieseltier, *Nuclear War, Nuclear Peace* (New York: Holt, Rinehart and Winston, 1983).

3 Carl von Clausewitz, *On War* (Princeton, NJ: Princeton University Press, 1976), p. 95.

PART I
Taking War Seriously

The fact that slaughter is a horrifying spectacle must make us take war more seriously, but not provide an excuse for gradually blunting our swords in the name of humanity. Sooner or later someone will come along with a sharp sword and hack off our arms.

Carl von Clausewitz, *On War*

1
A Glance into the Abyss

Turn around and go back down,
Back the way you came.
Can't you see that flash of fire,
Ten times brighter than the day?
Stand before a mighty city,
Broken in the dust again.
O God, the pride of Man,
Broken in the dust again.

Hamilton Camp, "Pride of Man"*

The Effects of Nuclear Weapons

One ton of TNT can completely demolish a multi-storey reinforced-concrete building. A bomb containing one ton of TNT is among the most powerful of conventional warheads: a World War II medium bomber could carry one or two tons of bombs; the heaviest conventional artillery in service today, the 16-inch guns on the battleship *New Jersey*, fire a shell weighing slightly more than one ton.

A very small nuclear warhead, the atomic artillery shell fired by an 8-inch howitzer, has the yield (explosive power) of 1,000 tons of TNT – one kiloton. This one shell, therefore, has a yield equal to that of all the bombs dropped in a very large World War II bombing raid or all the shells fired in an artillery barrage employing several thousand guns and lasting for several days. Atomic bombs and warheads range in yield from a minimum of one one-hundredth of a kiloton (ten tons) up to a maximum of 500 kilotons. The bomb that destroyed Hiroshima and killed 80,000 of its inhabitants had a yield of somewhere between 12 and 15 kilotons. By present-day standards the Hiroshima bomb would be a small, "tactical" nuclear weapon, suitable for destroying a dozen or so enemy tanks. If nuclear weapons were to be used on the battlefield in a major war, weapons as powerful as the Hiroshima bomb would be used by the hundreds.

The yield of hydrogen bombs and warheads is usually measured in megatons; a megaton is the yield of one million tons of TNT. A small intercontinental ballistic missile (ICBM), the United States's Minuteman II, carries a one-megaton warhead. The largest ICBM in service today, the

One second after detonation

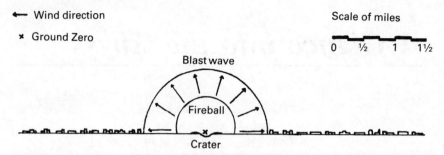

One second after detonation the fireball is nearly a mile in diameter, close to its maximum size, and has generated a blast wave travelling at twice the speed of sound.

One minute after detonation

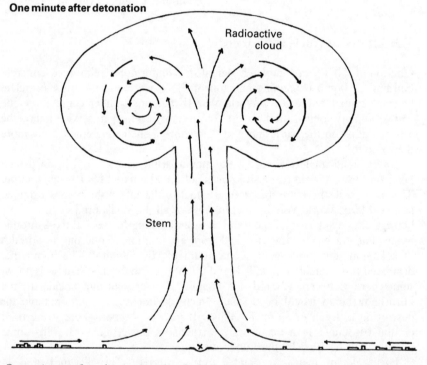

One minute after detonation the fireball has risen more than four miles from Ground Zero and has ceased to glow. The blast wave has passed over the city and been followed by strong winds being drawn in towards the center by the ascending radioactive cloud from the explosion.

Figure 1.1 *A one-megaton nuclear surface burst in a city*

Soviet Union's SS-18, can carry a 20-megaton warhead, and the Soviets have tested a 57-megaton hydrogen bomb. One SS-18 warhead has 13 times the yield of all the bombs dropped on Germany and Japan during World War II.

The difference between the yield of hydrogen bombs and atomic bombs is as dramatic as that between the yield of atomic bombs and conventional TNT-filled bombs. The heaviest conventional bomb ever used in combat weighed 11 tons, the atomic bomb dropped on Hiroshima had slightly more than 1,000 times the yield of that largest conventional bomb, and the SS-18 warhead has more than 1,000 times the yield of the Hiroshima bomb. It can, in fact, be argued that the invention of the hydrogen bomb changed the world far more than the atomic bomb. The destruction inflicted on Hiroshima and Nagasaki, horrible as it was, was not without precedent in the history of war. The conventional bombing campaign against Germany during World War II killed twice as many people as the two atomic bomb attacks put together. The destruction inflicted on a major city by even a single hydrogen bomb would, however, be quite without precedent in the history of war.

Nuclear explosives differ from conventional explosives in more ways than just their vastly greater destructive power. As it results from a fundamentally different physical process, a nuclear explosion has a number of effects which a conventional explosion, such as would be produced by TNT, does not have. Almost all the energy of a conventional explosion is released in its explosive blast, while the energy of a nuclear explosion is released in five different forms:

1　The explosive blast similar, but not identical, to that produced by a conventional explosion.
2　The thermal flash (or thermal radiation), an intense, seconds-long burst of light and heat.
3　The direct nuclear radiation from the explosion.
4　The radioactive fallout – minute particles of material, most of them scooped up from the ground, made radioactive by the explosion.
5　The electromagnetic pulse (EMP).

Any nuclear explosion produces all five of these effects, but their impact depends very much on where the explosion occurs. Nuclear explosions high in the atmosphere, a short distance above the surface of the earth, at the surface of the earth, underwater, and deep underground all have radically different impacts on their surroundings. What follows is a description of the detonation of a one-megaton nuclear warhead at the surface of the earth and at the center of a major city (see Figure 1.1).

At the instant the explosion occurs, the temperature at its center rises to several tens of millions of degrees Fahrenheit, hotter than the interior of the sun. A fireball of extremely hot, luminescent gas forms and begins to rise. As it rises, it spreads out at the top and trails behind it an ascending column of radioactive dust, thus forming the mushroom-shaped cloud of our nightmares. "The fireball from a one-megaton nuclear weapon would appear to an observer 50 miles away to be many times more brilliant than the sun at noon."[1]

The blast wave from the detonation expands outwards from its center with tremendous force and for many miles. It is followed by winds as high as 700 miles an hour a mile from the center of the explosion, diminishing to "merely" hurricane force five or ten miles out. These winds blow outwards from the center and then, after a moment of deathlike stillness, are succeeded by strong winds blowing in towards the center as air is sucked in by the rising fireball.

At Ground Zero, the point where the warhead exploded, there would be a deep crater on the surface of the earth. In dry soil the crater left by a one-megaton surface burst would be about 1,000 feet in diameter and 200 feet deep, a hole wide enough and nearly deep enough to bury the United States Capitol Building in. All around the crater would be a ring of soil and debris, twice as wide as the crater itself, flung out by the explosion.

The blast wave would demolish every man-made structure within a mile of the crater except for bridge abutments and building foundations. Further out it would leave the shells of reinforced-concrete buildings standing but with their interiors completely blown out. Virtually everybody less than 2 miles from Ground Zero would be killed by the blast and its secondary effects, regardless of anything else inflicted on them by the explosion. They would be crushed under falling debris, blown out of the windows of multi-storey buildings, flung against solid objects at 60 miles an hour, or have their lungs ruptured.

The blast wave would demolish ordinary single-family homes 3 miles from Ground Zero and severely damage stronger buildings. It would destroy most vehicles at that distance and leave the streets clogged with debris. It would tip over mobile homes 5 miles from Ground Zero and break windows at a distance of 7 miles.

The counterpart of the blast wave above the surface of the earth would be a shock wave below its surface, shaking the earth much as an earthquake does. Heavy underground structures, such as subway tunnels, are very resistant to shock damage, but whether people in them would survive a nearby nuclear explosion is another question. Presumably, if their access to the surface is blocked by debris they die, even if the tunnel does not collapse. It is possible, at considerable expense, to build underground shelters which can withstand a very heavy shock indeed. The underground silos for the United States's Minuteman ICBMs are designed to withstand the enormous overpressure of 2,000 psi and could probably survive a one-megaton surface burst 1,000 feet away. (Five psi overpressure will demolish an ordinary brick building.) However, even the strongest structures right in the crater would be destroyed.

All of the destruction described above would be caused by half of the energy released in the explosion. Another 35 per cent of its energy would be released in the form of light and heat: the thermal flash. This is a blindingly intense burst of light and heat from the fireball, lasting for about 10 seconds in the case of a one-megaton explosion. The light given off by the explosion is so intense that it temporarily blinds people, even if they are not looking towards it. They may be flashblinded as far away as 13 miles on a clear day or 53 miles on a clear night. (After several minutes the victim recovers completely, but if he is driving a car or flying an airplane at the time he is almost certain to crash.) A person unlucky enough to be looking in the direction of the explosion when it occurs could, in addition to the flashblinding, suffer severe and permanent burns of the retina.

The thermal flash is strong enough to inflict severe burns on people caught out in the open even several miles away. The amount of damage the flash does depends on the weather at the time of the explosion, the time of day, the time of year, and other factors. Moisture in the air absorbs thermal radiation so there would be less damage done on an overcast, drizzly day than on a clear day. Fewer people would be outside and thus vulnerable to the thermal flash at night than during the day. People would be wearing more clothing and would be better protected during the winter than during the summer. (Clothing provides some but not complete protection. A photograph of one of the victims of the Hiroshima bombing shows a neat checkerboard pattern of burns on her skin corresponding to the dark areas of the kimono she was wearing. Thick, light-colored clothing provides more protection than thin, dark-colored clothing.)

The possible variations in all of these factors lead to a very great uncertainty about the possible death toll from the thermal flash of a nuclear explosion. One study estimates that the number killed would be 190 times as great under the worst conditions possible (a clear summer weekend afternoon) as under the best conditions (a cloudy winter night). Nevertheless, under all but the most benign circumstances the thermal flash of a large nuclear explosion in an urban area would kill and injure a great many people. If visibility is good, a one-megaton ground burst can cause first-degree burns as far as 7 miles away. It can cause third-degree burns, those in which the skin tissue is destroyed all the way through, as far as 5 miles away. People who suffer third-degree burns over a quarter or more of their bodies generally die unless they receive prompt and extensive medical treatment, which would rarely be available in a nuclear war.

The thermal flash is likely to cause widespread fires in the ruins of a bombed city and in the area around it. The heat from a one megaton burst is sufficient to set newspapers on fire 8 or 10 miles away. It can ignite curtains, bedding and furniture inside buildings, and thus set the buildings themselves on fire, over a large area. And it can ignite dead leaves on the ground as well as living vegetation, thus setting forests, parks, and backyard gardens ablaze. This is probably the greatest threat posed by a nuclear explosion to the outlying suburbs of the typical American city with their tens of thousands of wooden homes and widespread wooded areas. The blast wave, traveling at somewhat more than the speed of sound, would arrive a few seconds after the thermal flash, which travels at the speed of light. Thus many of the fires started by the thermal flash might be blown out almost immediately by the blast wave. This effect is not, however, to be counted on; scientific studies and tests remain inconclusive.

Many fires would also be started by the blast wave: fuel tanks and fuel lines would be ruptured, wood stoves overturned, high voltage transmission lines blown down, etc. It is possible that the fires started by the blast wave and those set by the thermal flash would coalesce into a firestorm, a massive fire covering entire square miles of the city and driven by high winds. Hamburg and Dresden in Germany and several Japanese cities, including Hiroshima, suffered firestorms as a result of World War II bombing attacks. A nuclear explosion at the center of a city would set fires all around its outskirts and then generate extremely high winds blowing in towards the center, which is exactly what is required to start a firestorm. American cities, however, have far less inflammable material per acre than did the German and Japanese cities in World War II, so whether they would suffer firestorms in a nuclear war is uncertain.

Still another effect of a nuclear explosion is the direct nuclear radiation emitted from the center of the explosion during the first second after it occurs. This accounts for 5 percent of the energy released in the explosion. The direct nuclear radiation given off by a large nuclear detonation does not kill very many people, simply because it has a comparatively short range. Almost everybody close enough to feel its effects would already have been killed by the blast or the thermal flash. In the detonation of a very small nuclear weapon (one kiloton or less) the lethal effects of the direct radiation reach further than those of the blast and the thermal flash and so kill people whom the blast and thermal flash had spared. It is possible to design a nuclear weapon so as considerably to increase the lethal range of its direct radiation and at the some time reduce its blast effect; this is the principle of the neutron bomb.

A nuclear explosion can kill by its blast, its thermal flash, its direct nuclear radiation, and also – perhaps months or years after it goes off – by the radioactive fallout it creates. Radioactive fallout accounts for 10 percent of the energy released in a nuclear explosion but for far more than 10 percent of the casualties. In a surface burst, the explosion sucks up great quantities of microscopic debris from the crater and into the mushroom-shaped cloud. This debris, and the remnants of the bomb itself, become heavily radioactive as a result of the nuclear reactions which occur within the fireball. They constitute the radioactive fallout, a material harmful and frequently lethal to all life in its vicinity. Some of these particles, those in the stem of the mushroom, very soon fall to earth again in the immediate vicinity of the explosion and harm only those who are already dead from its other effects. A small amount of the material at the top of the cloud is carried into the stratosphere and blown all around the globe, not to fall to earth again until months or years later. Some of the fallout from nuclear weapons tests carried out during the 1950s is still up in the stratosphere. Fortunately the longest-lived radioactive materials are the least virulent, so the material falling to earth now from the 1950s tests has lost almost all its lethality.

Most of the radioactive material in the mushroom cloud is picked up by local winds in the troposphere, drifts away, and falls to earth again within several days and over a long, plume-shaped area extending downwind of Ground Zero. After a one-megaton surface burst, assuming a wind speed of 15 miles an hour, fallout heavy enough to cause significant harm to human beings would be distributed as far as 250 miles from the site of the explosion. Stronger winds would spread the fallout over a larger area, but there would be less of it on each square mile affected.

Fallout kills or injures people who are directly exposed to it, with those closest to the site of the explosion being subjected to the heaviest doses and thus suffering the greatest injury. After a week people within 30 miles of the site of a one-megaton surface burst would receive enough fallout to kill them even if they were to stay indoors. People from 30 to 90 miles away would become very sick if they stayed indoors and would probably die if they were to venture out for very long. People 90 to 170 miles away would become fairly sick; those more than 170 miles away would suffer few immediately visible ill effects but would run an increased risk of developing cancer. People can also be killed or made sick by drinking water or eating food contaminated by fallout. Rainwater falling after a nuclear explosion would be a deadly poison, dangerous not only to drink but even if it touches human skin. (Well water would be fairly safe to drink.) Fallout can be lethal to animals and plants, as well as to human beings – more or less lethal according to their species. Mammals and trees are the most vulnerable, insects and grasses the most resistant. Jonathan Schell writes that after a nuclear war had killed all of its human population, other land vertebrates, woody plants, and cultivated crops, the United States would be "a republic of insects and grass."[2] He may exaggerate the damage, but the insects and the grass would have the best chance of survival.

There is one more effect of a nuclear explosion, one that was not clearly recognized until about 1960 – the electromagnetic pulse (EMP). (The failure to recognize this phenomenon for 15 years after the first nuclear weapons test suggests the disturbing, although remote, possibility that there are still other effects of nuclear explosions we have yet to find out about.) The electromagnetic pulse is a very large burst of energy in the form of radio waves, radio waves at the same frequency as those broadcast by radio and television stations. It is powerful enough to destroy or temporarily paralyze many types of electrical equipment, particularly those attached to antennas, overhead power or telephone lines, and other long exposed wires or other metal objects. The EMP produced by a high-altitude nuclear detonation can knock out computers, power grids, radio stations, radios, and telephone systems over a very large area. Communications systems can be shielded against the EMP, and the most vital military communications lines are protected against it, but one or two nuclear detonations at an altitude of 200 miles could knock out unprotected electrical equipment over most of the United States. The EMP does not harm human beings or other living things.

A nuclear explosion a short distance above the surface of the earth has an impact somewhat different from that of a surface burst. As the air burst does not produce a crater it therefore creates little nuclear fallout, as almost all the radioactive particles put into the air by a surface burst come from the crater. On the other hand, its thermal flash reaches further and with greater intensity than that from a surface burst. The air burst also delivers a considerably more powerful blast wave at all but very short ranges, while the surface burst delivers the stronger shock wave and blast wave at close range. Therefore, an air burst is the more devastating against a large, unprotected target such as a city, while a surface burst is more effective against a small, heavily-protected target such as an underground command post or a missile silo.

A Nuclear Attack on an American City

It is possible to get a fairly good idea of what a nuclear warhead could do to a typical American city. The blast, thermal flash, direct radiation, and fallout-inducing effects of various sizes of nuclear detonations vary according to well-understood physical laws. The damage done to Hiroshima and Nagasaki was carefully studied after World War II and the damage that would be done to larger cities or by larger bombs can readily be calculated from those studies. Damage calculations cannot be absolutely precise because cities differ in terrain, in the types of buildings which predominate in them, and in other relevant ways, but they can show approximately what would happen. Several years ago the Office of Technology Assessment made for the United States Senate Committee on Foreign Relations a careful study of the effects of possible nuclear attacks on the United States and the Soviet Union. That study, published as *The Effects of Nuclear War*, includes descriptions of the damage inflicted on Detroit, Michigan by three possible nuclear attacks: a one-megaton surface burst, a one-megaton air burst, and a 25-megaton air burst. It was assumed in these descriptions that there would be no warning of the attack so people would not have taken shelter or left the city, that the attack would take place at night, and that there would be clear skies.

The one-megaton surface burst would do the least damage. It would kill "only" 230,000 of the Detroit metropolitan area's 4,300,000 inhabitants, injure another 420,000, and leave behind 70 square miles of ruins. The center of the city, where the warhead went off, would simply disappear. Within a circular area 3.4 miles in diameter there would be hardly anything left but rubble, except for the shells of a few strongly-constructed buildings around the outer edge of the circle. There are over 200,000 people in this area during the day and about 70,000 after working hours; very few of them would survive. In the ring extending from 1.7 to 2.7 miles from the explosion, steel-framed buildings would have their walls blown away and the steel framework left standing, while single-family homes would be totally destroyed. The streets would be covered with debris, in some places to a depth of 10 feet or more. About half of the people in this area would be killed, another two-fifths injured, and a lucky one out of ten uninjured. Farther out, from 2.7 to 4.7 miles away from the explosion, large buildings would be severely damaged and many single-family homes destroyed. About one-sixteenth of the people in this area would be killed by the blast and the thermal flash and nearly half of the survivors would be injured. This is where numerous fires or even a firestorm would be most likely. Tens of thousands might be burned to death in this area, particularly because most of those injured by the blast and the thermal flash would be unable to flee and would die in the flames. The Hamburg firestorm killed 40,000 people in an area of 4 square miles. More than 4.7 miles out there would be few deaths, by nuclear war standards, but about a quarter of the people in the area would be injured. Most homes would remain habitable.

Air bursts, the reader will recall, do more damage to cities than surface bursts. A one megaton air burst over Detroit would kill 470,000 people and injure another 630,000, nearly twice as many as a surface burst of that magnitude. Its effects on buildings and other structures would be similar to those of a ground burst, but extending over larger areas; the air burst does as much damage 7 miles from Ground Zero as the ground burst does 4.7 miles away. A 25-megaton air burst would kill almost half the inhabitants of Detroit and injure more than a quarter of them. With as many injured to be cared for as uninjured survivors, and few buildings left habitable, it would probably be necessary to evacuate all survivors from the city and perhaps to abandon the site permanently.

The types of nuclear attack that might be made on a city and the conditions under which they might be made are extremely variable, and so therefore is the amount of damage they might do. One of the virtues of *The Effects of Nuclear War* is that it employs a judicious mixture of optimistic and pessimistic assumptions. On one hand, it assumes that the attack is made without warning and that therefore all the inhabitants of Detroit are in the city and unprotected by any form of civil defense. On the other hand, it assumes that the attack is made at night, which greatly reduces the number of casualties caused by the thermal flash. The one-megaton scenarios also assume a much smaller attack than Detroit would be likely to suffer in a nuclear total war. A city the size and importance of Detroit would probably be hit with one or two dozen 500-kiloton weapons. This would kill or injure about as many people as the attack with a single 25-megaton weapon.

A Nuclear Attack on the United States

There have been several attempts to estimate the damage that would be inflicted by a strong Soviet nuclear attack on the United States. Arthur Katz's *Life After Nuclear War* offers a recent, balanced, and fairly detailed estimate. Katz describes the effects of a Soviet attack on the United States's 71 largest metropolitan areas and its most vital manufacturing plants outside those metropolitan areas. His study assumes that those targets would be hit by 700 or 800 Soviet warheads: 500 one-megaton warheads on the metropolitan areas and 200 or 300 100-kiloton warheads on the vital manufacturing plants outside the metropolitan areas. The Soviet Union actually has, on its ICBMs and the missiles carried on its missile submarines, nearly 10,000 nuclear warheads, most of them with an explosive power between 500 kilotons and one megaton. Probably most of those warheads would be used against the United States's ICBMs, bomber bases, and other military targets: the Soviet government's first priority in a nuclear war would not be to kill as many

American civilians as possible, but to protect the Soviet Union against American retaliation. Much of this tremendous arsenal might also be used against the Soviet Union's other enemies in Western Europe, China, Japan, and elsewhere. The Soviets have at their disposal plenty of short-range nuclear weapons capable of hitting targets all over Eurasia, but they might wish to supplement them. Some, perhaps many, Soviet missiles would malfunction on their launching pads or after launch. Nevertheless, taking all of these factors into account, in a nuclear total war the Soviet Union almost certainly would visit on American civilian targets the level of devastation Katz envisages. It would require fewer than one-ninth of the strategic nuclear weapons the Soviet armed forces have available.

Almost the entire area of the United States's great urban centers would be blanketed with shock, fire, and radiation. Sixty one-megaton warheads would descend on the New York metropolitan area, 40 each on Chicago and Los Angeles, 17 on Boston, 16 on St Louis, and so on down to Albuquerque, New Mexico and Fresno, California, places with populations of about 300,000 which would each suffer a single one-megaton blast. This attack would kill or injure two-fifths of the population of the United States. Two-thirds of the casualties would be dead and one-third injured: 50 or 60 million Americans dead and 20 or 30 million injured.

Most physicians and nurses would be dead or injured and most hospitals wrecked. Physicians and other health care specialists tend to congregate in urban areas; about two-thirds of all physicians in the United States live in the 71 largest metropolitan areas, which would be the primary targets of this Soviet attack and in which the vast majority of the population would be killed or injured. Katz estimates that each surviving unhurt physician would have to care for 200, 600, or even as many as 1,000 patients, in makeshift or tremendously overcrowded facilities. There would be 30 or 40 patients for every remaining hospital bed. Caring for the severely burned would be hopeless. The United States's existing specialized facilities for severely burned patients can handle about 200 patients at a time; a nuclear attack would leave millions of severely burned casualties. Almost all of those casualties would have to be left to die with little or no medical care.

The basic economic effect of the attack would be "the virtual elimination of the 71 largest SMSAs [standard metropolitan statistical areas] as productive economic units."[3] About two-thirds of the manufacturing capacity of the United States would be destroyed. The oil industry would be particularly hard hit. Petroleum refineries are large, extremely vulnerable targets which can take years to rebuild, even in an otherwise intact economy. Oil storage tanks, oil wells, and pipelines are also vulnerable to attack. Katz estimates that a major attack on United States cities and industry would destroy 98 percent of the country's petroleum-refining capacity. Coal and other energy sources can be substituted for oil to a large extent, although not completely, but the coal mining industry and the means for transporting coal would also be severely damaged.

Life After Nuclear War looks also at the impact of a Soviet attack on the United States's strategic nuclear force: ICBM silos, bomber bases, and missile submarine bases. As most of the ICBM silos and bomber bases are located in the more sparsely settled parts of the country, the casualties caused by this attack would be far fewer than those caused by an attack directed against cities. Katz estimates that it would kill 16.3 million people and injure 10 or 20 million – still a horrendous toll. However, an all-out Soviet attack would almost certainly hit both American cities *and* the United States's strategic nuclear force, so the casualties and damage caused by the attack on the strategic nuclear force should be added to that caused by the attack on the cities to give the full picture. This gives a total of about 70 million dead and 30–50 million injured, about half the population of the country.

The attack on the strategic nuclear force would have very serious effects on American agriculture. The radioactive fallout from this attack, drifting eastwards from ICBM silos and bomber bases in Montana, Wyoming, North and South Dakota, Arizona, Kansas, Missouri, and Arkansas, would blanket the country's most productive agricultural areas, with catastrophic effects on grain production and livestock raising. The destruction of the oil industry would also harm American agriculture severely. Without fertilizer, the production of which depends on the oil industry, the production of corn, wheat, and other major crops would be halved. Without gasoline and diesel fuel, which would be in extremely short supply after a nuclear attack, American agriculture would be almost unable to function at all. There would, of course, be far fewer people to feed after a nuclear war, but even so, many of the survivors could starve to death.

Perhaps the casualty rate could be substantially reduced by nationwide civil defense measures. Blast shelters strong enough to protect very many people in the urban areas would be prohibitively expensive and perhaps ineffective. The best way to protect people would be to evacuate the cities, move as much of the city population as possible into the rural areas. The warheads that had been aimed at the cities could be redirected at the rural areas, but they would kill far fewer people.

People in the rural areas – both evacuees and the original inhabitants – would need fallout shelters as protection against the radioactive fallout which would cover most of the country. They would also need stocks of food, water, and other supplies sufficient to permit them to stay in their shelters for several days to a month or more, depending on the level of fallout in their area. It is possible to convert the basement of an ordinary single-family home into a crude fallout shelter by piling dirt on the floor above it and around the outside, installing a toilet and an airtight hatch, and stocking it with provisions. T. K. Jones offers the assurance that if people take these simple measures the United States can survive a nuclear war substantially intact. "If there are enough shovels to go around, everybody's going to make it."[4]

Maybe. There are, however, several problems with this rosy scenario. It is totally dependent on several days' warning time so as to move people out of the cities and build or occupy fallout shelters. A nuclear total war is most likely to develop from a desperate crisis or a limited war, and a Soviet nuclear attack would most likely be preceded by the evacuation of Soviet cities, so probably there would be several days' warning of the attack. However, there

probability that the United States government would not issue the order in time. Governments have been known not to heed what retrospect to have been very clear warning of an impending attack; both tne Japanese attack on Pearl Harbor and the German attack on the Soviet Union in June of 1941 achieved complete surprise, despite ample evidence in both cases that an attack was coming. Even if the order were to be issued in time, many people probably would not heed it, either because they refused to believe the attack would happen, or because they thought that nothing they might do would save them. To the extent that an evacuation took place, it would be chaotic to the point of disintegration. The reader will recall that a large nuclear explosion temporarily blinds people over a wide area; should the warheads explode while the highways out of the cities were still clogged with vehicles moving out, there would be some very extensive multiple-car accidents. Also, few of the evacuees who did get out would have the time, or be able to find the materials, to build and stockpile their own fallout shelters. Living conditions in the shelters, for those lucky and foresighted enough to obtain them, would be extremely grim – several dozen people trapped in a dark, filthy basement, perhaps several of them dying of radiation sickness, all worried about the fate of close relatives and friends, and with little information about what was going on outside. Not everybody would have the self-control and endurance to remain calmly in a shelter for a week or several weeks waiting for the radioactivity to subside. Those who did survive would emerge into a world devastated almost beyond our ability to imagine and perhaps incapable of supporting human life.

There is a case to be made for civil defense (as well as a case to be made against it). Extensive and well-conceived civil defense measures taken a long time before a nuclear attack could, with luck, save many lives. However, even the best civil defense system would not prevent the deaths of tens of millions of Americans in a nuclear total war.

A Global Nuclear War

A nuclear total war would not, of course, involve only the United States. It would devastate the Soviet Union also and almost certainly the United States's allies in Western Europe and the Soviet Union's in Eastern Europe. Very likely every major country in the world would be attacked by one superpower or the other, because the devastation inflicted on the superpowers would be so great that neither could afford to leave a potential enemy unscathed. If the Soviet Union were to suffer a major nuclear attack and China did not, then China would be left militarily and economically far stronger than the Soviet Union. Furthermore, even countries that were not attacked directly with nuclear weapons would suffer from the radioactive fallout and other global effects of a nuclear total war; winds and fallout do not stop at national boundaries. A nuclear total war between the United States and the Soviet Union would also be a nuclear total war for the entire human community.

The effects of a global nuclear war have been measured in a study sponsored by the Royal Swedish Academy of Sciences and published in the United states as *The Aftermath: The Human and Ecological Consequences of Nuclear War*. For the purposes of this study it was assumed that a global nuclear war would involve the use of 14,737 nuclear warheads with a total explosive force of 5,750 megatons. This is somewhat less than the striking power available to the Soviet Union alone, so the assumption seems plausible. It was further assumed that slightly fewer than half the warheads would be used against military targets and slightly more than half against cities and industrial plants. This would be enough to inflict one megaton or more of nuclear destructive power on every city of more than 100,000 population in North America, Europe, the Soviet Union, North and South Korea, Vietnam, Australia, South Africa, and Cuba. Every city of more than 500,000 population in China, India and most of the rest of Asia would also be hit with one megaton or more.

The major cities in the Northern Hemisphere would cease to exist. Of the 1.3 billion people in those cities, more than half would be killed by the blast, thermal flash, and direct nuclear radiation of the attack and more than a quarter would be injured. Perhaps another couple of hundred million would be killed by the fallout. There would be fewer casualties if the nations attacked had blast and fallout shelters for their people, but most do not. Switzerland, which has the best protection against nuclear attack of any country in the world, can shelter most of its population. Sweden, China, and the Soviet Union also have devoted considerable resources to civil defense, but other countries have little or no protection for their peoples. Even leaving out of account the deaths from epidemics, starvation, civil disorders, and the widespread fighting with conventional weapons that would almost certainly accompany a global nuclear war, the war would kill about a billion human beings, one-quarter of the world's population.

All of this discussion of the impact of a nuclear war has been confined to the direct effects of nuclear detonations on human beings, and the work of human hands. If these direct effects were the only effects of a global nuclear war then such a war would be the greatest catastrophe in history, comparable only to the Black Death, but the human race would survive it. However, a global nuclear war might also inflict catastrophic, world-wide damage on the ecosphere: damage severe enough to make our planet temporarily incapable of supporting human life. That this would happen seems much less likely now than it did a few years ago, but the possibility cannot be entirely discounted.

In the 1970s there was much concern about the danger that a global nuclear war might greatly diminish the amount of ozone in the stratosphere. Ozone is one of the forms in which oxygen occurs and as the ozone layer in the stratosphere it shields the earth against ultraviolet radiation coming from the sun. If this shield were greatly weakened, the ultraviolet radiation getting through could cause sunburns so severe that nobody would be able to work outdoors for more than a few minutes a day. It would also blind land animals and any human being not wearing protective goggles, and kill both plants and the aquatic microorganisms on which life in the sea is based. Nuclear explosions set off large-scale chemical reactions in the atmosphere which destroy stratospheric ozone; a global nuclear war might reduce the amount of ozone in the stratosphere by 70 percent or more.

However, scientific opinion now is that large-scale loss of stratospheric ozone would be caused only by numerous very large nuclear explosions or explosions very high above the surface of the earth. Explosions of one megaton or less at or near the surface of the earth have little effect on the amount of ozone in the stratosphere. Almost all of the nuclear warheads in existence today have yields of a megaton or less and in a nuclear war would be detonated at or near ground level. A few warheads might be detonated at very high altitudes in order to exploit their EMP effect, but not enough of them to damage the ozone layer. Thus a nuclear war fought today would probably not greatly damage the ozone layer in the stratosphere. But, there are two points worth remembering on this subject;

1 A nuclear war fought in the 1960s, in which many very large nuclear weapons would have been used, might have damaged the ozone layer very severely. We might have rendered our planet uninhabitable by an effect of nuclear explosions we did not know existed.
2 The effect of nuclear explosions on stratospheric ozone is not yet fully understood; further research might well change what looks now to be a mildly reassuring picture into an alarming one.

Soon after the "ozone depletion" threat had been assuaged the "nuclear winter" threat came into view. In 1983 a group of scientists published their study of the global climatic effects of a nuclear total war. This "TTAPS" study pointed out that the war would cause massive conflagrations spreading over tens of thousands of square miles. Soot, smoke, tar, and ashes from these fires would rise into the atmosphere and, together with the radioactive fallout from nuclear explosions, obscure the light of the sun for months to come. "For more than a week in the northern mid-latitude target zone, it might be much too dark to see even at midday."[5] Temperatures all over the world would fall catastrophically and a "nuclear winter" would set in. The TTAPS study asserted that a nuclear war on the scale of that considered in *The Aftermath* would, for a short period, reduce the average land surface temperature in the Northern Hemisphere to −10 degrees Fahrenheit, which is the average temperature *in the winter* of Northern Alaska. (The climate would start warming up again after a week or so.)

More recent work has exposed some serious defects in the TTAPS study. (Among a number of other peculiarities, the TTAPS computer model of the earth's surface assumed that the oceans do not exist, that the earth is all dry land.) It seems now that after a nuclear total war Northern Hemisphere temperatures would drop 10 to 20 degrees, but not the 65 degrees predicted by the TTAPS study.

A temperature decline of this magnitude would not *by itself* imperil the survival of the human race. It would, though, be a major environmental disaster. If there were to be a nuclear total war in the summer, the temperature decline alone could destroy most of that year's agricultural output in the

Northern Hemisphere. This disaster would fall upon a world rendered greatly more vulnerable by the other effects of thousands of nuclear detonations.

Notes

1 Samuel Glasstone and Philip J. Dolan, *The Effects of Nuclear Weapons* (Washington, DC: US Government Printing Office, 1977), p. 27.
2 Jonathan Schell, *The Fate of the Earth* (New York: Avon Books, 1982), p. 169.
3 Arthur M. Katz, *Life After Nuclear War* (Cambridge, Mass.: Ballinger, 1982), p. 102.
4 Quoted in Robert Scheer, *With Enough Shovels: Reagan, Bush, and Nuclear War* (New York: Vintage Books, 1983), p. 18.
5 Carl Sagan, "Nuclear War and Climatic Catastrophe: Some Policy Implications", *Foreign Affairs*, 62 (Winter 1983/84), p. 267. The "TTAPS" study is named after the five scientists who conducted it – Turco, Toon, Ackerman, Pollack, and Sagan.

2
The Nuclear Predicament

The Balance of Terror

The holocaust described in the preceding chapter could happen at any moment. While you are reading these words the missiles sit on their launching pads ready to be launched, the gyroscopes spinning in their warheads, the paths they are to follow to their targets programmed into their computers, the launch-control officers standing by ready to launch them. Wherever the president of the United States goes he is accompanied by one of the symbols of the Nuclear Age, a military officer carrying a locked briefcase. Inside that briefcase are the computer codes that the president would use to order a nuclear attack on the Soviet Union. Apparently the government of the Soviet Union has made similar arrangements: a man with a similar briefcase has been seen accompanying the Secretary General of the Communist Party of the Soviet Union. A few hours from now half of the inhabitants of the United States and the Soviet Union could be dead in a nuclear war; a year from now, most of the population of the world.

This is "the balance of terror": more than 2,500 Soviet ICBMs, bombers, and submarine-launched missiles poised to strike the United States whenever the word is given; nearly 2,000 United States missiles and bombers, poised to strike the Soviet Union. The horrendous power of the strategic nuclear forces deployed by the two antagonists and the approximate balance between them does ensure that the chances of a nuclear war happening in any given year are extremely small. Each of these great arsenals is balanced and constrained by the other. An all-out Soviet attack on the United States or an American attack on the Soviet Union would trigger an equally-devastating retaliatory attack, probably within the hour; aggressor and victim would perish together. The political and military leaders of both countries are well aware of this and none of them would deliberately start a nuclear war under any but the most desperate circumstances. Also, much thought and effort have gone into ensuring that a nuclear war could not be started by an accident, such as a malfunctioning radar or the actions of an individual bomber pilot. The balance of terror has prevented a nuclear war for nearly 40 years. It may be the best way attainable, at least in the near future, to ensure that there never will be a nuclear war. Nevertheless, as long as nuclear weapons exist it is possible that they will be used, and however small the chance of a nuclear war being fought in any given year, within a long enough period of time any event that is not impossible will probably happen.

The balance of terror not only menaces the lives of half of the people of the United States, but also it is based on the United States's being prepared to kill half of the people of the Soviet Union. To be sure, an American attack carried out in retaliation for a Soviet attack on the United States might be considered a justifiable act of revenge. However, those who did the United States the injury would not be the ones who suffered its revenge. Those who planned and ordered the Soviet attack would be a small group of the highest-ranking leaders in the Soviet Union, and they no doubt would be inside the strongest blast shelters the Soviet Union has, awaiting the United States's retaliatory blow. The victims of American retaliation would be the people of the Soviet Union – men, women, and children – hardly any of whom would have had any influence on the decision to launch the Soviet attack. They would die in their millions. John C. Bennett asks, "How can a nation live with its conscience and know that it is preparing to kill 20 million children in another nation if the worst should come to the worst?"[1] The people of the United States have lived that way for nearly 40 years. Surely it is not a good way to live.

Red or Dead?

As long as the Soviet Union has nuclear weapons, the United States cannot preserve its national independence and the freedoms of its citizens without also having nuclear weapons. However, some would argue that no political system is worth a nuclear war and if we go on the way we have up until now eventually there will be a nuclear war. If the price of American national independence is for the United States to have nuclear weapons, and if having them inevitably will lead to their being used in a nuclear war, then that is too high a price to pay. It would be better for the United States to carry out unilateral nuclear disarmament – destroy all its nuclear weapons regardless of whether or not the Soviet Union does so, even if this should lead to a Soviet conquest of the United States. That is the position taken by the psychologist Erich Fromm and the historian H. Stuart Hughes, among others.[2] It is summed up by a slogan popular in the United Kingdom some years ago: "Better red than dead."

It is also argued that the use of nuclear weapons, even in self-defense or in retaliation for any enemy attack, cannot be morally justified. This issue has been addressed in a national pastoral letter issued by the bishops of the Roman Catholic Church in the United States. Their letter holds that:

> Under no circumstances may nuclear weapons or other instruments of mass slaughter be used for the purpose of destroying population centers or other predominantly civilian targets. . . . Retaliatory action which would indiscriminately

take many wholly innocent lives, lives of people who are in no way responsible for reckless actions of their government, must also be condemned.[3]

The Catholic bishops do not completely condemn all possible use of nuclear weapons. They have, though, imposed such stringent conditions on the permissible use of nuclear weapons that it is almost impossible to frame a strategy that satisfies them.

The bishops' position reflects a Catholic just-war doctrine going back to St Augustine in the fifth century. It may not be the only position that can be derived from Catholic just-war doctrine but it must be taken seriously by the 50 million Roman Catholics in the United States, by Catholics in other countries, and by many non-Catholics. It also stands in direct contradiction to the long-established national security policy of the United States. Ever since the late 1940s United States national security policy has indeed been based on the threat that "nuclear weapons ... [would be) used for the purpose of destroying population centers" in the Soviet Union in retaliation for a Soviet attack on the United States.

A case can be made for unilateral disarmament and it is, in some respects, a strong case, stronger than can be made for any other alternative to the balance of terror. That is one reason for discussing the possible results of the unilateral disarmament of the United States. Another is that looking at what might happen if the United States did *not* have nuclear weapons may show why the United States has them and why it needs to keep them, at least for the near future.

The only country Americans can disarm directly by their own actions is the United States. There are four other countries known to have nuclear weapons: the Soviet Union, the United Kingdom, France, and China. Even if the United States were to destroy all its nuclear weapons, that action would not necessarily induce other countries to destroy theirs. The United Kingdom might follow the United States's lead because of the strong political ties between the two countries. Also, the British government might feel that because the British nuclear weapons program depends so heavily on the United States it would be impossible to maintain a British nuclear deterrent if the United States were to disarm. On the other hand, the British might feel that without the United States's nuclear umbrella to shield them from Soviet aggression, it would be more important than ever to have a strong British nuclear deterrent force. France would be less likely than the United Kingdom to follow the United States's example, and China even less likely than France.

Even if all the other nuclear powers were to disarm, the Soviet Union would not. It would have no reason to. The leaders of the Soviet Union do fear nuclear war and fear it very deeply. They are well aware of the death and destruction it would cause. Unfortunately, they fear it because of what it would do to the Soviet Union, not because of what it would to other countries. Nuclear war is a threat to the Soviet Union only because other countries have nuclear weapons. If the Soviet Union were the only nuclear power in the world, then nuclear weapons would not menace the Soviet Union; they would be instead the great bulwark of Soviet national security and very likely the means by which the Soviets could impose their will on everybody else.

Probably not a single one of the present nuclear powers would give up nuclear weapons if it were the only nuclear power in the world and could count on remaining so. The United States did propose a world-wide ban on nuclear weapons in 1946 when it was at that time the only country in the world which had them, but at that time United States leaders were well aware that the Soviet Union would also have nuclear weapons in the near future.

If the United States were to destroy all its nuclear weapons, perhaps the Soviet Union and one or more of the other nuclear powers would keep theirs; that is one possibility. In that case the balance of terror would still prevail: there would still be several nuclear-armed countries, their interests would still be in conflict, and their governments might be tempted to use nuclear weapons to resolve those conflicts. It would simply be a much less stable balance than before. Today the leaders of the Soviet Union must know that if they attack the United States the Soviet Union will suffer a devastating retaliatory blow from the United States. They also know that an attack on China, France or the United Kingdom, even if the victim were unable to retaliate, would be a hideously risky adventure, because the United States might intervene in the conflict. If the United States were disarmed then a Soviet attack on any one of the three lesser nuclear powers would have a greater chance of being successful and there would thus be a greater chance of its being made. The Soviets might well calculate that their attack could destroy most or all of a lesser nuclear power's retaliatory force before it could be used, and thus that the Soviet Union would have little to fear from any retaliation. Therefore they might, in a crisis, carry out such an attack.

If no other country had nuclear weapons, perhaps the Soviet Union would not use its nuclear monopoly to impose its rule on the rest of the world. Nothing in the behaviour or world view of the Soviet Union's leaders suggests that this would be the case, but it is conceivable. A nuclear war would still be possible; the nuclear sword of Damocles would still hang over the human race, and it would hang by a thinner thread than ever. There would still be nuclear weapons in the world, they would be in the hands of a sovereign state, and they could be used, without fear of retaliation, whenever the rulers of that sovereign state decided. The only time nuclear weapons have been used in war was against a country which did not have nuclear weapons and could not retaliate. The nuclear disarmament of every country in the world but one would mean that every country in the world but one would be as vulnerable to nuclear attack as was Japan in August of 1945.

Perhaps it would be better for humanity if unilateral disarmament *did* lead to a Soviet conquest of the United States and the rest of the world. That might well mean the loss of a great deal that makes life meaningful for Westerners, the extinction of human freedom perhaps forever, but life would go on. And, as Jonathan Schell reminds us, "Life comes first. The rest is secondary."[4] Some would see giving up one's liberty and that of generations of one's descendants in return for life as rather an ignoble bargain. Others might see it as the noblest sacrifice the present generation of Americans could make to ensure the survival of the human race.

It would be a great sacrifice, greater than some unilateral disarmament advocates acknowledge. The conductor Leonard Bernstein, pondering a possible Soviet occupation of the United States, asks:

> What would they do with us? Why would they want to assume responsibility for, and administration of, so huge, complex and problematical a society as ours? And in English, yet! . . . how can they fight when there is no enemy? The hypothetical enemy has been magically whisked away, and replaced by 200-odd-million smiling, strong, peaceful Americans.[5]

The people of Afghanistan could tell Mr Bernstein what a Soviet army of occupation might do with us, as could the peoples of Hungary, the Baltic states, and a number of other countries. Alexandr Solzhenitsyn and other Russians could also shed some light on what the Soviet regime can do to those whom it has in its power. Communist-ruled regimes have perpetrated atrocities comparable only to the actions of Nazi Germany or the results of a small-scale nuclear war. The Pol Pot regime in Cambodia killed one-third of the population of that country in three years. Joseph Stalin, according to some careful estimates, caused the deaths of over 40 million of his fellow Russians. A third of the population of Afghanistan was driven from the country by the anti-guerrilla war waged there by the Soviet Union. Not all Communist-ruled regimes have been as brutal as Stalin's or Pol Pot's, and one can hope that a Soviet occupation regime in the United States would not be. However, as with nuclear war, policy should be guided by the worst that can happen.

The sacrifice of freedom for survival would not even ensure survival. The people of the United States might well find that they had given up their freedom and submitted to all the evils of a Soviet conquest and yet still not have saved the world from the menace of nuclear war. The Soviets probably could impose a Communist-ruled government on the United States but they probably could not control it once it had become well-established. The People's Republic of the United States, as it might style itself, would be distant from the Soviet Union and have immense resources at its disposal. Without ceasing to be a Communist dictatorship, it would have national interests different from those of the Soviet Union and might well become solidly-enough based to defy the Soviet Union, just as China, Yugoslavia, Albania and to a degree Rumania, have done. If so, it would be the most formidable enemy the Soviet Union has ever faced. Our grandchildren might die in a nuclear war between the People's Republic of the United States and the Soviet Union. It would be a tragically ironic, but all too likely, end to the search for peace through unilateral disarmament.

In the realm of abstract arguments, "red or dead" sounds like a real choice – or at least a question that can be debated. Perhaps one would prefer to live under Communist totalitarianism, or perhaps one would prefer to be dead. Perhaps a world in which the light of human freedom had been snuffed out by a universal totalitarian regime would be preferable to one in which the human race had been snuffed out by a global nuclear war. Perhaps not. In the real world a sane mind recoils from the question and refuses to make the choice. To anybody who knows that totalitarianism is like, and what a nuclear war would be like, the only possible answer to the "red or dead" question is "neither red nor dead." The national security policy of the United States must be to do everything possible to preserve American freedom and independence and everything possible to prevent a nuclear war.

The people of the United States and other free countries confront two great enemies: the aggressive imperialism of the Soviet Union and the threat of a nuclear war. Perhaps because they are both so menacing, people tend to minimize the threat posed by one in order better to deal with the other. Those who feel very strongly about nuclear war tend to soften down the threat posed by Soviet imperialism and urge their fellow citizens to behave as if the nature of the Soviet government were different from what it is. George Kennan, for example, urges that the United States halt all economic sanctions against the Soviet Union, "exercise restraint in the tragic question of human rights and national independence" (i.e. do not object to what the Soviet government has done in Poland and Afghanistan), and "put an end to the often systematic condemnation of another great people and its government – a condemnation which if not stopped will really make war inevitable by making it seem inevitable."[6] On the other hand, those who feel very strongly about the threat posed by the Soviet Union tend to minimize the devastation that would be caused by a nuclear war. They sometimes argue that if Americans fear nuclear war too much, the United States will not resist Soviet aggression strongly enough. This is one of the arguments for the various "damage-limitation" nuclear war-fighting strategies discussed in chapter 9.

Both these tendencies are understandable and both must be resisted. Americans and other free peoples cannot defend themselves against either of the two great threats facing them by yielding to or minimizing the other. Unilateral disarmament, or refusing to defend the United States and other countries against Soviet expansionism, would not prevent a nuclear war; it might even make a nuclear war more likely. Conversely, fighting a nuclear war would destroy, not save, democracy in the United States.

The American people's first priority after a nuclear total war would not be to maintain the Bill of Rights; it would be to survive. Liberal democracy, as the most complex and demanding of all forms of government, requires very favourable conditions. Those conditions have been met in only one-fifth of the countries in the world today, generally the most-favoured, economically-developed, best-educated countries. It is hard to see how they could be met in the United States after a nuclear total war. The most probable form of government of a post-nuclear war United States would be dozens or hundreds of small dictatorships dividing up the country among them; the next most probable would be a centralized dictatorship. The least likely would be any form of democracy.

Disarmament, World Government, and Arms Control

Unilateral disarmament does have its virtues, at least in comparison with other alternatives to the balance of terror. It is a radical solution for a radical evil. It is also the only alternative which could be carried out immediately, by the United States alone, and which holds out any hope of eliminating the balance of terror. That is why it has been examined at some length, even though most opponents of current United States national security policy advance other alternatives – multilateral disarmament, world government, or various forms of arms control. All of these other alternatives require the cooperation of other countries. Multilateral disarmament and world government will take a long time to attain if they are attainable at all, and any readily-attainable form of arms control would leave the balance of terror essentially unchanged.

Multilateral nuclear disarmament and the establishment of a world government will be discussed in chapter 20 as long-term alternatives to the balance of terror. In the short term they are not alternatives at all. It is almost impossible to conceive of either being attained before the end of this century, and hard even to have much confidence that either could be attained before the end of the next century.

At the end of World War II there was a widespread conviction – at least in the United States and Western Europe – that nuclear weapons and unfettered national sovereignty were too dangerous a combination and that one or the other or both had to go. There was a sense of urgency in the search for nuclear disarmament and in the efforts to build a world government on the foundation of the United Nations. That sense of urgency faded rather quickly. Multilateral nuclear disarmament was seriously proposed by the United States in 1946, half-seriously proposed by the Soviet Union in 1955, advanced for propaganda purposes by the Soviet Union in 1959 and then by the United States in 1961. Since then it has been generally forgotten and meanwhile the nuclear arms race has gone on almost unabated. Instead of one nuclear power with two or three nuclear weapons, as there was in 1945, there are now five nuclear powers with tens of thousands of nuclear weapons among them. The effort to build a world government has made no greater progress. The United Nations is even further today from being a world state than it was in 1945. It has become less and less a means of resolving conflicts between countries and more and more a weapon which they use against each other in those conflicts. The sovereign state, far from being superseded, has become more important than ever before in history. Peoples all over the world have attained their own sovereign states since World War II and are not about to give up the national sovereignty they have fought to attain.

Multilateral nuclear disarmament would constitute a radical change in the way the world is organized, but the human race has carried out radical changes in its institutions before: Schell cites the abolition of slavery as a change equally as radical in its day. However, in one absolutely crucial respect the abolition of slavery was much easier of attainment than multialteral disarmament would be. Slavery could be, and was, abolished one country at a time. Britain could abolish slavery within its Empire in 1830, even though the United States and Brazil were not yet ready to do so. Then the abolition of slavery by Britain and other European states helped move the United States and Brazil towards taking the same action. Global nuclear disarmament, however, cannot be achieved one country at a time. Every nuclear power must disarm at the same time or there will be no disarmament at all. Even if one country were to discard its nuclear weapons unilaterally, it would not induce all the others to follow suit. It is always much more difficult for a number of countries to act simultaneously than for one of them to act alone, or even for all of them to act one after the other. Nothing can be done until the most reluctant is ready to act. Not only must they all be ready to act, but also each of them must be certain that all of the others will carry out whatever has been decided on. A multilateral nuclear disarmament treaty can only be made and carried out if every nuclear power is certain that every other nuclear power will destroy all its nuclear weapons and not build any more, and that no non-nuclear power will acquire nuclear weapons. This is the problem of "verification" – locating every one of the tens of thousands of nuclear weapons in the world, making sure they are destroyed, and making sure that no more are built. This verification problem has bedeviled every attempt at nuclear disarmament since 1946 and it seems no closer to being solved in 1988 than in 1946.

The same considerations apply *a fortiori* to efforts to establish a world government. World government would be the most radical political innovation ever achieved, and much more difficult to attain even than multilateral nuclear disarmament. While there are only five nuclear powers, there are 165 sovereign states. Perhaps not every one of those 165 sovereign states would have to join a world government initially for it to work, but the vast majority would have to, including every nuclear power. Also, a world government would require far more of a sacrifice from every country that joined it than nuclear disarmament would require of the nuclear powers: it would require the sacrifice not just of a particular type of weapon, but of the cherished right of every people to choose its own institutions and chart its own destiny without outside interference. Finally, while there is perhaps emerging a world-wide consensus that nuclear weapons serve no country's purpose and all would be better off without them, there is no world-wide consensus that national sovereignty is obsolete.

Various arms control measures have been concluded by the United States and the Soviet Union, most notably SALT I of 1972 and SALT II of 1979, limiting their strategic nuclear weapons, and the 1987 INF Treaty banning their intermediate range nuclear missiles from Europe. More far-reaching measures may be attainable. Strategic nuclear force reductions as great as 50 percent are at least to be discussed at the START (Strategic Arms Reduction Talks) negotiations, and some commentators have proposed that the superpowers reduce their nuclear arsenals by as much as 90 percent. Arms control is clearly more feasible than complete nuclear disarmament or world government. On the other hand, it does not put an end to the balance of terror. Even the elimination of 90 percent of their nuclear arsenals would still leave the United States and the Soviet Union able to destroy each other in an hour.

It seems then that there are no real alternatives to the balance of terror. It does have its virtues; it has served two fundamental American objectives of containing the expansion of Soviet power and preventing a nuclear war. Nuclear weapons have not been used in war since 1945, most of the time the Soviet Union has been successfully contained, and where it has not been, the balance of terror has not been the reason. It has made direct Soviet military aggression, otherwise a most potent instrument of Soviet expansionism, very dangerous and hence rare. This is not to say that living under the balance of terror is a desirable way to live. Plainly it is not. But it *is* to say that for any alternative to be preferable it must better serve the two major purposes which are served by the balance of terror, however imperfectly.

The balance of terror does not maintain itself automatically. It persists because both the United States and the Soviet Union have built and maintain powerful strategic nuclear forces and each has taken measures to ensure that its force could not be destroyed in a surprise nuclear attack by the other. It will continue to exist in the future only if both sides continue to maintain their strategic nuclear forces. The possibility of winning a nuclear total war may long since have disappeared, but the need to maintain strategic nuclear forces unhappily has not.

Wars fought with conventional weapons have not been prevented by the existence of nuclear weapons, as the history of the world since 1945 shows all too clearly. Conventional wars will continue to be fought, their outcomes will continue to affect the vital interests of the United States, and the United States must be prepared to fight them. To prevail in war, or to prevent war, requires understanding and study as well as effort and sacrifice. As the military scholar B. H. Liddell Hart used to say, "If you wish for peace, understand war."[7]

Notes

1　Quoted in Robert E. Osgood and Robert W. Tucker, *Force, Order, and Justice* (Baltimore, Md: The Johns Hopkins Press, 1967), p. 207.

2　Erich Fromm, "The Case for Unilateral Disarmament", in *Arms Control, Disarmament, and National Security*, ed. Donald G. Brennan (New York: Braziller, 1961); Sidney Hook, H. Stuart Hughes, Hans J. Morgenthau and C.P. Snow, "Western Values and Total War", *Commentary*, 32 (October 1961).

3　"The Third Draft of the Proposed National Pastoral Letter of the U.S. Bishops on War and Peace: 'The Challenge of Peace: God's Promise and Our Response'", *Origins*, 12 (April 14, 1983), p. 711.

4　Jonathan Schell, *The Fate of the Earth* (New York: Avon Books, 1982), p. 172. Schell does not advocate unilateral disarmament. He proposes multilateral nuclear disarmament, i.e. that all countries discard their nuclear weapons. How he would choose, between unilateral disarmament and the balance of terror, if multilateral disarmament were to prove unattainable, is not clear from his writings.

5 Leonard Bernstein, "Just Suppose We Disarm", *New York Times*, June 10, 1980, p. A19.
6 "George Kennan Calls on the West to End Sanctions Against Soviet", *New York Times*, October 11, 1982, p. A10.
7 Liddell Hart, *The British Way in Warfare* (London: Faber and Faber, 1932), p. 8.

3
The Study of War

This author is a defense professional, but he is also a husband, father, and taxpayer. In short, his vested interest in Soviet–American conflict is as nothing compared with his interest in East–West harmony.
Colin S. Gray, *Strategic Studies: A Critical Assessment*

Taking War Seriously

One way to find out how people feel about war is to announce in casual conversation that one is writing a book about the subject. It usually stops the conversation. War is a disconcerting topic. Very large numbers of people get killed in wars. As John Keegan notes, with characteristic English understatement, "Killing people, *qua* killing and *qua* people, is not an activity which seems to carry widespread approval."[1] War is not just mass slaughter. It is also an extremely important instrument of national policy and sometimes the only means by which a nation can survive. It has shaped the entire course of history. It has been the arena in which have been displayed some of the noblest virtues: heroism, endurance, loyalty to a cause and loyalty to comrades. And it has inspired great art and great literature, from *The Iliad* to Picasso's *Guernica*. For these and other reasons most people see war as more than just a means of killing people, and regard a professional soldier as different from a professional executioner. However, it is inescapable that killing people is a major part of war.

Just as war itself has its repulsive aspects, so does the study of war. Nuclear war is the most gruesome topic of study imaginable; there is, as the journalist Thomas Powers says, "a period of intense melancholy, common to people when they first begin to think hard about nuclear weapons."[2] Certain aspects of research into conventional warfare also require a strong stomach. For example, one of the important subspecialties in the design of military small arms is "wound ballistics", i.e. the behaviour of bullets and other projectiles after hitting a human body. Students of wound ballistics have determined that a small bullet, such as that fired by the M-16 rifle, actually does more damage and hence is more desirable than a large bullet. Whereas the large bullet goes straight through its victim, the small bullet flips end over end inside the body, gouging the flesh and leaving a large exit hole.

It is possible to question the motives and moral sensitivity of people who devote their lives to such studies. The motives and moral sensitivity of so-called "defense professionals" have been questioned, particularly when they discuss nuclear war for any reason other than to condemn it and find ways to prevent it. In 1960 Herman Kahn published a book, *On Thermonuclear War*, in which he undertook to describe the course and outcome of a nuclear total war and point out measures which the United States could take to survive it. Kahn's book aroused a storm of condemnation. One reviewer described it as "a moral tract on mass murder: how to plan it, how to commit it, how to get away with it, how to justify it." Another wrote of Kahn, "To be blunt, his book makes me ashamed that we are fellow countrymen."[3] Some of the critics concentrated on the very considerable weaknesses in Kahn's argument, but others were simply outraged that anybody should discuss how to survive a nuclear war at all, particularly in the calm, clinical manner which Kahn adopted. While few subsequent books about nuclear strategy have aroused such passion as *On Thermonuclear War*, the topic remains disreputable in some circles.

If some people choose to know nothing more about war than that it is a horrible slaughter, others contrive almost to forget that fact. The latter is a common failing among professional students of war and of national security policy. The literature is full of works on national security policy which fail to suggest that anybody might be killed in pursuit of it, and equally of military histories written in the flat, laconic, emotionless jargon affected by professional military officers. War has been studied and described as if it were a business, a poker game, or an athletic contest. In some respects, it does resemble each of these. Like a business, it consumes certain inputs (men, munitions, and supplies) and produces certain outputs (enemy casualties and territory captured). Success in war, like success in business, depends partly on securing the maximum output from the minimum input. Like a poker game, war is a contest of intellect, nerve, and experience, the outcome of which depends heavily on chance. Like an athletic contest, war is a test of the strength, skill, endurance, and strategies of two opposing sides. Lessons learned from these other fields of endeavour can aid in understanding war and conducting it effectively. However, they obscure more than they illuminate if they are used without remembering what is unique and absolutely central to the nature of war: people get killed in wars. The constant presence of violent death affects every other aspect of war and its conduct.

To understand war one must approach it in the spirit which von Clausewitz advocates in the quotation that opens Part I of this volume. "Taking war seriously" means to acknowledge war's cruelty and destructiveness, but also to acknowledge the tragic necessity for a country to prepare for war and sometimes to fight if it is to retain its sovereignty. Von Clausewitz's awareness of both the tragedy and the necessity of war permeates *On War*, and is one of the reasons why, 150 years after it was first published, it remains the most important book on the subject ever written. It might seem that nuclear weapons have made both war and von Clausewitz's book obsolete. However, as long as the balance of terror prevails, to prevent nuclear war one must prepare to fight it. The way to prevent an enemy nuclear attack is to be able to launch a retaliatory strike if the enemy does attack. As von Clausewitz would put it, the balance of terror means precisely that we cannot afford "blunting our [nuclear] swords in the name of humanity. Sooner or later someone will come along with a sharp sword and hack off our arms." To prepare for war we must understand it and understanding begins with taking war seriously.

The Study of War in the United States

World War II changed the way Americans study war. It brought into the war effort scholars in many fields whose contributions proved valuable not only in the design of new weapons, but also in the development of new tactics. After the war this collaboration between the military and civilian scholars continued. Military men had learned to value the insights which civilian scholars could provide and some scholars were willing to work on national security problems.

Most of those civilian scholars had been trained in mathematics and the physical sciences or in the more mathematically-inclined branches of the social sciences. They were physicists like Herman Kahn, mathematicians like Albert Wohlstetter, or economists like Thomas Schelling. Few of them had studied military history, or any other aspect of history, or any of the traditional liberal arts in any depth. Few had seen military service of any kind, let alone combat. Bernard Brodie, the first and probably the most distinguished of the civilian students of war, was something of an exception; he majored in philosophy as an undergraduate and earned his doctorate in international relations. Brodie, who made a name for himself during World War II as an authority on naval warfare, was also the only one among the new group of scholars who had made a serious study of war before the Atomic Age.

There were sound reasons why people with this sort of background should play such an important role in the study of war at that time. The scientists had created the atomic bomb and in so doing changed the nature of warfare radically. It seemed reasonable to turn to them for counsel on the implications of their creation. Quantitative techniques had borne fruit in many other areas of military endeavor. Nor did ignorance of history and lack of military experience seem to be decisive handicaps to understanding wasr. People generally believed that nuclear weapons had so changed the nature of war that most of what had been learned about it before 1945 had become obsolete. If everything would have to be thought through again from the beginning, this could best be done by those who could approach the subject afresh, without the excess baggage of obsolete knowledge or experience. As a high-ranking civilian official in the Pentagon trained in economics and systems analysis said to a general in rejecting the latter's views on nuclear strategy, "General, I have fought just as many nuclear wars as you have."[4]

The quantitatively-oriented scholars who took up the study of war evolved a number of techniques for the rational analysis of their subject. One of these is "systems analysis," a means of choosing the most effective from among several different types of weapons. The question asked and answered in a typical exercise in systems analysis is: "How much damage could a particular weapon do to the enemy for every dollar spent on that weapon?" In other words, what is its "cost-effectiveness"? The best, the most cost-effective, weapon can then be seen to be the one that inflicts the most damage to the enemy, or the one that carries out some other mission, for a given amount of money. Presume that the cost of keeping one missile submarine in service for a year is equal to that of keeping 13 strategic bombers in service for a year. If the submarine could be expected to destroy 12 enemy targets during a nuclear total war, and the bombers could destroy only eight targets, then the missile submarine is the more cost-effective weapon. It should be built instead of the bombers. Another systems analysis exercise might answer the question of how many weapons of a particular type a country needs to have. Suppose that United States defense planners have decided that missile submarines are the best type of strategic nuclear weapon for the United States and each missile submarine could be expected to destroy 12 targets in the Soviet Union. If the United States needs to be able to destroy 1,200 targets in the Soviet Union during a nuclear total war in order to prevent such a war, then the United States strategic nuclear force should consist of 100 missile submarines – no more and no less.

Systems analysis can be a powerful analytic and administrative tool. It can help the Secretary of Defense decide how many weapons and of which types to buy, in the face of military men demanding that as many weapons as possible be bought and of the types used by their particular branches of the service. Equally important, the Secretary can use systems analysis to defend his decisions. The first of the systems analysis exercises sketched out above would help to decide whether bombers or missile submarines should be built. It would help explain to the air force generals and their supporters in Congress (who will take some persuading) why missile submarines are to be built instead of bombers. The second exercise could be used to explain to the admirals in the navy why fewer missile submarines than they want are needed. Robert McNamara, United States Secretary of Defense from 1961 to 1967, was very much taken with systems analysis for just these reasons and ordered it used extensively in United States defense planning.

Wargaming and game theory are two more techniques for studying war. Game theory derives the general principles of conflict, competition, and bargaining from thinking about them in a very abstract way. Rather than discussing an actual war, or even a hypothetical war, it discusses "A" and "B", two opponents placed in a hypothetical situation in which each can attain his best result only at the expense of the other. It assumes that both "A" and "B," unlike real decision-makers, are completely rational and attempts to determine what each would do in that situation. (The "prisoner's dilemma," which will be discussed in chapter 4, is a classic example of a game theory scenario.)

The most important insight offered by game theory is the "minimax principle": assume that your opponent will make what is the best possible move from his point of view, which means that your move should be whatever is the most effective response to his best move. To take the homely example of a ticktacktoe game:

```
        b │     │
   O      │     │
 ───────┼─────┼───────
        │     │  a
   ×    │  ×  │
 ───────┼─────┼───────
        │     │
        │     │
```

It is "O's" move.

Obviously "×'s" best move on the next turn is to put an "×" in square "a" and win. Therefore, "O" must assume that that is what "×" will do unless thwarted, and in the light of that "O's" best move is to put an "O" there first. This would seem obvious to an eight-year-old but in real life military planners in "O's" position surprisingly often put their "O" in square "b." That would be the best move if "×" did not have to be taken into account.

Wargaming attempts to simulate closely a possible war or battle, with the purpose of testing various proposed strategies and thus finding the best one for each side. A typical war game is played between two people, or two groups of people, each taking the role of the leadership of one side or the other. One player might take the role of President Reagan and another the role of Secretary-General Gorbachev; in another game, one might take the role of the commander of an infantry battalion defending a position and the other the role of the commander of a tank battalion attacking it. Some war games have had more than 200 participants and have gone on for several weeks and others have had generals and ambassadors participating in them. The board games sold by Avalon Hill and Victory Games are also war games, often as sophisticated as those used by military planners and researchers. Chess and Go can be considered war games, although very abstract and "unrealistic" ones, and perhaps the first to be invented.

Wargaming can be a very valuable tool for studying war, for learning what might happen in a future war and how various moves might be countered by either side, and for training decision-makers. It is particularly valuable as a means of impelling military planners to take the enemy into account and consider what moves and counter moves he might make. Playing the role of Gorbachev, or the commander of a Soviet tank battalion, forces the participant in a war game not only to look for the flaws in United States strategy and tactics, but also to contemplate what an intelligent, enterprising enemy might do to counter the planner's strategy, a factor which military planners do not always take into account. As Bernard Brodie writes about conventional planning studies, "The irrepressible tendency is to regard the enemy in the body of the study as rather dim-witted and passive, however respectful may have been the statements about him in the preface."[5] So, while game theory yields the minimax principle, wargaming impels military planners actually to observe it in their planning.

The new techniques for studying war and managing the military establishment reached the height of their influence in the United States during the early

1960s, when Robert McNamara was Secretary of Defense. Even then, however, they never played the role their practitioners promised they could and their detractors feared they might. Most national security policy decisions reflected Kennedy administration politics and Pentagon institutional interests as much as they did systems analysis logic. Since the McNamara years the new techniques and their practitioners have become less influential. These techniques have their uses, but they also have their defects.

One limitation common to systems analysis and wargaming is their dependence on the data used in them. As computer programmers say: garbage in, garbage out. If one does not know how much it costs to operate a missile submarine or a bomber for a year, and how many enemy targets each would destroy in a war, then one's systems analysis study of the cost-effectiveness of missile submarines and bombers will be faulty or cannot be made at all. Sometimes the analyst has available precise and fairly undisputable figures for what he needs to know, but too often he is reduced to quantifying what cannot be quantified with any precision.

How men behave in combat, and how different eventualities affect their behavior, are vital to understanding almost any aspect of war, but they are not readily quantifiable. One can know with some precision how many bullets a machine gun can fire in a minute but one cannot know how likely it is that the machine gunner will stay at his post and keep firing under enemy shellfire. There have been some studies of how likely soldiers are to fire their weapons in combat and the results are rather startling. In *Men Against Fire* General S.L.A. Marshall summarizes the results of a systematic study of the behavior in combat of United States infantrymen during World War II. He concludes that only about *15 percent* of the infantrymen participating in a battle would ever fire their weapons during that battle. (The systems analyst cannot, however, use this 15 percent figure in his calculations. Sometimes the rate of fire was as high as 25 percent, at other times much lower than 15 percent.) The gap between what fighting men are supposed to do in an engagement and what they actually do is commonly rather large, large enough to invalidate many quantitatively-oriented studies. To cite one example, an elaborate study of bomber versus fighter duels, done by American systems analysts in 1949, concluded that the fighter pilot should be able to destroy the bomber 60 percent of the time. Combat data from World War II show that in similar situations the fighter pilot actually did shoot down the bomber only 2 percent of the time. The reason for this tremendous gap between theoretical and actual performance was that most fighter pilots were reluctant to get close enough to a heavily-armed enemy bomber to shoot it down.

The fundamental difficulty facing any quantitative approach to the study of war is what General von Clausewitz called "friction." In war nothing ever happens quite the way it is supposed to. Battalions arrive at the front not when they were ordered to but minutes or hours later; they bring not their full strength but a portion of it; when they are ordered to attack some of the soldiers advance, some drop out and hide, and some never get the word; the enemy whom they are to attack is not where the high command thought he was but somewhere else, etc., etc. There is friction in the workings of any large organization – in a business, a university, a government agency, or a church – but more so in an army in combat than in anything else. This is because of all the ways in which war is different from other things that people do. First of all, as von Clausewitz keeps pointing out, war is very dangerous; people get killed in wars. Most soldiers fight bravely and carry out their orders as well as they can, but to few is it given to be not affected at all by the fear of death. Secondly, there is the sheer physical stress and effort of fighting a war: soldiers march 20 or 30 miles a day and live for months on end in debilitating conditions; generals make vital decisions having had four hours of sleep in two days; an entire armored division has to advance along a road which in peacetime accommodates a dozen cars a day, etc. Thirdly, commanding officers have to bear not only this physical stress, but also a paralyzing weight of responsibility. It was said of the commander of the British fleet during World War I that he could lose the war in an afternoon. Even before the Nuclear Age a single bad decision could lose a battle, and a single lost battle seal the fate of a nation. Today, a single bad decision could seal the fate of the human race. It is hard even for a very tough individual, as successful generals tend to be, to think clearly under such pressure. Finally, there is what von Clausewitz termed "the fog of war." In a battle nobody, from the commander-in-chief down to the newest recruit, ever quite knows what is going on. This fog of war envelops the battlefield partly because of the stress imposed by combat and the speed with which events can develop, but even more because war is waged against an enemy, and that enemy takes deliberate measures to conceal his strength, his intentions, and anything else of interest about himself; he attempts to act in ways that one won't expect. For these reasons, war on the battlefield is quite different from a war on paper or a computer simulation.

If men and machines performed at 100 percent efficiency in combat, systems analysis would be much more reliable than it is. However, it does not deal well with friction; not because its practitioners are not aware of the role friction plays in war, but because friction's effects are almost impossible to quantify. The systems analyst can calculate the firepower of an infantry company fighting at 100 percent efficiency (everybody fires his weapon and uses it properly). He may be able to calculate its firepower at 15 percent efficiency (15 percent of the soldiers fire). What he cannot do, at least before the fighting starts, is know how many of the soldiers will fire their weapons in combat and therefore what the company's level of efficiency will be. Only if there is a war going on when he does his analysis might he be able to get some data on how well various weapons are employed in combat in that war. World War II operations researchers were able to do this, but their present-day counterparts are dealing mostly with weapons that have never been used in combat and are trying to predict how they might perform in a war that may never be fought.

Wargaming and game theory suffer from similar weaknesses in dealing with friction and the human element in war. They can determine what a perfectly rational opponent would do and what would be the most dangerous move he could make, but they cannot determine what an actual opponent would do in a real war. Yes, a United States military officer, government official or student of war can play the part of Mikhail Gorbachev in a war game; and by doing so he may gain some insight into how Gorbachev sees the world. This is not to be despised. But really to discover what Gorbachev himself would do in a war or in a war-threatening crisis, we would have to have the role played by Gorbachev himself. We would need at least to have the role played by somebody who is a Marxist–Leninist and a Russian, who had worked his way up to Politburo rank through the Party hierarchy, who had lived through World War II in the Soviet Union, etc., etc. We would also have to put our role-player under the extreme stress felt by a national leader facing a nuclear total war. Too much American thinking about nuclear war and national security policy ignores the tremendous differences in character, world view, and objectives between American and Soviet leaders, the differences between political leaders in any country and American defense professionals, and the differences between participating in a war game and fighting a nuclear war. In many studies one seems to be dealing, not with actual leaders, but with "strategic man," an oddly affectless, totally rational being who conducts war as if it were a war game: a game in which today he takes Reagan's role, tomorrow he will take Gorbachev's, and only pieces of paper die.

The quantitative approaches do deal effectively with some aspects of warfare and do contribute to the study of war. They are deficient only in that there are other aspects of warfare which they deal with poorly or not at all. They need to be supplemented, not abandoned. Systems analysis can tell a planner which of two types of ICBM to procure and, with somewhat less confidence, whether to procure an ICBM or a strategic bomber or how many strategic nuclear weapons to procure. A war game can give those who participate in it a feel for how the world looks through an opponent's eyes and for what might happen in a war. Game theory helps to illuminate the nature of war as one among many forms of conflict. However, there are two vital aspects of war which the quantitative methods do not deal with and which need to be understood: the nature of the opponent, and the way that fighting men behave on the battlefield.

It seems obvious that to protect oneself against a threat one must first understand what the threat is. An effective defense of the United States against the Soviet Union requires the accurate assessment of the strength, equipment, and capabilities of the Soviet armed forces, and also of Soviet military strategy and political objectives. United States military planners, political leaders, and some members of the general public have studied Soviet military power attentively since the early days of the Cold War. They have observed carefully the Soviet Union's acquisition of the atomic bomb and then of the hydrogen bomb, the development of Soviet strategic nuclear forces, the growth of the Soviet navy, and similar trends. United States national security policy has been very largely a reaction to the strengthening, or in some cases the perceived strengthening, of various parts of the Soviet military machine. However, until the past decade or two, comparatively little attention was paid to Soviet military strategy.

When Raymond Garthoff and other scholars began in the 1950s the serious

y of Soviet military strategy, they soon found that Soviet views on nuclear
were quite different from those prevalent in the United States. Soviet
strategists were rather less hopeful that a nuclear war could be prevented and
rather more hopeful that the damage caused by it could be limited. The
Soviets thought that large land armies would play an important role even in
a nuclear total war, and they at least considered defeating an imminent
American nuclear attack by attacking the United States first. (Soviet nuclear
strategy is discussed at greater length in chapter 8.) This discovery raised
serious doubts about the United States's nuclear strategy, which was based
on the assumption that Soviet and American views on nuclear war more or
less corresponded. At first Soviet views on nuclear war were mostly ignored
by American nuclear strategists. As the evidence mounted and became harder
to ignore, a number of nuclear strategists ascribed the differences to the
"traditionalism" and intellectual backwardness of Soviet military leaders.
Presumably, once the Soviets had studied nuclear war as long and as rigorously
as the Americans had, they would come to hold much the same views as the
Americans. Arms control negotiations were seen as an opportunity to educate
the Soviets about the realities of the nuclear age. Earnest efforts were made
during the 1970s, at the SALT I and SALT II negotiations, to form a
Soviet–American consensus on nuclear strategy. They failed. It seemed that
Soviet military planners did not think they had anything useful to learn from
their American counterparts. In retrospect, considering the repeated failures
of United States national security policy during the past several decades, it is
hard to see why anybody ever expected the Soviets to take lessons from the
Americans.

The failure for so many years to take adequate account of Soviet views on
nuclear war reveals a serious defect in the quantitative approach to war. To
its practitioners – mathematicians, physical scientists, and economists – it was
natural to view nuclear strategy as akin to a problem in physics or a
mathematical equation whose solution is always the same in all countries. For
them the characteristics of nuclear weapons ineluctably dictated that there
would be one correct nuclear strategy: a strategy which intelligent, well-
informed people anywhere would inevitably identify and upon which they
would reach a consensus. A political scientist or a historian, especially one
who had made a serious study of some country other than the United States,
would probably be much more aware that what seems obvious to people in
one cultural setting is not at all obvious to those in another. That observer
would see a country's nuclear strategy as dictated primarily by the political
goals and world view of its leaders, which it is. Since the goals and perceptions
of American and of Soviet leaders are profoundly different, our observer
would expect their nuclear strategies to be different as well. Unfortunately,
most American nuclear strategists are not historians or political scientists, and
those who are seldom have studied any foreign country in depth.

The previous failure to take account of Soviet military strategy is being
remedied. Not only are there now a fair number of scholars studying this
subject, but also their conclusions are taken seriously in forming United States
national security policy. American understanding of Soviet military strategy is

far from complete and will inevitably remain so. Soviet writings on military strategy are quite opaque and may be deliberately misleading, at least on some issues. They are not designed for the enlightenment of those outside the Soviet military establishment. Also, one cannot really know what a government's military strategy is until that government implements it in fighting a war. As the intelligence experts say, you cannot photograph an intention the way you can a missile. Nevertheless, United States military planners now realize that they need to know as much about Soviet intentions and military strategy as they can find out.

In recent years there has been in the United States a strong surge of interest in war on the battlefield, in what happens during a battle, how soldiers react to it, and how they can be helped to endure the almost unendurable pressures battle imposes on them. The need to consider these things was brought home to many thoughtful military professionals by the Vietnam War. United States ground forces in Vietnam were seriously demoralized during the latter part of the war; officers were "fragged" (killed) by their men, soldiers went into combat high on drugs, some men deliberately refused to fight and large numbers of others deserted. The conclusion which many students of war reached was that one reason for all this was that the United States army had been run too much like a business and too little like an army.

Perhaps because of the great influence businessmen have traditionally enjoyed in the United States, the United States armed forces have long been more willing than those of other countries to use managerial techniques taken over from the business world. This was clearly evident in the way American forces fought World War II. Like systems analysis and wargaming, the managerial approach has its uses but it fails to encompass that crucial reality, that people get killed in wars. Nobody is ever called on to die for General Motors. Since soldiers are called on to risk their lives in combat, leading a military unit in combat is quite different from managing a business. The most important consideration is not to allocate men and resources as efficiently as possible, as it might be in running a firm. Instead, it is to sustain "the miracle of the spirit which takes thousands of young men, ties them together in strange self-forgetfulness, and enables them to walk steadfastly and without faltering into the certainty of pain and death."[6]

Among the reasons why the United States army lost the Vietnam War was that its personnel policies were better adapted to business management than to fighting a war. By far the most important of the factors which enable a soldier to endure the stress of combat is his loyalty to the dozen or so other soldiers in his unit and his respect for the officer who leads them. As the saying goes, "A soldier doesn't take a hill for his country. He takes it for his buddies." Loyalty to comrades and respect for a leader are best promoted by having the members of a unit serve together under the same leaders the entire time they are in combat. American personnel policy in Vietnam seemed almost as if designed to prevent that. Enlisted men served for a year, but the term of service of every man in a unit would end at a different time, so the personnel composition of the unit changed constantly, and officers served only for six months. This system may have been efficient and fair from a managerial point of view, but it was disastrous in its effects on combat effectiveness.

War and the Citizen

At the height of the Vietnam War *New York Times* columnist James Reston pointed out that there was something new about this war. For the first time in history television cameras accompanied United States soldiers on to the battlefield "recording daily for vast television audiences the most brutal and agonizing scenes of the struggle." On one hand, the average American had been told very little about Vietnam or why the United States was fighting a war there and had given no more than passive acquiescence to the war. On the other hand "the public [was] being invited to tune in on the eleven o'clock news and see Johnny killed."[7] Reston questioned whether it would be possible to sustain popular support for the war under such conditions. The course of events was to prove that it could not be sustained for long enough to win the war. The American people became well aware of the costs of the war, had no reason to believe it could be win quickly, and saw no compelling reason to fight it. They withdrew their support and the war was lost.

In the past civilians could and did remain quite ignorant of what a war was like as long as that war was being waged in another country. The long agony of World War I dragged itself out with the vast majority of the people back home in all of the countries involved blissfully ignorant of the horror being enacted in the trenches. This is still true in dictatorships; their leaders have told the people of the Soviet Union little of what their soldiers have endured in Afghanistan and nothing of what they have inflicted on the people of that unhappy land. It is no longer true in democracies. A democratic government can still fight a very short, small-scale war, as the United States did in Grenada, without the public finding out very much about the fighting until it is over. However, any war that a democracy fights for very long or on more than a small scale will be fought in front of the TV cameras. People back home will have some idea of the suffering it causes and their government will need to have a good explanation of why that suffering is necessary. This is true of the waging of conventional wars and it is equally true of the threat of a nuclear war. Most Americans now have at least some idea of how devastating a nuclear total war would be – enough to be thoroughly frightened by the possibility.

The problem that public awareness of the horrors of war poses for democracies is that it is hard to maintain public support for a war even when that war needs to be fought. Once the war starts, the costs of fighting it become very clear, while the costs of not fighting it, or of losing, even if equally great, are not so clear. The price of fighting the war has to be paid in the present, while the war is going on, and it has to be paid in the very tangible forms of lives lost, money spent, and perhaps a country devastated. The price of losing does not have to be paid until after the war is over, perhaps many years after. While the war is going on the price of losing can look rather vague and hypothetical, particularly to the people of a country that is fighting in defense of another country. For the United States fighting in defense of an ally, the price of losing a war is likely to be an adverse shift in the balance of power, a loss of credibility or prestige, perhaps the loss of access to vital raw materials. These are important considerations, American national security depends on them, but they don't touch anybody as directly as losing a son in combat. So, as Bernard Brodie puts it, "Those who talk abstractly of national prestige and national honour, and of other interests that inevitably differentiate the state from its people, find themselves matching their discourse with those who speak of dead bodies, burnt villages, My Lai massacres, and other ugly matters that are highly visible in the field, noted by reporters, and sometimes picked up on television screens."[8] The latter tend to win the argument, which is what happened in the United States during the Vietnam War. The American people acquiesced in a major war without really knowing why or foreseeing the full cost of it, and then withdrew their support when it became clear how long the war would be. The only antidote to this disastrous pattern is for the American people to be aware of the likely cost before the United States goes into a war and be prepared to pay it. If their government cannot convince them that fighting the war is necessasry, if the people of the United States do not give their informed consent before any war starts, then their government should not enter that war. This may embarrass the conduct of United States national security policy, but it is the only way a democracy can wage war in the Television Age.

The people should in fact be informed about and, through their elected representatives, control all important aspects of national security policy. The most important decisions a government can make lie in this area. A bad decision on economic or social welfare policy can waste money and wrong many people, but a bad decision on national security policy can destroy the country. No matter how openly and democratically a government adopts new taxes or manages its urban renewal program, if it makes its national security policy behind closed doors it is something less than a democracy. To be sure, many aspects of national security policy are very complex, very technical, and should be left to the experts. Often the data necessary to form an intelligent opinion are, for good reasons, classified information known only to few people. It would be folly to hold a national referendum about the proper design of a new ICBM or whether a particular target should be attacked by tactical aircraft or by naval gunfire. Many Congressmen should in fact be better informed than most of them are before attempting to decide such questions. But the basic choices between peace and war, or between strengthening the armed forces and disarmament, must be made by the people of the country, acting through their elected representatives. It is they who will pay the price if the wrong choice is made. As the slogan goes, "no annihilation without representation." It seems a reasonable enough demand.

Notes

1 John Keegan, *The Face of Battle* (New York: Vintage Books, 1977), p. 314.
2 Thomas Powers, *Thinking About the Next War* (New York: Alfred A. Knopf, 1982), p. 117.
3 James R. Newman, "Books: Two discussions of thermonuclear war", *Scientific American*, March 1961, p. 197; George C. Kirstein, "The Logic of No Return", *The Nation*, January 14, 1961, p. 35.
4 Quoted in Fred Kaplan, *The Wizards of Armageddon* (New York: Simon and Schuster, 1983), p. 254.
5 Bernard Brodie, *Strategy in the Missile Age* (Princeton, NJ: Princeton University Press, 1965), p. 246.
6 Bruce Catton, *Mr. Lincoln's Army* (New York: Pocket Books, 1964), p. 15.
7 James Reston, "The Press, the President and Foreign Policy", *Foreign Affairs*, 44 (July 1966), p. 554.
8 Bernard Brodie, *War and Politics* (New York: Macmillan, 1973), p. 8.

PART II
War and Politics

War is thus an act of force to compel our enemy to do our will.
                              Carl von Clausewitz, *On War*

4
The Condition of War

During the time men live without a common Power to keep them all in awe, they are in that condition which is called Warre; and such a warre, as is of every man, against every man ... For as the nature of Foule weather, lyeth not in a showre or two of rain; but in an inclination thereto many dayes together; So the nature of War, consisteth not in actual fighting; but in the known disposition thereto, during all the time there is no assurance to the contrary.

Thomas Hobbes, *Leviathan*

War and the Sovereign State

In the seventeenth century there was an elegant Latin phrase for the resort to war. It was said to be the "ultima ratio regum" (the final argument of kings), words which King Louis XIV of France had cast on the cannons in his army. Throughout history war has indeed been the final argument of kings, dictators, presidents, and peoples, of whoever has ruled a sovereign state and carried on its quarrels with other sovereign states. It has been the primary means by which each state has sought to impose its will on other states or to avoid having the will of others imposed on it. There are, of course, many other means a state can use to get its way in its dealings with other states – diplomacy, economic suasion, appeals to public opinion, etc. – but they are only decisive in the absence of war. If one state is powerful enough to attack another with whom it has a quarrel and is resolved to do so rather than yield the point at issue, then the victim of its aggression must fight. No degree of diplomatic, economic, or propaganda pressure the victim can exert will in itself save him from being conquered. These other means of struggle may help the victim fight more effectively, but they cannot save him from having to fight. Only if the resort to war is ruled out do other forms of force or persuasion become decisive.

War and the state go together: only a state or an entity that closely resembles a state can wage war. A corporation cannot, nor can a religion, a political party, an ethnic group, nor any other association unless it has control of a sovereign state. To wage war requires an army, and to form, equip, and maintain an army almost invariably requires the exclusive control over some inhabited territory which only a sovereign state has. Guerrilla war may seem an exception to this rule, but guerrillas rarely succeed unless they can create a quasi-state of their own, either on the territory of the state they seek to overthrow or adjacent to it. Without what Mao Tse-tung called "base areas," a guerrilla movement can build only a very weak guerrilla army and can prevail only against a very weak or irresolute opponent.

War and the state go together in another way; wars are fought to determine something about a state. They are fought to determine whether or not a particular state is to go on existing, or who is to govern it, or whether or not it is to control a particular territory, or what its policies are to be in some respect – whether, for example, it is to allow foreigners to carry on trade within its boundaries and on what terms. The American Civil War was fought over whether or not the Confederacy was to be a sovereign state, the English Civil War over whether the King or Parliament was to govern England, the conflict in the Falkland Islands was over whether Argentina or the United Kingdom was to possess those islands, the Opium War to determine China's foreign trade policies, and so on. Even wars of religion are wars about state power, about which state is to control a particular territory so as to propagate the religion which its leaders favour and suppress other religions.

Because war and the state are so much bound together, and each is so dependent on the other, they have evolved together over the millennia. Step by step, as the level of organization of human societies has risen from the nomadic band to the sovereign state, the level of organization of violence between human beings has risen from family feuds to global wars.

Human beings are not inherently warlike. They are born capable of violence, perhaps even predisposed to violent behavior, but not all violence is war. War is a particular type of *organized, collective* violence, that which communities employ against each other. Without some kind of community there cannot be any kind of war. Man in the state of nature can kill, but he cannot wage war.

The organization and level of development of a community determine the kind of war it can wage. Nomadic bands and village societies wage what Harry Turney-High terms "primitive war." Sovereign states wage "true war", which requires:

1 Tactics – at the very least, employment of the line of battle in combat and the column of march to approach the battlefield.
2 Command and control – one commander of an army, able to give orders and have them obeyed, and with junior commanders under his authority.
3 Being prepared to conduct a campaign, not just one battle, without which only very limited war aims are attainable.
4 A group rather than an individual war aim – true wars are fought for the interests of the community, not the benefit of the individual warriors.
5 Supply, the ability to maintain an army in the field long enough to conduct a campaign.

Warfare among the American Plains Indians was primitive war in its classic form. An Indian tribe rarely had much of a chain of command. Frequently the commander of an Indian war party was whoever could assemble one by persuading a number of other braves to join it. Nobody was compelled to join the war party and anybody who did participate could leave and go home if he had had an ill-omened dream or for whatever else he regarded as an adequate reason. Sometimes the commander himself did not join the war party; he was essentially a priest who kept the sacred war-bundle and stayed behind to pray for the expedition's success. Secondly, the Plains Indians fought for characteristically "primitive" war aims. They did not seek to conquer territory or to subjugate the enemy. They went to war so that individual warriors could assuage their griefs, gain prestige, and capture horses or slaves. War was an opportunity for individuals to earn coups, not an instrument of politics. Finally, Plains Indians war parties could not conduct real campaigns, partly because they did not take with them enough food for more than a few days.

In contrast, the Zulus of southern Africa, who were otherwise at the cultural and economic level of the Plains Indians, waged true war. In the early nineteenth century they had the most formidable military machine in Sub-Saharan Africa. Their armies employed well-conceived tactics, enjoyed a strong chain of command, and had on strength porters to supply the fighting men with food and beer. All Zulu warriors had to serve when summoned, they could not drop out after the campaign started, and any who showed cowardice in battle were put to death. The Zulus fought, not to assuage their griefs or to count coup, but to destroy their enemies. They waged war, not just primitive war.

The Zulus were able to wage true war because they lived under the authority of a rudimentary sovereign state, a state which itself had been built by making war. A series of able, ruthless Zulu leaders made themselves kings and their community a state by organizing powerful armies and conquering neighbouring clans. In other cases war did not play a major role in building the state, but once a state existed it was able to wage true war. However, whichever appears first, true war or the sovereign state, the other is sure to follow. War and the state make each other possible.

International Anarchy

Only a sovereign state can fight a true war, and it can fight it only with another sovereign state; war is a transaction between sovereign states. Even a civil war is a conflict between a state and another political entity which behaves like one, seeks to become one, and must come more and more to resemble one in order to succeed. It is possible to conduct some forms of collective violence against a primitive tribe which has not yet organized itself into a state – a slave-catching raid or a punitive expedition, perhaps – but not to wage war against it.

It follows that for there to be a war there must be at least two sovereign states to fight it. If the entire world were controlled by a single government strong enough to prevent any regional authority or guerrilla movement from challenging it, there would never be a war. Instead there would be worldwide peace and order of the kind which a state normally maintains within its borders. The domestic order within a state is based on its government's monopoly of the right to employ violence, a monopoly which allows it to compel all parties under its jurisdiction to resolve their disputes peacefully and to obey its laws. The absence of such order within a state is known as anarchy.

As there is no world government to control them, the 165 sovereign states of the world today live with each other in a state of international anarchy. All but a few of them maintain armies in one form or another, every one of them can go to war, and any of them may have to go to war in self-defense or for some other reason. Neither the United Nations nor any other international organization has ever been able to prevent states from waging war when they choose to do so. The international laws prescribing the rights and duties of states are observed most of the time, but every state must be prepared on occasion to defend its rights by force of arms. The sovereign state has established peace within its borders only to wage war outside them.

The English political theorist Thomas Hobbes wrote a gloomily realistic description of life in a state of anarchy. He pointed out that if there were no governments of any kind any man would be able to kill his neighbour and would know, therefore, that his neighbor could kill him. Each would be in competition with his neighbor for the available food and shelter and therefore would derive some benefit from killing that neighbor, just as the neighbor would from killing him. Therefore, every man would have to look upon every other man as a threat to his life, a threat which he could eliminate only by killing the other man first. Life in a state of anarchy would be a continual and universal battle, "a warre . . . of every man, against every man." There would be no cultivation of the earth because nobody could count on harvesting the crops he had planted; no industry because there would be no security for property; no science, scholarship, or art because nobody would have the leisure or the security to carry them on – nothing but "continuall feare, and danger of violent death." Human life would be, in Hobbes's famous phrase, "solitary, poore, nasty, brutish, and short."[1] He himself advocated a totalitarian dictatorship in preference to this state of affairs.

Hobbes's description of the anarchy in which men would live if there were no governments rather resembles the international anarchy in which sovereign states lives in the absence of a world government. International anarchy does not, of course, mean a perpetual "war of every state against every state." No country is at war all the time. Usually a government deals with the threat of being attacked by a neighbor by maintaining an army rather than by destroying the neighbor, and international conflict is moderated by treaties, alliances, and international law. International anarchy *does* mean, however, that countries frequently go to war with each other, that every country may at some time have to go to war, and that most countries must in time of peace devote much of their resources to preparation for a possible war.

Hobbes makes the important point that men living in anarchy act aggressively not necessarily because they are aggressive by nature, but because the condition in which they live compels each one to act aggressively to ensure his own survival. If "a generall inclination of all mankind [is] a perpetuall and restless desire of Power after power," it is not necessarily because every man wants unlimited power for its own sake. Rather it is "because he cannot assure the power and means to live well, which he hath present, without the acquisition of more."[2] So it is with governments. Often they engage in arms races, launch aggressive wars, and expand their territories not because they are aggressive and expansionist by nature, but because they wish to retain what they already have and feel they cannot do so without being more powerful. The trouble is that in becoming more powerful in order to ensure their own security, they threaten the security of their neighbours, who then have to build up their own strength. If no governments harbored aggressive intentions towards their neighbors, and all knew that none did and would not in the future, perhaps there would be no wars. The reality is, however, that some governments are very aggressive, no government can be certain of the intentions of its neighbors, and there have been more than 6,000 wars in the course of history.

The situation in which international anarchy places sovereign states is elegantly demonstrated by a little drama which students of game theory call the "Prisoners' Dilemma." Two men, "A" and "B", are accused of committing the same crime and held in jail awaiting trial, kept in separate cells so neither can talk to the other. The District Attorney has no evidence against either one except for the testimony of the other. Also, he does not care who actually committed the crime but merely wants to get a conviction; his objective is to persuade either prisoner to testify against the other. He meets with each one separately and offers him a deal: "If you testify that the other man did it, and he doesn't testify against you, then he will be convicted and get a ten year sentence and you will not only go free but get a reward for helping the prosecution. If you testify against him and he testifies against you, the judge will find that the two of you share the guilt and give each of you five years. If you remain silent and he testifies against you, then you get ten years in prison and he goes free and gets the reward. If you both remain silent, then you both go free." (Probably the District Attorney does not mention this last possibility, but both prisoners know it exists.) The possible outcomes of what each prisoner decides to do are shown in the matrix below:

	If Prisoner B accuses Prisoner A	If Prisoner B remains silent
If Prisoner A accuses Prisoner B	Both get five years	B gets ten years; A goes free, gets reward
If Prisoner A remains silent	A gets ten years; B goes free, gets reward	Both go free

The best move from the point of view of the two prisoners would be for both to remain silent. However, each of them is thinking only of his own interests and neither knows or can control what the other will do. So A

	Country B goes to war	Country B refuses to fight
Country A goes to war	WAR	Country A wins; Country B conquered
Country A refuses to fight	Country B wins; Country A conquered	PEACE

thinks, "Either B will accuse me or he won't. If he accuses me and I remain silent I will get a ten year sentence, whereas if I accuse him too I only get five years. If he remains silent and I do too I go free, but if he remains silent and I accuse him I get a reward as well as my freedom. So, whatever he does, my best move is to accuse him; I'll tell the judge B did it." B, who has thought through his options the same way A did and come to the same conclusions, will tell the judge A did it, and they both go to prison instead of going free. Each has acted rationally from his point of view, but the outcome is what neither of them wants, which shows that behaving rationally and selfishly will not keep either one out of prison. Unfortunately, neither will altruism. If A is thinking of what is best for B he will remain silent, but if B is also thinking about what is best for B he will accuse A; the altruistic A will go to prison for ten years while B goes free and gets a reward. Only if both of them are altruistic and remain silent will both go free.

The prisoners' dilemma is a good simulation of several of the dilemmas faced by sovereign states living in international anarchy. For example, the decision matrix of two states who find their vital interests to be in conflict looks like this:

	Country B builds more weapons	Country B doesn't build weapons
Country A builds more weapons	ARMS RACE	Country A attains military superiority
Country A doesn't build weapons	Country B attains military superiority	NO ARMS RACE

It would be best for both of them to keep the peace, but each consulting its own interests, will find them best served by going to war, so there will be a war. The same mechanism leads to arms races between potential enemies: In each case, the prisoners deciding whether or not to accuse each other, governments deciding whether or not to go to war or whether to engage in an arms race, each does what seems to be in its self-interest and gets what it does not want – imprisonment, war, or an arms race – because of the situation in which each is placed. The prisoners are trapped in a dilemma structured for them by the District Attorney and governments are trapped in the international anarchy in which they have lived throughout history.

The Prisoners' Dilemma and the condition of international anarchy resemble each other in another respect; in both situations the results of individual altruism are likely to be disastrous. The prisoner who does what is best for the other prisoner while the other prisoner does what is best for himself ends up serving ten years in prison. A country that refuses to resort to war in a dispute with another country, while the other country does go to war, loses whatever is at stake in the dispute, which may be its existence. War is a costly and usually unprofitable adventure, which is why countries are reluctant to resort to it, but it is costly only when both parties to a dispute are prepared to fight. War against a country that refuses to defend itself is neither costly nor dangerous. There is much to be gained from it, little at risk, and hardly any reason not to wage it.

Perhaps the conflicts that lead to war could be settled by some less brutal, less destructive means. In *All Quiet on the Western Front* a German soldier suggests one alternative:

> Kropp ... proposes that a declaration of war should be a kind of popular festival with entrance-tickets and bands, like a bull-fight. Then in the arena the ministers and generals of the two countries, dressed in bathing-drawers and armed with clubs, can have it out among themselvs. Whoever survives, his country wins.[3]

It is rather an attractive idea and no doubt has occurred to many soldiers in many wars. Some wars have been settled this way, by each side choosing a champion to fight for it, with victory going to the side whose champion wins. The duel between David and Goliath was a famous example of champion combat and there are other examples in the histories of Ancient Greece, Ancient Rome, and Medieval Europe, and in the Hindu epics. Legal arbitration is another means of resolving a conflict between two countries, one which somewhat resembles champion combat, with each side hiring a lawyer to argue for it rather than a champion to fight for it. Both of these alternatives are much less costly than war. Their great weakness is the difficulty of making the loser accept the result. In the absence of any international authority strong enough to compel him to accept defeat, the loser can always go to war, probably with a good chance of winning; there is no correlation between the strength of a country's champion or of its legal case and the strength of its army. Champion combat or legal arbitration may settle an issue that is not of great importance to the contestants, but not one that either or both regard sides as important enough to fight a war about.

Why Countries Go to War

Every war is fought for at least two reasons. It takes two sides to fight a war and each must have a reason, or a number of them, for going to war. Even when an aggressor sends its army across the border of another country, that aggression will not lead to war unless the victim fights back. When the Soviet Union invaded Czechoslovakia in 1968 or Estonia, Latvia, and Lithuania in 1940, there was no war because the victims did not resist.

The most common reason why a country goes to war is to protect its national security. This is most clearly and justifiably the case when a country that has been invaded fights to repel the invader. Any country will defend itself against an invader unless its cause is absolutely hopeless, and often even when it is. Almost anybody will concede it the right to do so.

Going to war with an invader is only questionable when the invader seizes only a small part of the victim's territory and makes it clear that he intends to take no more. Even then nobody doubts the victim's right to resist, only whether what is at stake is worth fighting a war over. This was the question raised by the United Kingdom's decision to go to war with Argentina in 1982 in order to take back the Falkland Islands which Argentina had just seized. In the words of Lieutenant David Tinker, a young officer in the British task force that retook the islands, "Here we are ... 28,000 men going to fight over a fairly dreadful piece of land inhabited by 1,800 people."[4] The islands had belonged to the United Kingdom for nearly 150 years and their inhabitants strongly preferred to remain under British rule. On the other hand, there were only 1,800 Falkland Islanders, the territory has little economic value, and it cost nearly 1,000 British and Argentine lives to enable those 1,800 people to continue living under the government of their choice. Had the several billions of dollars spent on the war been given to the inhabitants of the Falkland Islands as compensation for their land, it would have made them all millionaires.

But even in this extreme case resistance to aggression was justified. It may have averted another and larger war elsewhere in the world. There are several other such remnants of the British Empire which are coveted, against the will of their inhabitants, by other countries – Gibraltar comes to mind – and the British willingness to fight for the Falklands may well deter a future attack on one of them. The war was also for the British a matter of that elusive but very real consideration, national honor. One of the promises a nation makes to its members is that it will come to the defense of any of them who are attacked, even if they are only a few and the cost of defending them is great. When a nation is no longer willing to fulfill that promise it begins to cease being a nation. If the British government had abandoned 1,800 loyal Britons in their hour of need, the meaning of being a member of the British nation would have been diminished for all of its members. Another of the servicemen in the British task force put it this way: "We want to sort this out so that our children can walk about in the world with their heads held high."[5]

If all that were ever necessary to ensure a country's national security were to fight when the country was invaded, there would be fewer wars and no doubt as to whom to blame for a war. The guilty party would always be whoever struck the first blow. However, simply repelling any invasion may not be enough to ensure a country's security; it is possible for one country to pose a serious threat to another without actually invading it. This is one of the reasons why the rights and wrongs of international relations are not as clear as international law tries to make them. Often the country that strikes the first blow does so in order to survive and the victim of its attack has been threatening the aggressor's survival. (Or so it seems to be to the leaders of the country which launches the attack.) On the other hand, whatever the victim of the attack is doing usually is not a violation of international law, while attacking another country certainly is.

A country can menace another country by encouraging the overthrow of the latter's government. It may direct propaganda appeals to the people of the other country; it may provide sanctuary, funding, or arms for a subversive movement within the other country; and it may even control and direct that subversive movement. It may also be, or be seen as, a threat to the government of the other country simply by being such an attractive alternative that its very existence undermines the loyalty of the people of that country to their government. The threatened government may decide that the only way to ensure its national security is to invade and destroy the source of the subversion. This was one of the reasons why Iraq attacked Iran in 1980 – the Iraqi government feared Khomeini's propaganda appeal to the Shi'ite Moslem minority in Iraq. This was also the main reason why Austria–Hungary attacked Serbia in 1914. Serbia had sheltered those plotting to undermine Austrian rule, including the assassins of Archduke Franz Ferdinand, and simply by being a Slav-ruled sovereign state inspired unrest in the Austro-Hungarian empire. The Austro-Hungarian government, blaming its domestic unrest on Serbia, made demands on Serbia which it expected to have rejected. When they were, Austria–Hungary attacked Serbia, thus beginning the conflict that developed into World War I.

Sometimes one country attacks another country because it fears that the other will attack it unless forestalled. In some cases launching such a preemptive attack, rather than waiting for the enemy to mobilize all his forces for an attack, seems the only way to ensure national survival. On the other hand, if every country has the right to eliminate by force of arms any threat to its existence, no matter how remote, then every country has the right to conquer the world given that the existence of any other country is to some extent a threat to its security. To determine whether a pre-emptive attack is necessary, and hence justified, one must ask how great is the threat which is to be averted by striking first. How likely is it that the enemy will attack unless forestalled? How likely is it that he will win the war if he does strike first?

When the threat is both imminent and extreme it is hard to blame the government of a threatened country for striking first. Just such a threat confronted Israel in June of 1967. Egypt, Syria, and Jordan, proclaiming their intention to destroy Israel, had mobilized their armies, moved them to Israel's borders, and had sent packing the UN forces which had been stationed along those borders to prevent another Arab–Israeli war. The Israeli army was also mobilized, but it could not stay mobilized for more than a few weeks without the Israeli economy collapsing. All the three Arab countries had to do was wait several weeks until Israel was forced to demobilize most of its army, and then attack with an overwhelming numerical advantage. Under these circumstances, the Israeli government felt the only way to ensure the survival of Israel was to launch a preemptive attack on the three Arab countries, which it did. While one can see the justification for the Israelis doing what they did, there will always be some doubt as to whether the Arabs *would* eventually have attacked, and thus whether the Israeli attack was necessary.

Sometimes countries have gone to war to prevent another becoming too powerful, even though that country was probably not about to attack them immediately. They feared that if the other country became too powerful it would attack them eventually and they would be unable to defend themselves; their only real defense therefore was to prevent it from becoming more powerful. In other words, countries have gone to war in such situations to preserve a balance of power between themselves and a potentially overwhelming enemy. Many wars have been fought to preserve a balance of power. Thucydides, the world's first great student of the causes of wars, pointed out that the Peloponnesian War was caused primarily by "the growth of the power of Athens, and the alarm which this inspired in Lacedaemon [Sparta]."[6] Similarly, the early eighteenth century War of the Spanish Succession was caused by the growth of the power of France and the alarm which this inspired elsewhere in Europe. Britain, Austria, and Holland went to war to prevent France from gaining control of Spain, because they feared that, with Spain under her control, France would be powerful enough to dominate Continental Europe. Many other European wars have been fought for a similar reason. Repeatedly over the past five centuries a European country has threatened to become powerful enough to exert supremacy over the continent, but when it sought to increase its power, it was contained or destroyed by the united efforts of the countries which it had threatened to dominate. The empire of Charles V, Spain under Philip II, France under Louis XIV and then under Napoleon, Germany under Wilhelm II and again under Hitler – each has in turn threatened to dominate the continent and has been contained or destroyed by its neighbors.

The necessity and justifiability of a war to maintain the balance of power depends very largely on the nature and intentions of the country whose growing power threatens the balance. If that country has given real proof of its aggressive intentions, its neighbors may well be justified in going to war to curb its power. On the other hand, for a country simply to be or to threaten to become powerful does not justify other countries in waging war on it. The countries of Latin America would not have been justified in invading the United States in the early nineteenth century and dividing it up among themselves to prevent it from becoming too powerful (although there may be Latin Americans who regret that this was not done). It is, though, both inevitable and understandable that the neighbors of a powerful country regard it with apprehension and mistrust and take measures to protect themselves against it, regardless of the policies the powerful country follows. A family of mice which shares its living quarters with an elephant inevitably feels apprehensive and mistrustful no matter how benevolent are the intentions of the elephant.

Wars of self-defense, wars to forestall a threatened attack, and wars to maintain a balance of power are all made necessary by the condition of international anarchy which requires that sometimes a country must fight to preserve its freedom. Other wars are made possible by international anarchy, by the freedom to go to war which every country has.

Sometimes a country goes to war to recover an *irredenta*, a territory which is ruled by another but which those who wage the war feel should be part of their country because it had been so once before and was taken away unjustly, because its people are related to them by blood and language, or because it is adjacent to their country. Argentina went to war for the Falkland Islands to take back a territory which had belonged to Argentina before 1833 and had been taken from her by force. Part of the reason why North Korea invaded South Korea and why North Vietnam went to war against South Vietnam was to reunite the sundered halves of countries that had been split up. The country that goes to war to recover an irredenta can usually justify its actions as a remedy for past injustice. Most national borders are the result of a war and every country in the world today, with the possible exception of Iceland, occupies a territory which its inhabitants took from somebody else. Most invaders can argue that the territory they are invading once belonged to them or ought to belong to them and can portray their invasion not as an act of aggression but as the remedy for a previous act of aggression. There is usually some justification for a war to reclaim an irredenta, and sometimes considerable justification. On the other hand, if every country which has ever had territory taken from it by force were to go to war to get it back, the world would never be at peace.

In the past a country could greatly increase its economic and political power by seizing more territory and many countries went to war for this reason. By the middle of the twentieth century this was becoming a much weaker reason for fighting wars as industry replaced agriculture as the primary source of wealth, thus causing the economic importance of land to lessen. West Germany and Japan since World War II have shown that it is possible for a country to lose much of its territory without losing its economic strength, in fact to become stronger than before. In the late twentieth century, however, going to war to seize territory just might make sense again. Certain raw materials, particularly oil, have become extremely important economic assets, possession of which may sometimes be considered a vital national interest. The struggle for raw materials may be a major cause of war in the twenty-first century. Even today it seems to some respected authorities justifiable to invade another country to secure access to a vital raw material. Robert W. Tucker, for example, has argued that if the Arab oil-producing countries were in the future to impose an oil embargo on the United States, the United States would be justified in invading them to seize their oil fields.[7] The primary objective of the Soviet invasion of Afghanistan may have been to move the

Soviet armed forces within striking distance of the Persian Gulf and its oil fields. Soviet control of the Persian Gulf's oil would both avert a possible future Soviet oil shortage and deny a vital natural resource to the United States and its allies.

A few countries have made war because they wished to conquer the entire world, or as much of it as they could. Once in a while there comes a national leader who seeks to conquer everything he can; one who will make no permanent peace with his neighbors – an Alexander the Great or a Genghis Khan, a Napoleon or a Hitler. Their careers of conquest have been rare, brilliant, and appalling episodes in history. Napoleon and Alexander the Great fought for glory and Genghis Khan for booty, neither perhaps a motive likely to inspire future wars of conquest. But Hitler fought to impose on the world his ideology and his monstrous vision of the future, and this *is* all too likely a motive for future attempts to conquer the world. The modern world is full of ideologies, some of them the ruling creeds of governments which are quite prepared to impose them on others by force of arms. Another distinctly modern motive for seeking to rule the world is paranoia, a regime's conviction that it is not safe as long as there are in the world sovereign states which it does not control. If the leaders of the Soviet Union today continually seek to increase the power of their country at the expense of the security and freedom of every other country, it is paranoia far more than the dream of a world Communist state that impels them.

Any regime which has set itself the objective of conquering the entire world may reasonably be described as inherently aggressive. Not only will it attack any of its neighbors when it can do so safely, but also its aggressions are dictated by the very nature of the regime itself, not by anything others have done to it or any legitimate grievances it may have. When its legitimate grievances are satisfied, it invents others. The more powerful and the more secure it becomes, the more expansively it defines its security needs. Every conquest it makes leads to an attempt to conquer more. Its neighbors cannot appease it except by ceasing to exist; all they can do is be as strong as possible, be prepared to fight, and perhaps destroy the aggressor nation before it destroys them.

One of the great dilemmas of statecraft is that it is very hard to distinguish in time between an inherently aggressive, permanently expansionist regime and one that is simply ruthless and violent in the pursuit of limited objectives. A country can commit unjustified aggression against another, as Argentina did against the United Kingdom in 1982 or as Syria did against Israel in 1948 and 1973, without being inherently aggressive. Nobody believes that if Argentina had been peacefully ceded the Falkland Islands she would have gone on to attack other British possessions or that if Syria had conquered Israel she would have gone to to attack Turkey. Usually, in fact, an aggressor has only limited objectives. He may wish to seize a particular territory which he, but nobody else, regards as rightfully his but that does not mean he wants to conquer the world. An aggressor with limited objectives can be appeased, perhaps at some cost in justice and national honor, by ceding to him what he demands. This is not to say that he should be appeased – that depends on the circumstances of the particular case – but it can be done. An inherently aggressive regime cannot be appeased, as the French and British governments found to their sorrow in dealing with Hitler. Every concession to an inherently aggressive regime strengthens it, weakens its victims, and whets its appetite for more. One of the most important and difficult tasks for those who determine national security policy is to tell the difference between a ruthless, assertive, unpleasant regime with limited aims and an inherently aggressive, expansionist regime with unlimited aims. The difficulty arises because early in his career of conquest the would-be world conqueror always claims to be, and often seems to be, seeking only limited objectives, as Hitler always claimed he was. One can only know that his objectives are unlimited when he is still asking for more after his initial set of demands has been met. By then it may be too late to stop him.

Wars are caused by two factors: the condition of international anarchy in which sovereign states live and the nature and policies of those states, their willingness to resort to war to achieve their objectives. A few states are inherently aggressive, many will aggress for certain limited ends, and almost all will go to war in self-defense. If all states were inherently peaceful there would never be a war, but also if there were only one sovereign state in the world there would never be a war. International anarchy helps to make states warlike because as long as it prevails the peace-loving are either conquered by their warlike neighbors or become warlike themselves, at least warlike enough to defend themselves. Thus international anarchy permits the warlike states to decide how much war there will be. If a state is prepared to go to war with its neighbor, the peaceful intentions of its neighbor do not count very much; either the neighbor rises to the aggressor's level of violence and fights or it must give up whatever is at stake in the conflict – which may be its national existence.

Notes

1 Thomas Hobbes, *Leviathan* (New York: E.P. Dutton, 1950), pp. 103–4.
2 Ibid., pp. 79–80
3 Erich Maria Remarque, *All Quiet on the Western Front* (New York: Fawcett World Library, 1967), p. 28.
4 Max Hastings and Simon Jenkins, *The Battle for the Falklands* (New York: W.W. Norton, 1983), p. 133. Lieutenant Tinker was killed in the war.
5 Ibid.
6 *The Complete Writings of Thucydides*, The Modern Library (New York: Random House, 1951), p. 15.
7 "Oil: The Issue of American Intervention", *Commentary*, 59 (January 1975).

5
Military Power as an Instrument of Politics

The first, the supreme, the most far-reaching act of judgement that the statesman and commander have to make is to establish ... the kind of war on which they are embarking; neither mistaking it for, nor trying to turn it into, something that is alien to its nature.

Carl von Clausewitz, *On War*

Deterrence and Defense

The best kind of war is the one that does not happen. Even the most ruthless and predatory governments prefer not to fight if they can obtain their objectives without fighting. As von Clausewitz wrote, "the aggressor is always peace-loving ... he would prefer to take over our country unopposed."[1] The only exceptions to this rule are the few regimes which see war as desirable in itself, as a means of strengthening the nation, disciplining and uniting its citizens, or as the arena in which the warrior can display his heroism. This was a common attitude before World War I but it has become quite rare since. Probably Hitler's Germany was the only post-World War I regime ever to proclaim war to be desirable in itself, and even Hitler managed to gain many of his objectives without war.

Very often, by threatening to fight, a regime can achieve most or all of what it might hope to get by actually fighting. It can do so because its opponent also wants to avoid a war, particularly one that he might lose. One's own fear of war makes it desirable to attain one's objectives without fighting, but it is the opponent's fear of war that makes it possible. As Samuel Johnson once said:

> Mutual cowardice keeps us in peace. Were one half of mankind brave and one half cowards, the brave would always be beating the cowards. Were all brave, they would lead a very uneasy life; all would be continually fighting; but being all cowards, we go on very well.[2]

As war has become more and more costly governments have become more and more reluctant to resort to it, particularly in its most devastating forms. In eighteenth-century Europe war was an accepted and frequently used instrument of policy – European governments resorted to it often and sometimes for rather minor reasons – precisely because wars were fought for limited ends and consumed only a limited proportion of a country's resources. A country that lost a war might have to give up a province or a few colonies, nothing more. In the first half of the twentieth century war became a much more serious and more costly business, to be entered into only with great reluctance. In the two world wars the belligerents mobilized all of their human, economic, and moral resources and the price of defeat was not the loss of a province but the overthrow of the loser's government and the military occupation of part or all of his territory. Today, the price of defeat or even "victory" in a nuclear war is likely to be national annihilation. It is hard to conceive of anything, short of the starkest possible threat to its own survival, that would induce any regime to launch a nuclear war. However, as war itself has become less and less usable as an instrument of national policy, the threat of war has become more and more important. Precisely because a nuclear war would be so destructive, a credible threat to launch such a war is the most powerful instrument of coercion a country can use. The United States and the Soviet Union dominate the world very largely because they can and do make that threat in defense of their interests.

The threat of war serves usually as a deterrent, a means of preventing another country from doing something, such as invading one's territory. When Americans talk about "deterrence" or "nuclear deterrence" they usually refer to the United States's threat to launch a nuclear war against the Soviet Union if the Soviet Union attacks the United States. The threat of war may also be used to deter an attack upon one's allies – a Soviet invasion of Western Europe or Soviet intervention in a future Arab–Israeli war. It might even be used to dictate another country's domestic policy, perhaps to prevent the Soviet government's sending a dissident such as Andrei Sakharov to a concentration camp. However, unless the deterrer is very powerful and very determined and the other country is weak and irresolute, using the threat of war to dictate domestic policy is much more dangerous and less effective than non-military forms of suasion. The threat to launch a nuclear retaliatory attack is the best deterrent of a Soviet attack on the United States, but the threat to end exchanges of scientific information between American and Soviet scientists is probably the best protection for Sakharov.

Deterrence works on the mind of a potential enemy; it works by affecting the way he sees things and through that, what he does. He must be persuaded that the cost to him of doing what one does not want him to do is greater than what he would gain from doing it. One must send him a message, he must receive it, and he must believe it. The message must make clear what it is one wants him not to do and what will happen to him if he does it. One's message must also assure him that the threatened evil consequence will not befall him if he does not do it. No government will change its policy to avert an evil that will befall it regardless of what it does. It is very important that the United States government make clear to the Soviets that a Soviet nuclear attack on the United States would trigger an American nuclear attack on the Soviet Union. It may be equally important, particularly in a desperate crisis such as might lead to a nuclear war, to make it clear that the United

States will *not* attack the Soviet Union unless and until the United States is attacked.

A deterrent threat must be both powerful and credible to be effective. The opponent must believe that he would be hurt badly if the threat were carried out, badly enough to offset whatever he might gain from what he contemplates doing. He must also believe that the deterring power is prepared to carry out the threat. To deter a Soviet nuclear attack on the United States the United States must be able to devastate the Soviet Union even after being attacked, and the Soviets must know that it can. Weapons which the opponent does not know exist do not have any deterrent effect. In addition, the United States government must firmly intend to strike back if the United States is attacked, and the Soviets must be made aware of that intention. Firm intentions which have never been communicated to the opponent have no more deterrent value than powerful weapons which he does not know exist.

That is why the United States government should not adopt as policy the position taken by the American Catholic bishops in their pastoral letter (see chapter 2, p. 19–20), that the United States should never, under any circumstances, launch a nuclear attack on the Soviet Union. Adopting that policy would weaken the credibility of American nuclear deterrence. Soviet leaders might be tempted to believe that the Soviet Union could attack the United States with impunity. Admittedly the Soviets would place little reliance on any American promise never to use nuclear weapons. They would judge, rightly, that any such commitment would be forgotten once the United States had been attacked with nuclear weapons. Nevertheless, anything that lessens the credibility of American nuclear retaliation destabilizes the balance of terror, and anything that destabilizes the balance of terror makes a nuclear war more likely.

As Thomas Schelling points out, "some threats are inherently persuasive, some have to be made persuasive, and some are bound to look like bluffs."[3] The threat to launch a nuclear attack on the Soviet Union as retaliation for a Soviet nuclear attack on the United States is inherently persuasive. That threat might not be carried out – the attack would be, as the Catholic bishops point out, immoral – but the possibility that it would be is strong enough to deter any Soviet leader. A threat to attack the Soviet Union if Andrei Sakharov were to be imprisoned is bound to look like a bluff. It might make the Soviets hesitate, because even the slightest threat of a total catastrophe will make people hesitate, particularly if what they hope to gain by ignoring the threat is not important to them. (Few of us would go to work in the morning if there were a one-in-a-thousand chance we would be killed when we got there.) Even so, it would be an extremely reckless threat. No government should try to defend objectives of minor value to it by threatening nuclear war, because the danger is too great that its bluff will be called. If its bluff *is* called it faces a very unattractive choice between launching a nuclear war over a minor issue and backing down. Launching the war would be criminally insane, but backing down would weaken the credibility of whatever deterrent threat it might make in any other situation.

The threat to attack the Soviet Union if it invades Western Europe can perhaps be made persuasive. It would not be rational for the United States to carry out that threat, but execution of the threat is not inconceivable. Schelling and other game theorists have devoted much thought to how such an irrational but not incredible threat can be made persuasive. Some governments can make an irrational threat persuasive by themselves appearing irrational and reckless. Even the most irrational threat will be taken very seriously if it is made by a heavily-armed madman. This option, however, is not open to the government of a liberal democracy because it cannot afford to look irrational to its constituents. If the government of a democracy is to appear reckless and irrational to an opponent who is to be deterred, it cannot help but appear reckless and irrational also to its constituents. They will vote it out of office the first chance they get.

The only way a democratic government can make an irrational threat persuasive is deliberately to deprive itself of its freedom of action. It must leave itself no alternative to taking the action it threatens, or at least no alternative that is not extremely costly. One way to do this is to make a solemn and public pledge to carry out the threatened action if the occasion arises. It can, for example, make with an ally a mutual security treaty pledging it to go to war if the ally is attacked.

A commitment can also be strengthened by stationing troops on the territory of a threatened ally. If the ally is attacked those troops will be attacked as well and it will be hard for the protector to stay out of the war; some of his soldiers will have been killed in action, others will be locked in combat and perhaps unable to disengage without suffering heavy casualties, and his national honor will be at stake. In principle a very small force can strengthen a commitment. Before World War I a British general asked a French general how many British soldiers France would need to have on her territory as protection against a German invasion. The Frenchman replied, "One single private soldier, . . . and we would take good care that he was killed."[4] The one British soldier could do very little by his own efforts to stop the German army, but his presence in France would be a warning to the Germans and a promise to the French. It might bring Great Britain into the war on France's side – particularly if he were killed in the fighting. As the French general implied, the soldier would be much more useful to France dead than alive. In reality, of course, the fate of one soldier has never brought a country into a war it would not have fought for other reasons. A military force designed to strengthen a commitment must be fairly large for the commitment to be credible. However, it does not have to be large enough to defend the territory on which it is stationed. That is not its purpose.

The great defect of deterrence is that it can fail. Deterrence exists only in the mind of the opponent; it is a limit on what he intends to do imposed by his fear of what will happen to him if he does it. This makes deterrence dependent on the opponent's rationality and his ability to assess risks and rewards correctly; one cannot deter a madman, a masochist, or a child too young to connect cause with effect. It also means that no government can ever be sure that its deterrent is adequate, no matter how strong its force is or how firmly it intends to use it. No government can ever be sure what its opponent is thinking. The opponent may fail to realize how strong the deterrent force is, he may think it would not be unleashed against him no matter what he does, or he may value his objective so highly that he does not care what he suffers to attain it. Policy-makers must assess as well as they can the impact of their deterrent threats on the minds of potential opponents. This is one reason why it is important to understand the political culture of the Soviet Union as a guide to formulating United States national security policy. The test of the adequacy of an American retaliatory threat is not whether or not it would deter Americans but whether or not it deters the leaders of the Soviet Union with their distinctive perceptions of risks and rewards. Even so, the impact of a deterrent threat can never be assessed with absolute precision. Even those who have spent their lives studying the Soviet Union find the Kremlin mind hard to penetrate.

If deterrence fails then one must either surrender or wage a defensive war. Defense, the waging of a defensive war, is the only alternative to deterrence which protects a country against an enemy attack. Deterrence seeks to avoid a war while defense seeks to conduct that war successfully if it cannot be avoided. The primary objective in fighting a defensive war is to limit the damage to one's country, prevent any territory being lost to the enemy, and limit the number of citizens killed and the amount of property destroyed. One may also wish to punish or even destroy the aggressor so that he will not aggress again, but the basic objective is to limit damage to oneself. The best defense therefore is that which allows the least damage to the defender. On the other hand, the best deterrent is that which would cause the most damage to the aggressor if he were to start a war.

Inflicting the most damage possible on the enemy and suffering the least possible damage oneself are two fundamentally different military objectives. Strategies and forces designed to achieve one often are not well adapted to achieving the other. The best deterrent may be a very poor defense, and vice-versa. The United States's nearly 2,000 ICBMs, strategic bombers, and submarine-launched missiles are an excellent deterrent to any Soviet attack on the United States but would prove a poor defense if the United States were to be attacked. Even sending every single one of them against the Soviet strategic nuclear force would not prevent the destruction of the United States. If the United States did not have any strategic nuclear weapons and had instead invested the money it has spent on them in civil defense, anti-bomber defenses and anti-ballistic missiles (ABMs), it would perhaps have a much stronger defense against nuclear attack but only a weak deterrent. A strong defense with no deterrent effect limits the damage an attack would do but leaves it an ever-present possibility. A strong deterrent which offers no defense makes an enemy attack very unlikely but extremely devastating, should it ever occur.

Ideally a country should have both a strong, credible deterrent and an effective defense. It should be able, as much as possible, to deter any enemy attack, but it should also be prepared to defend itself if deterrence should fail. Unfortunately, this is not always possible. The country may not be able to afford both types of forces and the existence of an effective defense often makes a deterrent threat less credible and thus less effective. The military planner must choose between deterrence and defense. United States military planners must make just that choice in selecting strategies for the protection of the United States and Western Europe. Of the two alternative strategies for protecting the United States against Soviet attack, one, "Mutual Assured Destruction," offers an excellent deterrent but a very weak defense. The other, "Damage Limitation," offers a somewhat stronger defense but a much less sure deterrent. Similarly, of the possible strategies for the protection of Western Europe, "Massive Retaliation" is primarily a deterrence strategy and "Flexible Response" a defense strategy. This choice between deterrence and defense is one of the most difficult dilemmas United States military planners face.

Political Goals and Military Strategy

War is, in von Clausewitz's famous phrase, "the continuation of policy by other means."[5] It is a means of achieving certain political goals. This requires of any government which resorts to war a clear understanding of what its political goals are, why it is fighting that war, and what it hopes to achieve as a result. It should know how the military effort which it is making will lead to the political result which it wants. It should have measured the resources it can mobilize for the war and determined that they will be sufficient. Finally, it should have soberly judged the importance of its political goals: are they important enough to justify the military effort necessary to attain them?

Any government needs to have thought through these questions before it gets into a war, but it is particularly important for the government of democracy to do so. A democratic government must know the answers to these questions because it must give them to the people it leads when it asks for their support. It must convince them that the war needs to be fought, that it is being conducted intelligently, and that it can be won at a justifiable cost. Otherwise, as was discussed in chapter 3, the people will not support it, at least not for long.

The reader will recall from chapter 4 that there are a variety of political goals which governments seek by fighting wars. These are:

1 To prevent an invader from seizing one's territory and imposing his authority and his political system.
2 To prevent an invader from taking away a part of one's territory – a province, perhaps, or a colony.
3 To prevent another country from stirring up trouble within one's borders and perhaps encouraging the overthrow of one's government.
4 To prevent another country from becoming so powerful as to menace one's security.
5 To take away part of the territory of another country.

6   To conquer another country and thus impose one's authority and political
    system on it.

For a country to attain its political goal in a war its armed forces must
reach a certain military objective. Political goals dictate military objectives.
Once a government has decided what its political goal is, it then must decide
what military objective its armed forces must reach. A particular political
goal does not always dictate the same military objective. In both the
Spanish–American War and the Mexican War the United States had the
same political goal: to take certain territories away from its opponent. In the
Spanish–American War the goal was to seize Cuba, the Philippines, and
Puerto Rico, and in the Mexican War it was to seize Texas, California, New
Mexico, and Arizona. However, the military objectives that had to be attained
to secure the political goal were different in each case. In the Spanish–American
War, once the United States had conquered the territories it wanted the
Spanish government was prepared to make peace and give up those territories.
There was no need to invade Spain. The Mexican government was not
prepared to make peace after it had lost Texas, California, New Mexico, and
Arizona. The United States had to send an army into the heart of Mexico
and capture the capital city before the Mexican government would accept the
United States's peace terms.

As a general rule how far-reaching the military objective must be is
determined by the magnitude of the political goal and the obstinacy of the
enemy. If a government's political goal is limited but it faces a very determined
enemy, then it must attain a very far-reaching military objective, as the United
States had to in the Mexican War. Also, if its political goal is very ambitious
then it must attain a very far-reaching military objective even if it faces a
moderate and peace-loving opponent. If it wishes to take away another
country's national independence then it must occupy all of the victim's
territory, as Nazi Germany did to Belgium, Norway, the Netherlands and
other small countries which sought desperately to avoid war. Only if a
government's political goal is limited and its opponent not very obstinate can
it set for itself a limited military objective.

The military objective dictates the military strategy which must be pursued
to attain it. If the military objective is to occupy the enemy's entire territory,
the appropriate strategy usually is to destroy his army on the field of battle.
Presumably he will keep fighting as long as he can, that is as long as he has
an army in the field. In the American Civil War the North had to destroy
the Confederate army to win the war, as both General Grant and President
Lincoln realized. Similarly, in World War II the Allies had to destroy the
German army because as long as Hitler had armed forces he would keep on
fighting. In such cases, capturing territory is useful only as it contributes to
destroying the enemy army.

If the military objective is to seize a particular piece of land, usually the
best strategy is to overrun that piece of land rapidly at the beginning of the
war and then defend it against enemy counterattack. This was the strategy

employed by Egypt in the Yom Kippur War: cross the Suez Canal rapidly, seize the hills on the east bank of the canal, and then hold off Israeli counterattacks.

If the military objective is to prevent being conquered by a more powerful enemy, the best strategy may be to keep one's army intact and prolong the conflict as long as possible until the enemy tires of the war or allies come to one's aid. This is a particularly appropriate strategy for a weak country which is totally committed to the war and is facing a powerful country which is less committed. This strategy was employed successfully by the United States during the Revolutionary War and unsuccessfully by the Confederacy during the Civil War.

A country facing two or more opponents may find it best to set itself a different military objective and pursue a different strategy *vis à vis* each one, thus exploiting the differences between them. For example, during the Vietnam War North Vietnam faced two very different opponents, the United States and South Vietnam. The United States is a very powerful country whose commitment to the war was rather weak because no possible outcome threatened it fundamentally; North Vietnam had no intention or prospect of invading the United States. South Vietnam was a very weak country whose commitment (or at least the commitment of its government) was total; North Vietnam intended to conquer and annex South Vietnam. North Vietnam's military objective *vis à vis* the United States was to get the United States out of the war, and the strategy calculated to achieve that was to prolong the war until the United States decided to pull out. Then South Vietnam could be dealt with. The military objective *vis à vis* South Vietnam was to conquer the country and the appropriate strategy was to destroy the South Vietnamese army on the battlefield. The North Vietnamese achieved their political goal, a reunited Vietnam, by achieving both their military objectives.

Having decided what it wishes to do a government must then decide whether it can do it. Will the military effort which the country can exert be sufficient to attain the political goal of the war? Is that goal worth the price that must be paid to attain it? If not, then there is a difficult choice to be made. Sometimes a country goes to war knowing that its total defeat is unavoidable, as Belgium and the Netherlands did in World War II, because it has been invaded and prefers to fight rather than accept subjugation without resistance. In other cases the goal can be scaled down to make it attainable by the means available. During the Korean War the United States government concluded, after China had entered the war, that North Korea could not be conquered and reunited with South Korea at an acceptable cost. The United States's political goal was scaled down from reuniting the two Koreas to protecting South Korea. In still other cases, when national survival is not at stake and victory seems unattainable or attainable only at too high a price, it may be best not to go to war.

Because wars are fought to attain political goals the political leaders of a warring state, not its military high command, must make the decision to go to war and direct the war effort. Only they, knowing their political goal, can set the appropriate military objective. Only they can judge whether attaining that goal will be worth the price that must be paid. The role of military officers is to advise the political leaders as to whether the military strength available is sufficient to attain the objective. They must be consulted, but the decision to go to war and the determination of the objective are not theirs to make. Civilian control of the military is necessary in a democracy to ensure the people's control over how their armed forces are used. It is also necessary for the intelligent use of war as an instrument of policy by any government, autocracy or democracy. Von Clausewitz, the servant of an absolute monarchy, advocated civilian control of the military as strongly and cogently as any democratic leader.

Escalation and Escalation Control

What I have just said about political goals and military strategy may have made the conduct of war seem more rational and controlled than it really is. Governments don't always go to war to attain clearly perceived political goals, they do not always choose the right military objectives, and quite often they do not ask what victory will cost or whether it is attainable at any price. Even when they attempt to do these things they quite often miscalculate, particularly by underestimating the opponent's commitment to his cause. It is all too easy to regard one's own motives for fighting as so obviously just and important that one will make any sacrifice for victory, and the enemy's motives as so weak and unjust that he will not fight at all or will give in easily. One reason why wars occur is that governments contemplating a war commonly overestimate their chances of winning it. Many times the aggressor, if he had known how the war would end, would not have started it. Sometimes the victim, if he had foreseen how futile his resistance would be, would have given in. Usually both sides in a war expect to win it: one of them has to be wrong. As Mark Twain says, "it is difference of opinion that makes horse races"[6] – and wars.

The difficulty of making war a rational and controlled instrument of policy is reflected in the tendency of wars to escalate, to become bigger and more destructive the longer they continue. A war very often begins as a strictly limited conflict, using only a fraction of the belligerents' resources, because what is at stake is not greatly important to either side. Each belligerent devotes to its war effort only as much as its political goal justifies. Then, as the war drags on, stalemated, one side decides that it must escalate to win. It may start using powerful weapons which it has hitherto held back – nuclear weapons or poison gas, for example. Or, it may escalate by attacking targets which had been held inviolate before – cities, oilfields, or merchant shipping. It may also escalate by inducing an ally to join the war. The opponent may not be able to escalate (in which case he probably loses the war), but usually he can. He may not be able to escalate in exactly the way his opponent has done, but he finds some other course of action. Soon the war is stalemated again, but at a higher level of violence. Then one of the belligerents raises the ante again, the other responds – and a war which may have started over a relatively minor matter has become a total war.

The dynamics of escalation are well illustrated by an exercise known as "the dollar game." Some sly person offers to give a dollar bill to whoever bids the most for it, but making two stipulations:

1 Each bid must be at least five cents more than the previous bid.
2 *Both* of the two highest bidders must pay the auctioneer the amount they have bid, even though only one of them gets the dollar.

It seems a reasonable kind of game, up to the point at which the two highest bidders, call them "A" and "B", have bid $1.00 and $0.95 respectively. At this point "B" must raise his bid to $1.05 or drop out. If he raises his bid the best he can hope for is to lose $0.05, pay $1.05 for the dollar bill, but if he drops out he loses $0.95. He raises the bid, hoping that "A" will see reason and drop out. Instead, "A" raises his bid to $1.10. There is no logical reason why the bids should not go higher and higher forever, but in practice they tend to top out at around $3.50. It is a good way to make some money if one has a couple of obstinate and rather dim friends whose friendship one does not mind losing.

The dollar game is a good simulation of a war escalating out of control. The reason why the bidding goes beyond $1.00 is that both the loser and the winner have to pay what they have bid, so that it makes sense to pay more than it is worth to get the dollar bill rather than lose. Even if the bidding goes above $1.00 the "winner," the one who gets the dollar, loses less than his opponent does. So it is in war. Even if winning a war is not worth what it costs, it is less costly than losing. The vanquished loses not only the resources he devoted to fighting the war – the lives of his war dead, the money spent on his war effort, the damage done to his territory – but also whatever the war was fought about.

It is even harder to control escalation in a war than in the dollar game because at least in the game the stake ($1.00) remains the same throughout, whereas in a war the stake tends to become larger the longer the war goes on. Perceptions of the enemy change in the course of a war. After he has killed many of one's young men on the battlefield, committed atrocities upon one's people (there are atrocities in most wars), and perhaps bombed one's cities or invaded and devastated one's territory, he begins to resemble evil incarnate rather than just an opponent with whom one has a conflict of interest. After one has done these things to him one is tempted to rationalize them by thinking him to be evil incarnate. One's war aims escalate from self-defense to weakening the enemy, and then to total victory and the imposition of one's domination upon him. Wars do not always generate such hatred but a prolonged war generally does, at least between the civilian populations of the warring countries if not between the armies on the battlefield. Another reason why war aims escalate, particularly in a democracy, is that as a war becomes more and more costly the only way to justify it may be to set more and more expansive war aims. A president or a prime minister who offers his people a compromise peace after millions of their young men have been killed will be asked why so much was sacrificed to gain so little. He can hardly expect to retain his high office. This was one reason why World War I was not ended by a compromise peace in 1917 or earlier.

Wars, especially long wars, do tend to escalate both in what is at stake and in the means used to fight them. In the American Civil War the Union's war aims escalated to include the abolition of slavery as well as the restoration of the Union. In World War I the war aims of most of the belligerents escalated and the means employed to attain them came to include poison gas, unrestricted submarine warfare, the bombing of cities, and the starvation by naval blockade of entire peoples. In World War II the Allies escalated from dropping propaganda leaflets on German cities in 1939 to bombing German industry in 1940, then to bombing German residential areas in 1942, and finally to dropping atomic bombs on Japanese cities in 1945. In the war which Iraq launched against Iran in 1980, Iraq escalated to the use of poison gas and Iran to human-wave attacks carried out by 12-year-olds. However, even a long war does not necessarily escalate to total war, neither the Korean War nor the Vietnam War did, and it is at least desirable to prevent it from doing so. In wars involving the United States or the Soviet Union it is more than just desirable to prevent escalation, it is absolutely vital. The last step on their escalation ladder is a nuclear total war.

Wars can be kept limited only by the exercise of deliberate restraint by both sides. "Limited war . . . is like fighting in a canoe. A blow hard enough to hurt is in some danger of overturning the canoe."[7] The belligerents must not do everything they can to defeat each other lest, perhaps, they destroy the world in which they both must live. Instead, they must reach and keep while the war is going on a series of tacit agreements as to what is and is not permitted in that war. The war started because they could not resolve their differences in peacetime and fighting the war probably exacerbates those differences. Nevertheless, they must conclude what is in effect an arms control treaty while the war rages.

Each side proposes a desired limitation by observing it himself as long as the opponent does likewise. For example, by not bombing enemy cities at the beginning of the war one side in effect says to the other, "Let's not bomb cities." If the other side also refrains from bombing cities then perhaps the message was received and a tacit agreement reached not to bomb cities. This works only if the proposed limits are very clear and important to both sides. A national boundary may be a feasible limit, particularly if it follows some conspicuous geographic landmark, a large river or an arm of the sea perhaps. During the Korean War the belligerents were able to limit the fighting to the Korean peninsula and prevent it from spreading to China or to United States bases in Japan even though both China and the United States were in the war. They were able to do this for several reasons, chief among them being the conspicuous landmarks separating Korea from China and Japan – the Yalu River between China and Korea, and the Korea Strait between Japan and Korea.

The difference between one type of weapon and another, particularly between nuclear and conventional weapons, may also be a feasible limit. Nuclear weapons are not only far more powerful than conventional weapons but also easy to distinguish from them in other respects; only a nuclear explosion produces radiation, a thermal flash, and an EMP. Because it is so easy to tell if nuclear weapons have been used in a war, a tacit agreement not to use them should be relatively easy to reach and observe. However, once even small nuclear weapons have been used, it would be very difficult to prevent escalation to a nuclear total war. The difference between, say, 10-kiloton and 50-kiloton nuclear weapons is not nearly as clear or important as that between nuclear and conventional weapons.

"Escalation matching" also plays an important role in keeping a war limited; neither side should be able to seize a decisive advantage over the other by escalating. Neither should be able to win, or to avoid losing, by resorting to a higher level of violence. Each should be about as strong in conventional weapons as it is in tactical nuclear weapons or strategic nuclear weapons. Otherwise, the side which is weaker at a lower level of violence but stronger at the next higher level will be strongly tempted to escalate. If NATO is much weaker than the Warsaw Pact in conventional forces but stronger in tactical nuclear weapons, NATO will be tempted to escalate a war between them into a nuclear war. If, on the other hand, NATO is strong enough in conventional weapons but weaker in tactical nuclear weapons, the Warsaw Pact is likely to escalate. It may be desirable for the United States to maintain "escalation dominance" over any potential enemy, to be able to defeat him at any level of violence to which he may resort. It is absolutely vital that the United States be capable of "escalation matching" with any potential enemy. That enemy must not be able to defeat the United States by resorting to a higher level of violence nor to force the United States itself to escalate to avoid defeat.

Notes

1 Carl von Clausewitz, *On War* (Princeton, NJ: Princeton University Press, 1976), p. 370. Lenin, who read von Clausewitz carefully and learned quite a bit from him, was very much struck by that point.
2 Quoted in James Boswell, *The Life of Samuel Johnson*, The Modern Library (New York: Random House, 1931), p. 828. The reader will note a certain contrast between Hobbes's and Johnson's views of human nature. Perhaps Hobbes explains best why nations sometimes go to war, while Johnson shows why they are not always at war.
3 Thomas C. Schelling, *Arms and Influence* (New Haven, Conn.: Yale University Press, 1966), p. 36.
4 C.E. Caldwell, *Field-Marshal Sir Henry Wilson: His Life and Diaries* (2 vols, New York: Charles Scribners Sons, 1927), vol. 1, pp. 78–9. The two generals were Sir Henry Wilson, later to be chief of staff of the British army, and Ferdinand Foch, the future Commander-in-Chief of all Allied forces in World War I.
5 von Clausewitz, *On War*, p. 87.
6 Mark Twain, *Pudd'nhead Wilson*, (New York: New American Library, Signet Classic, 1964), p. 138.
7 Schelling, *Arms and Influence*, p. 123.

6
United States National Security Policy

Life so long untroubled, that ye who inherit forget
It was not made with the mountains, it is not one with the deep.
Men, not gods, devised it. Men, not gods, must keep.
Rudyard Kipling, "The Islanders"*

To guide the use and development of its military power over the long term, just as to fight a war, a government must know the political purpose of its military efforts. It must decide: What does it seek to attain by the use or threat of military power? What are its most likely or most dangerous enemies? What are their political goals and how might they use their military power to attain them? Given the enemies it faces and their military capabilities, how can it use its military power to attain its political goals? A government's answers to these questions constitute its national security policy.

The objectives of United States national security policy are primarily defensive: to prevent or repel any attack upon the United States or other countries which are important to it, and to prevent other developments which would adversely affect United States interests. This does not mean that the United States is completely satisfied with the world as it is or would never attack another country. It does mean that the United States is primarily a status quo power, less interested in changing the world to its advantage than in preventing the world from being changed to its disadvantage.

The Soviet Adversary

By far the most important military threat to the United States is the Soviet Union. The Soviet Union and the United States are the two "superpowers" in the world today: two countries which in military power and political influence are approximately equal to each other and far stronger than any others. The Soviet Union could, in a nuclear total war, destroy the United States as the United States could the Soviet Union. No other countries, not even the nuclear-armed ones, could do such damage to either superpower.

The Soviet Union is deeply hostile to the United States. Its leaders see

---

* *Rudyard Kipling's Verse: Definitive Edition* (Garden City, NY: Doubleday and Company, Inc., 1940), p. 301. Reprinted with permission.

the two superpowers as the champions of two antagonistic social systems, "socialism" and "capitalism," and the "struggle between the two social systems" as the fundamental reality of world politics today. So it has become.

Each superpower is the center of a network of military alliances, the leader of an assortment of allies, satellites and clients, and the spiritual home of an ideology. Each encourages and defends its allies with military and economic aid, with diplomatic support, and sometimes by going to war on their behalf. In Europe NATO, led by the United States, confronts the Warsaw Pact led by the Soviet Union (see map 6.1, p. 70). In the Middle East Israel, supported by the United States, confronts Syria and other Arab states supported by the Soviet Union. In Central America the United States resists what its government sees as an attempt by the Soviet Union's clients, Cuba and Nicaragua, to subvert El Salvador. The same pattern prevails in Southeast Asia, in the Persian Gulf, in southern Africa, and elsewhere around the world (see map 6.2, p. 71). Not all of the United States's enemies are allied with the Soviet Union – Iran is not – but most of them are. From the American point of view the Soviet Union is, if not the "focus of evil" which President Reagan once called it, at least the focus of an enormous amount of trouble.

Why are the United States and the Soviet Union enemies? One reason is the very fact that they are the two most powerful countries in the world. Perhaps this alone would make them enemies, regardless of whatever form of government each had or any other issue which might divide them. In a bipolar world, where there are just two great powers, each is almost compelled to regard any increase in the power of the other as a threat to its own security. Neither can afford to let the other become substantially more powerful than itself for fear that that greater power might some day be used against it.

Bipolarity helps to make the United States and the Soviet Union enemies, but the primary reasons for their conflict are the Marxist–Leninist ideology to which the leaders of the Soviet Union adhere and the way the Soviet government behaves in conformity with that ideology. Marxism–Leninism holds that Communist-ruled and capitalist-ruled governments are natural enemies. Capitalist-ruled governments, such as that which rules the United States, are inherently hostile to the Soviet Union and inherently aggressive even when they are forced to behave peacefully. As Secretary-General Brezhnev said in 1971, at the height of detente, "US imperialism . . . seeks to dominate everywhere, interferes in the affairs of other peoples, high-handedly tramples on their legitimate rights and sovereignty, and seeks by force, bribery and economic penetration to impose its will on states and whole areas of the world."[1] This implies that, as the Soviet Union is the primary obstacle to American domination of the world, the United States inevitably seeks to weaken the Soviet Union and, if possible, to destroy it.

The Soviets fear the United States's military power but only as one of many weapons the United States employs in its struggle against the Soviet Union. Among the others are United States diplomacy, government economic policies, scientific and technological research, and the actions of American private enterprise at home and abroad. Perhaps the greatest threat is the power of Western ideas and the subversive example of Western ways of life.

Iceland

North
Sea

Norway
Sweden
Finland

Denmark

Baltic Sea

Ireland

United
Kingdom

Netherlands

Soviet
Union

East
Germany
Poland

Belgium

West
Germany

(Lux.)

Czechoslovakia

France

(Sw.)
Austria
Hungary

Italy
Rumania

Portugal

Yugoslavia

Spain
Bulgaria

Mediterranean Sea
Greece
Albania
Turkey

Morocco
Algeria
Tunisia

| | | | | | NATO Countries

Scale of miles

0 100 200 300 400

Warsaw Pact countries

Neutral countries

Sw. Switzerland
Lux. Luxembourg

~ Boundary between NATO and Warsaw Pact

Map 6.1 *Superpower alliances in Europe*

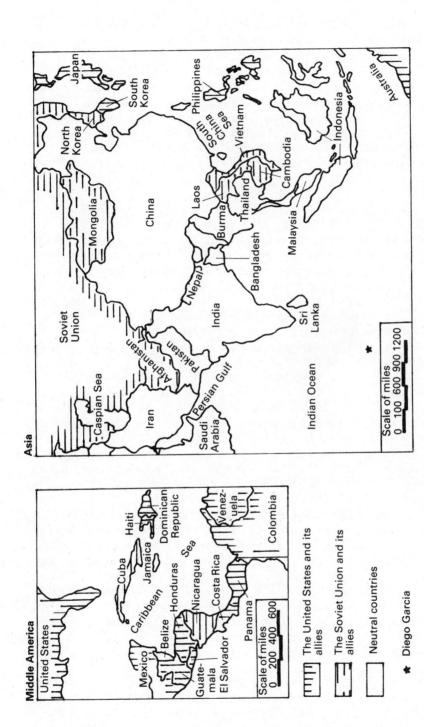

Map 6.2 *Superpower alliances in the Third World*

Soviet leaders fear that the people of the Soviet Union, learning about the freedom, prosperity, and other attractions of life in the liberal democracies, will demand the same for themselves and turn against a regime that cannot provide them.

Even if the United States and the other liberal democracies were to disarm completely and thus pose no military threat whatever to the Soviet Union, the Soviets would still have reason to fear and hate them and wish to subjugate them. The mere existence of attractive alternatives elsewhere seems to the leaders of the Soviet Union to threaten their control over those they rule. That is why they sent the Soviet army into Czechoslovakia in 1968. In 1981, by threatening another invasion, they forced the government of Poland to declare martial law and suppress the free trade unions in Poland. Nothing that was happening in either country posed any military threat whatever to the Soviet Union. The leaders of the reform movements in Czechoslovakia and Poland were very careful not to challenge their countries' political ties to the Soviet Union, the military alliance with the Soviet Union, or the Soviet Union's right to station troops in Poland and Czechoslovakia. They asked only for a certain degree of intellectual and political freedom in Czechoslovakia and free trade unions in Poland; that was more than the Soviets would allow them. Even a very limited amount of popular freedom in Poland and Czechoslovakia seemed to the rulers of the Soviet Union an unacceptable threat to their rule, a threat great enough to justify invading those countries. Surely they must find the much greater freedom enjoyed by the peoples of the United States and other democracies to be at least as menacing.

The people of the United States have in their turn become thoroughly hostile to the Soviet Union, mostly because of how the Soviet government has behaved. American leaders have done and said things which have increased Soviet hostility towards the United States, but nevertheless, the basic reason why the Soviets hate and fear the United States is because it exists. It is a very powerful country which they do not control and which could do the Soviet Union a great deal of harm. The Soviets will probably continue to hate and fear the United States as long as it exists and is free of their control.

The United States is not the Soviet Union's only enemy. Wherever the Soviets look, the entire length of their long, vulnerable borders, they see powerful enemies: NATO to the west, China and Japan to the east, and the United States across the North Pole. Even Iran to the south is almost as anti-Soviet as it is anti-American. Most of the other powerful countries in the world are indeed hostile to the Soviet Union, not because they are ruled by capitalists (which China is not), but because of how the Soviet Union has behaved in the past and continues to behave. The Soviet regime may have a paranoid view of the world, but even paranoiacs have enemies. In fact, paranoiacs have more enemies than other people because they treat others in such a way as to make them enemies.

A regime which feels itself to be surrounded by enemies, enemies who can never for very long be its friends because their mere existence threatens it, will above all else seek power. Believing itself to be engaged in a Hobbesian "war of all against all," it carries out a Hobbesian "continual and restless search of power after power." The fundamental objective of Soviet national security policy is to make the Soviet Union as powerful as possible. The Soviets feel somewhat secure today only because they could probably defeat or deter any attack which might be made against them by all of the other major countries of the world combined. They will feel entirely secure only when they control all of the other major countries.

The Soviets are not, of course, the only ruling elite ever to seek power. The leaders of the United States also seek power; they seek to maintain the power of the United States relative to that of the Soviet Union. The difference is that while American leaders seek a balance of power, Soviet leaders seek preponderant power, and perhaps some day total power. Americans tend to think peace is most secure when there is an approximately equal balance of power between the United States and the Soviet Union. The Soviets think that the more powerful is the Soviet Union, and the weaker the United States, the more secure is world peace. The Soviet Union is a threat to other countries not because its leaders seek to ensure their country's national security. All governments do that. It is a threat because the Soviet Union's national security, as defined by its leaders, requires a degree of Soviet predominance that threatens every other country. The Soviets seek absolute security and, as Henry Kissinger has pointed out, "absolute security for one country must mean absolute insecurity for all others."[2]

The search for power takes many forms, but territorial expansion is the surest and most permanent way to increase a country's power. That is why the Soviet Union is an aggressive, expansionist country. Its leaders do not and cannot long accept the global status quo. Theirs is a peculiar kind of "defensive expansionism," motivated primarily by a fear that the status quo does not serve the Soviet Union's long-term security needs, but it is expansionism all the same. It may be defensive in intent, but "there has never been anything more offensive than a Russian on the defensive."[3]

The Soviet government, then, is committed to a cautious but persistent effort to make the Soviet Union as powerful as possible: to attain military superiority over the United States and its allies, to bring more and more countries under Soviet control or influence, and to isolate its potential enemies from each other. Because the Soviet drive for power is motivated primarily by the desire for security, it has always been cautious. It is intended to ensure, not risk, the regime's survival. Because it arises from the fundamental character of the regime, it is bound to be persistent. As long as the Communist Party controls the Soviet Union it will continue. The United States's defense against the Soviet threat must also be both cautious and persistent.

The United States's Vital Interests

A country's "vital interests" are those it is prepared to go to war for. The most vital interest of the United States, as of any country, is to prevent an armed attack on its territory. As far as conventional warfare is concerned this is probably easier for the United States than for any other country. There is not now and has not been for more than a century any possibility of an overland invasion of the United States. It has land borders with only two countries: one of them, Canada, is a very close ally and the other, Mexico, is strongly disinclined ever again to go to war with its powerful neighbor. A seaborne invasion is conceivable, but only to those with very strong imaginations. A seaborne invasion force would have to cross thousands of miles of ocean and in much greater force than any other seaborne invasion in history. Even if the invaders were to make a successful landing the United States could use nuclear weapons to destroy them. Geography has protected the United States from attack to a remarkable degree. It is one of the many ways in which that country has been very fortunate.

The United States can be attacked only through the air, by strategic nuclear weapons, and the only country which might conceivably do so is the Soviet Union. The strategic nuclear forces of France, China, and the United Kingdom are intended almost entirely for use against the Soviet Union. China is the only one of the three which might conceivably go to war with the United States and only a few of her strategic nuclear weapons have sufficient range to hit any part of the continental United States. For all practical purposes defending the United States from direct attack amounts to preventing a Soviet strategic nuclear attack on American soil.

It can be argued that the only function of the United States's armed forces should be to protect the country against a direct attack, as it was for most of American history. Perhaps the fate of Western Europe or the Middle East is not of vital importance to the United States or the armed forces of those countries can defend themselves without American help; that is to say the United States should pull back its forces to American soil. This policy was advocated by the pre-World War II isolationists and is advocated today by "neo-isolationists" such as Earl Ravenal.[+]

This is not the policy of the United States government today and has not been since the beginning of World War II. The United States has military commitments all over the globe. It has military alliances with 43 foreign countries, gives military training to students from over 80 countries, and has more than half a million of its servicemen stationed abroad. In fact, most of the country's military effort is devoted to defending other countries rather than directly to the defense of American territory. The United States spends on its strategic nuclear forces, air defense forces and civil defense, which protect the United States itself, less than a quarter of what it spends on the general purpose forces (ground forces, tactical and transport aviation, and naval forces) designed primarily to defend other countries.

The United States has committed itself to defend these foreign countries out of a conviction that United States national security would be fatally compromised were all the rest of the world, or even most of it, to fall into the hands of its enemies. The United States is a large, rich, and powerful country but it has only about one-twentieth of the world's population, generates about a quarter of its economic output, and is dependent on other countries for many vital raw materials. An enemy who had brought most of the world under his control might well be able to conquer the United States. With the economic resources at his command he might be able to build such a powerful strategic nuclear force and such strong defenses against nuclear retaliation as to be able to carry out a successful strategic nuclear attack on the United States. If all of Eurasia were to fall under Soviet control the Soviet government would have at its command three times as much economic output as the United States. This would make it nearly as far superior to the United States in that respect as the Soviet Union is to China today. The Soviet Union probably could carry out a successful strategic nuclear attack against China. An enemy controlling all of Eurasia could build naval and other forces powerful enough to seize control of part of Central or South America. From that bridgehead the enemy could launch a seaborne invasion of the United States, employing the threat of nuclear retaliation to prevent the United States's resorting to nuclear weapons to stop it. Or he might be able to cripple the United States's economy by denying it imports of essential raw materials. The best the United States could expect would be the fate President Franklin Roosevelt predicted for it were the Axis to conquer the rest of the world: "the nightmare of a people lodged in prison, handcuffed, hungry, and fed through the bars from day to day by the contemptuous, unpitying masters of other continents."[5] For the past 50 years that nightmare has shaped United States national security policy. The United States fought World War II to prevent the Axis from conquering Eurasia, and today it resists Soviet expansionism for the same reason.

United States national security policy today strikingly resembles the balance of power policy pursued by Britain in European power struggles from the sixteenth to the twentieth centuries. Just as the United States today seeks to prevent all of Eurasia from falling under the control of the Soviet Union, so the British for centuries worked to prevent any one country dominating Continental Europe. Both the United States today and Britain in the past have organized and led alliances of other powers against whoever threatened to dominate Eurasia or Continental Europe. The United States today is the center of an anti-Soviet alliance system which includes its NATO allies, Japan, and other countries. Britain was the center of the alliances which defeated France under Louis XIV and Napoleon, and Germany under Wilhelm II and Hitler.

The United States and the United Kingdom are similar in one more important respect: both are essentially islands, separated from the Eurasian mainland by bodies of water. Their insular condition has been both a blessing and a curse upon them. It has been a blessing by shielding them against overland invasions; to attack the United Kingdom an invader would have to cross the English Channel or the North Sea; to attack the United States he would have to cross the Atlantic or Pacific Ocean. It has been a curse by requiring that they themselves cross a body of water to reach the seat of any likely war and bring help to a continental ally. For these reasons both the United States and Britain have, throughout most of their histories, been primarily sea powers. They have had to control the seas around them as their primary line of defense against enemy attack, as the highway for the trade on which their economies depend, and as the communications link between them and their allies.

One of the most important national security policy decisions the leaders of the United States must make is how much of the world to protect against Soviet expansionism. The United States must defend more than its own territory. On the other hand, it cannot defend the entire world. Not only to stop direct Soviet military aggression against every potential victim but also to prevent any pro-Soviet group seizing power anywhere in the world is probably more than the United States can do, no matter how hard it tries. Defending the entire world is certainly more than it can do with the limited share of its resources the American people are willing to devote to national security.

The vast majority of Americans agree that it is essential to prevent Western Europe and Japan falling under Soviet control. There is less consensus on the importance of defending any part of the Third World, the underdeveloped countries of Asia, Africa, and Latin America. Many would agree with John Kenneth Galbraith: "The Third World consists, by definition, of poor rural societies. . . . Even by the crudest power calculus, military or economic, such nations have no vital relation to the economic or strategic position of the developed countries."[6] Ironically (and much to Galbraith's dissatisfaction), the United States has twice since World War II gone to war in defense of such "poor rural societies," in Korea and in South Vietnam. One of the peculiarities of the Cold War is that, although Western Europe and Japan are largely what is at stake in the conflict, neither superpower has gone to war for them, and only on a few occasions has either even seriously threatened war. Precisely because the stakes for both sides in a European war would be so high, both have felt that launching such a war would be too dangerous, too likely to escalate to a nuclear total war. Instead, they have confined their military strokes and counterstrokes to the Third World where the risks of escalation are much smaller.

The United States has interests in the Third World which it must defend, but it cannot protect every part of the Third World against every Soviet or Soviet-supported threat. Instead, United States policy-makers must decide where in the Third World their country's vital interests lie, what they would go to war over. Some Third World countries supply the United States with raw materials important enough to fight for, but most do not. The only Third World countries for whose raw materials the United States is likely to go to war are the oil-producing states around the Persian Gulf, particularly Iran, Iraq, and Saudi Arabia. They are important to the United States and its allies and also very vulnerable to Soviet attack. Other parts of the Third World are important because of their location, the Caribbean and Central America, for example, because they are so close to the United States. In 1983 the United States fought a micro-war in the Caribbean to overturn the pro-Communist government of Grenada. It has been involved for years in guerrilla wars in two Central American countries, El Salvador and Nicaragua. South Korea is also important because of its location; it has been described as "a sword pointed at the heart of Japan." The United States went to war once to defend South Korea and maintains a garrison there today. Finally, the United States probably would fight to defend Israel. Israel is not vital to United States national security but most Americans feel a moral obligation to its inhabitants. Thus the United States probably would go to war to defend the Persian Gulf, the Caribbean, Central America, South Korea, and Israel. Elsewhere in the Third World it gives its allies and others who ask for its help military and economic aid and diplomatic support, but it probably would not go to war for them.

There is, though, no Third World country which the United States certainly would never fight to defend. The loss of a country which is not important in itself may make difficult or impossible the defense of one that is; it may be necessary to defend the former in order to protect the latter. That is why United States military forces have to be prepared to fight almost anywhere in the world. However, United States policy-makers must weigh very carefully both the cost and the necessity of any military intervention in the Third World.

How the United States Defends Its Vital Interests

The United States armed forces have, then, three major missions. The first is to prevent a Soviet nuclear attack on the United States. The second is to preserve the balance of power in Eurasia by protecting Western Europe and Japan from Soviet aggression. The third is to protect United States interests and allies in the Third World. Each of these three missions makes different demands on the United States armed forces and requires a different set of arms of the service.

Preventing a Soviet nuclear attack on the United States is the most important mission the armed forces have; failure to carry it out could result immediately and directly in the destruction of the entire country if not of the human race. It is also in most respects the easiest, partly because it is so important. Virtually all Americans agree that the United Statest must be defended and that strategic nuclear weapons are necessary for this purpose. This task is made easier also by the United States's ability to perform it almost entirely by its own efforts. The United States asks of its allies for its own defense only the use of a missile submarine base in the United Kingdom and radar installations in Canada, Iceland, Greenland and the United Kingdom. All the rest of the United States's far-flung network of military bases abroad exists primarily for the defense of other countries. Planning for this mission is easier than for the other two because only the Soviet Union could carry out a serious strategic nuclear attack on the United States and there is essentially only one strategy the Soviets could employ to do this. They do not have (as they would if they were to invade Western Europe) a choice among several different strategies. Even the financial cost of defending the United States is less than that of executing the other two missions. Individual strategic nuclear weapons are very expensive, but the destructive power of each one is so awesome that comparatively few are needed. The personnel who man them must be highly trained and absolutely reliable, which requires that they be well paid, but, again, comparatively few are needed.

To deter a Soviet nuclear attack on itself, the United States relies on its strategic nuclear offensive weapons: the Strategic Air Command's ICBMs and bombers and the navy's missile submarines. It also makes a weak attempt to defend itself against attack should deterrence fail. To this end it maintains a small force of interceptors under the North American Air Defense Command and a modest civil defense program. None of these strategic forces, except for the bombers, could be used to fight any kind of war other than a strategic nuclear war. (Some Strategic Air Command B-52s were used in Vietnam and B-29s in Korea but they did not play a major role in either war.) The strategic nuclear force does, to some degree, deter Soviet aggression against Western Europe and Japan but not reliably enough so that other forces to protect them are not needed. It does not deter challenges to United States interests in the Third World because no threat to employ it in response to such challenges would be credible enough to be useful.

Protecting Western Europe is not as important as protecting the United States, but it is an important task and in some ways a difficult one. Any war fought in Europe could escalate into a nuclear total war and destroy the United States. Initially it would very likely be a limited nuclear war, employing the thousands of tactical nuclear weapons deployed in Europe, a war that would devastate Europe almost as completely as a nuclear total war would the superpowers. Even fought with conventional weapons alone it would call on all the military force the United States could muster to win it and all of the United States government's prudence and judgement to prevent it becoming a nuclear total war.

There is not, among the American people, the same unanimity about the need to defend Western Europe as there is about defending the United States. Many Americans feel that Soviet domination of Western Europe would not so fundamentally threaten the United States as to justify going to war to prevent it. Many feel that the Europeans should be able to defend themselves without American help. During the late 1960s three-quarters of the citizens opposed going to war if the Soviet Union were to invade Western Europe, and even in 1982 over a third were against doing so.[7] Another difficulty is that the United States can defend Western Europe only in cooperation with its NATO allies and it is always harder for a group of countries to act together than for one country to act alone. Also, planning for the defense of Europe is more difficult than it might otherwise be because it must meet a variety of contingencies. Even though a Soviet overland invasion is the only likely threat, the Soviets could begin and conduct that invasion in a variety of ways.

To protect Western Europe the United States relies primarily on the ground forces and tactical air forces it has stationed on that continent, others stationed in the United States but earmarked for dispatch to Europe in the event of war, the airlift and sealift needed to get them there, and anti-submarine forces to protect them in transit. The ground forces in or earmarked for Europe have many tanks and other armored vehicles, tremendous firepower, and great tactical mobility (the ability to move rapidly on the battlefield). These are the qualities they must have to resist the extremely powerful armored forces the Soviet Union would use in a European war. On the other hand, because they have so much firepower and so many vehicles, they require tremendous quantities of ammunition and fuel, and thus can fight at their best only where there are excellent roads, ports, airfields, and other transportation facilities. They are also very difficult to transport to the battlefield, i.e., have little strategic mobility. For these reasons they are not well adapted to fight in a typical Third World country. They can do so, but not very effectively, as the Vietnam War showed.

The defense of Japan is a vital task, but comparatively easy because Japan is an island country. As long as the United States navy commands the seas around Japan, even the small army Japan has should be able to deal with whatever force the Soviet Union could land on its territory. The same applies to Canada, Australia, and New Zealand, important allies fortunate enough to have large bodies of water between them and the Soviet Union.

Protecting United States interests in the Third World is the least important of the three missions and, in one respect, the least demanding. Any war the United States wages in the Third World almost certainly will be a limited war, fought with conventional weapons alone and with only a part of the country's conventional forces. Winning that war, if it can be won at all, will require only a limited effort and be worth only a limited effort. Furthermore, the United States dares not commit so much of its military power to a Third World conflict as to weaken the forces needed to defend Europe. Winning a war in the Persian Gulf or Southeast Asia would be little consolation for the Soviet conquest of Western Europe.

In other respects, however, the Third World mission is a very difficult one. There is much less popular support for fighting in defense of Third World countries than there is for defending Western Europe. Recent polls, the same ones that show majority support for defending Western Europe, show three- or four-to-one majorities opposed to sending United States combat troops to defend El Salvador, South Korea, Saudi Arabia, or even Israel. The Third World countries which the United States might go to war for generally do not seem greatly important to the average American and usually have rather unattractive governments: weak, unstable, repressive dictatorships. Threats to these countries often take morally ambiguous forms which make it even harder to rally popular support for American intervention. The threat probably will not be an overt invasion crossing a legally established boundary. Instead, it may be a guerrilla insurgency supported and directed from outside but waged primarily by natives of the threatened country, or a combination of invasion and guerrilla war. Countering the threat, particularly if it is a guerrilla insurgency, may require a low-level but extended American commitment rather than the intense but short-lived mobilization American popular opinion favors. It may, for example, require committing a relatively small force to combat for a decade or more. It is not impossible to sustain American popular support even for a long-term military effort in a Third World country, but it is difficult.

The United States cannot defend a Third World country without the cooperation of that country's government and perhaps the cooperation of its neighbors. Often they are difficult governments to work with, much more difficult than the United States's allies in Europe. There tends to be a considerable cultural distance between their ruling elites and the Americans who must deal with them, and sometimes they may be deeply hostile to the United States before (and even after) they ask for help. It could become necessary, for example, for the United States to defend Iran against a Soviet invasion, but the Ayatollah Khomeini would not be a very comfortable ally for the United States – nor the United States for the Ayatollah.

Sometimes it is necessary to go beyond cooperation and for the United States to involve itself deeply in the internal affairs of its ally. A guerrilla insurgency rarely can be overcome without the threatened government making fundamental reforms which the United States may have to compel it to make. Even defeating an invasion may require rebuilding the victim's army, which often cannot be done without offending its ruling elite. It is difficult for the United States to intervene in the internal affairs of a Third World country: its right to do so is questionable, the intervention is almost certain to be resented by the country's people and government, and few Americans know enough about any Third World country to intervene effectively in its internal affairs.

Planning for Third World conflicts is complicated by the inability of the planners to know where United States forces will have to fight, whom they will fight, or what kind of military action they will have to take. The United States could conceivably come to the aid of any one of the 110 or so Third World countries scattered across three continents. It might have to fight the Soviet Union, or more likely a Soviet client such as North Vietnam, North Korea, or Cuba, or perhaps a country which is not in the Soviet camp. It might have to repel an overt invasion, stop the covert infiltration of troops across a border, or defeat an indigenous guerrilla insurgency supplied and directed from outside. Its military action might comprise supplying military equipment to a country at peace, sending military advisers to a country at war, providing air and naval combat support, on up to the commitment of a large land army to combat for several years. To be sure, some threats are more evident than others, evident enough so that the planners can concentrate American resources on countering them. The threat to the Persian Gulf is evident enough to justify establishing a major military base on Diego Garcia island in the Indian Ocean, and sending a carrier task force to that area to counter it. The threat to South Korea is evident enough to justify stationing the Second Infantry Division there. Most potential threats, however, are not clearly evident until they lead to a war, so United States planning for Third World conflicts must be flexible.

United States military forces intended for use in the Third World are not likely to be stationed where they will have to fight. Not only is it difficult to predict where that will be, but also the Third World countries which might need the United States's help rarely permit United States military bases on their territories. Therefore, those forces have to be able to move rapidly thousands of miles to where they are needed. They may have to fight in extremely inhospitable terrain, in desert, jungle, or Arctic tundra. They almost certainly will have to fight where there are few good roads or other transportation facilities. The enemy they fight may be very hard to find. Chances are that he will know the terrain much better than the United States forces, be well adapted to move on it, and will use it to shield himself against United States firepower. The key to victory may be not to overwhelm the enemy but to find him. On the other hand, he is very unlikely to have the tremendous numbers and firepower that the Soviet forces in Europe have. Even in the remote event of a clash between United States and Soviet ground forces (perhaps near the Persian Gulf) the Soviets would probably be able to deploy only a small fraction of their total strength.

To defend its interests in the Third World the United States relies primarily on its surface navy, light infantry, and the sealift and airlift necessary to move military units to distant battlefields. Surface ships, as the most flexible form of military power, the best adapted to meeting unexpected contingencies far from home, and the one most suitable for shows of force, are ideal for this role. Light infantry are easier to transport to a Third World battlefield, easier to supply, and better adapted to the terrain than are the armored and mechanized units intended for the defense of Europe. The Marines and the army's airborne and air-portable units are especially useful because of their great strategic mobility. (If, however, the war were being fought in the Middle East, it might be necessary to use several armored divisions despite the considerable difficulty of getting them there; in desert warfare armored units are greatly superior to other types of ground forces).

Forces earmarked to fight in the Third World should be composed exclusively of volunteers. Popular support for an American war in the Third World is likely to be rather tenuous and it is easier to fight an unpopular war with volunteers than with draftees. Volunteers can also more easily sustain the psychological burden of fighting a prolonged small-scale war. Such a war requires a great deal from a few of a country's citizens and little or nothing from the rest. Some young men must fight it and endure all the stress and sacrifice of combat while others of their age never see military service at all and back in the United States life continues almost unchanged. As the Korean and Vietnam Wars showed, it is difficult as well as unjust to make draftees bear such an unequal burden. If the United States ever reinstitutes the draft, draftees and reservists probably should be called upon to fight only in defense of the United States and Western Europe.

Because the requirements of each mission are so different the United States armed forces can employ only a fraction of their strength in any particular war. Most of the United States military machine was ineffective, or not usable at all, in the Vietnam War. The Soviet armed forces face the same problem. The tremendous Soviet nuclear arsenal played no role whatever in the Soviet Union's war in Afghanistan, nor did the Soviet navy. Even the ground forces, equipped and trained primarily to fight in Western Europe, were not very effective in fighting guerrillas. Any country with global military commitments must be prepared to fight a variety of different kinds of wars and thus faces a similar problem. This is one reason why such small countries as North

Vietnam and Afghanistan have been able to give the United States and the Soviet Union so hard a fight. A small country can devote all of its military resources to its struggle against a superpower, while the superpower can utilize only a fraction of its great military might.

Notes

1 L.I. Brezhnev, *Following Lenin's Course* (Moscow: Progress Publishers, 1972), p. 339.
2 Henry A. Kissinger, *The Necessity for Choice* (New York: Harper and Brothers, 1961), p. 148.
3 Strobe Talbott, "What Ever Happened to Détente?", *Time*, June 23, 1980, p. 34.
4 Earl C. Ravenal, "Europe Without America: The Erosion of NATO", *Foreign Affairs*, 63 (Summer 1985).
5 Quoted in Bernard Brodie, *War and Politics* (New York, Macmillan, 1973), p. 346.
6 John Kenneth Galbraith, "The Plain Lessons of a Bad Decade", *Foreign Policy* 1 (Winter 1970–1), p. 37.
7 Graham Allison, "Cool It: The Foreign Policy of Young America", *Foreign Policy*, 1 (Winter 1970–1), p. 144; John E. Rielly, "American Opinion: Continuity, Not Reaganism", *Foreign Policy*, 50 (Spring 1983), p. 99.

PART III
Strategic Nuclear War

The spectre of extinction hovers over our world and shapes our lives with its invisible but terrible pressure. It now accompanies us through life, from birth to death. Wherever we go, it goes too; whatever we do, it is present. It gets up with us in the morning, it stays at our side throughout the day, and it gets into bed with us at night. It is with us in the delivery room, at the marriage ceremony, and on our deathbeds. It is the truth about the way we now live.

Jonathan Schell, *The Fate of the Earth*

It may well be that we shall, by a process of sublime irony, have reached a stage in this story where safety will be the sturdy child of terror, and survival the twin brother of annihilation.

Sir Winston Churchill

7
Strategic Nuclear Weapons

The Types and Characteristics of Strategic Nuclear Weapons

Strategic nuclear weapons are nuclear weapons[1] of intercontinental range, bombers or missiles able to hit targets in Eurasia from bases in the continental United States, or targets in the continental United States from bases in Eurasia. Submarine-launched ballistic missiles are also strategic weapons because the submarines which carry them can approach close enough to any target in the world to hit it. All others are tactical nuclear weapons. An intercontinental ballistic missile (ICBM), which has a range of 3,400 miles or more, is a strategic weapon; an intermediate range ballistic missile (IRBM), with a range of 1,690–3,400 miles, is a tactical weapon. The key distance of 3,400 miles is the shortest distance between the northeastern United States and the northwestern part of the Soviet Union.

The distinction between strategic and tactical weapons is clear and very important to Americans; only strategic nuclear weapons could be used to attack the United States. It is much less significant to the Soviets and to people in Western Europe, as their countries could be devastated by tactical nuclear weapons. A cynical United States army saying has it that "A tactical nuclear weapon is one that explodes in Germany" – which does not reassure Germans or other Europeans about tactical nuclear weapons. NATO IRBMs could hit much of European Russia from bases in Western Europe, and Soviet IRBMs could hit every part of Western Europe from European Russia.

In calculating the nuclear threat to their country Soviet leaders look not only at the United States's strategic nuclear weapons but also at all the other nuclear weapons which are close enough to the Soviet Union's borders to be launched against it: the missiles and bombers the United States has in Western Europe, nuclear-armed attack aircraft on United States's aircraft carriers, and the nuclear forces of France, China and the United Kingdom. From their point of view there is little difference between an ICBM 7,000 miles from Moscow and an IRBM 2,000 miles away. Both can hit Moscow. Similarly, China and the United States's allies in Western Europe must take into account

Table 7.1   *Strategic nuclear forces of the United States and the Soviet Union*

	United States	Soviet Union
ICBMs	1,000	1,418
SLBMs	640	928
Strategic bombers	317	165
Total strategic nuclear weapons	1,957	2,511
Total bombs and warheads carried by strategic nuclear weapons	13,873	11,044

Figures represent the military balance as at July 1, 1987.
Source: The Military Balance 1987–8 (London: International Institute for Strategic Studies, 1987), p. 225

all of the Soviet Union's tremendous arsenal of nuclear weapons, not just the strategic weapons that could be used against the United States.

There are three types of strategic nuclear weapon: ICBMs, SLBMs (submarine-launched ballistic missiles), and long-range bombers. The United States and the Soviet Union both maintain strategic nuclear "triads," i.e., large numbers of each of the three types of weapon. A bomber can carry several bombs or missiles and most of the ICBMs and SLBMs carry a number of warheads, so the total number of targets a strategic nuclear force can hit is several times the number of strategic nuclear weapons it has. Table 7.1 gives a statistical summary of the strategic nuclear forces of the two superpowers in 1987:

Probably the strategic nuclear weapons which each superpower has in any one category would be sufficient totally to destroy its opponent. The United States could destroy the Soviet Union using only its ICBMs without launching a single bomber or SLBM, and the Soviet Union could do the same to the United States. Having three different types of strategic nuclear weapon simply ensures that, even if the opponent were to find a way to destroy or neutralize weapons of one type, the other weapons would still be effective. Even if, for example, the Soviet Union were to destroy almost all of the United States's ICBMs by a surprise attack with its own ICBMs, American bombers and SLBMs would survive to strike back. Each type is different enough from the other two that whatever measures would be effective against one would probably be useless against the others. So a superpower has the same motive for having three types of strategic nuclear weapon as a householder in a high-crime neighborhood has for having three locks on his door: if one does not work, the other two will.

The four most important characteristics of a strategic nuclear weapon are its striking power, its accuracy, its survivability, and its penetrativity. The most obvious of these is its striking power, the amount of damage it can do to whatever it hits. It is the tremendous power of these weapons that would make any war in which they were used so radically different from any other.

Even a small strategic nuclear weapon can destroy all but a very large city, and the largest can carry 20-megaton warheads or bombs. A 5-megaton explosion is sufficient to destroy any city in the world; the only reason for using more powerful weapons is more reliably to destroy a very well-protected target such as an ICBM silo or an underground command post. Even for that purpose accuracy and numbers are more important than megatonnage, and in fact the very large warheads and bombs have now almost entirely disappeared from nuclear arsenals.

How important a strategic nuclear weapon's accuracy is depends on the target it is directed against. Dropped on a large city it can miss the aiming point by several miles and still do the same amount of damage. Directed against an ICBM silo it must hit within several hundred feet or less to destroy the ICBM. The most modern strategic nuclear weapons can come within a few hundred feet of a target thousands of miles away – in theory. However, the accuracy a missile or a bomber will achieve in combat can never be known for certain because it depends on the performance of those who use the weapon and those who prepare it for use. It depends on the performance of a bomber crew in combat or the accuracy with which the inertial guidance mechanism of a missile has been set. Any strategy which depends on strategic nuclear weapons performing with a very high degree of accuracy is, for that reason alone, very risky.

Accuracy and striking power determine how much damage a strategic nuclear weapon will do to whatever it hits, but it has to reach its target to do any damage. To reach its target the weapon must do two things:

1 Survive enemy attempts to destroy it before it is launched – "survivability."
2 Overcome enemy attempts to destroy or deflect it while it is in flight – "penetrativity."

A weapon's penetrativity is important no matter how it is used. The importance of its survivability depends on the nuclear strategy it serves. If that strategy is to deter any enemy attack by the threat of retaliation carried out after the attack, then the weapon must be able to survive a full-scale enemy attack. The weapon's survivability is extremely important, probably its most important characteristic. If, on the other hand, the strategy is to attack first, to hit the enemy before he launches an attack, then survivability is much less important.

Intercontinental Ballistic Missiles

Intercontinental ballistic missiles are the most important component of the Soviet Union's strategic nuclear force and continue to be an important part of the United States's force. An ICBM is a very large, very long range rocket with a nuclear warhead (figure 7.1). It consists of several sets, or stages, of rocket engines, usually three, each with its own fuel, and the payload which includes one or several nuclear warheads. When an ICBM is launched, the first stage motor ignites and drives it straight up towards outer space. After

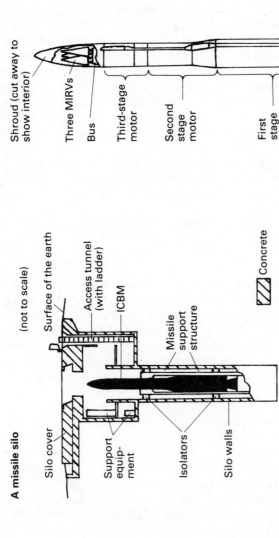

An ICBM – the United States's "Minuteman III"

Scale of feet

0 5 10 15 20

Launch weight: 39 tons

Range: 8,800 miles

Warheads: 3 MIRVs, each of 335 kilotons

Shroud (cut away to show interior)

Three MIRVs

Bus

Third-stage motor

Second stage motor

First stage motor

The Minuteman III is a three-stage, solid fuel ICBM. It entered service in 1970, and 550 of the type were deployed. They have been upgraded several times since their initial deployment.

A missile silo

(not to scale)

Surface of the earth

Access tunnel (with ladder)

ICBM

Silo cover

Support equipment

Isolators

Missile support structure

Silo walls

▨ Concrete

The isolators soften the shock to the missile of any nearby nuclear detonation, and the access tunnel allows technicians to enter the silo when necessary. Normally the silo is left unattended. The silo cover is shown rolled back. (This cutaway drawing shows the features of a typical missile silo but does not depict any particular installation.)

Figure 7.1 *An intercontinental ballistic missile (ICBM) and a missile silo*

a minute or two the first stage has exhausted its fuel and is jettisoned, and the second stage motor ignites. That one in turn soon drops away and the third stage takes over. As the missile rises its guidance mechanism swings it around on the path it must follow to hit the target. After 3–5 minutes of powered flight the missile is in outer space several hundred miles above the earth, arcing towards its target, and moving at about 1,600 miles an hour. At this point the payload separates from the third stage. Twenty-five minutes later the missile's warheads coast back to earth and explode.

Early-model ICBMs carried only a single nuclear warhead. Many of these remain in service today but most modern ICBMs are equipped with multiple independently targetable reentry vehicles (MIRVs). A MIRVed missile is one that carries several warheads, from three to ten on currently-deployed models, and can launch each warhead accurately at a separate target. The individual warheads are known as MIRVs and the missile's payload, which carries and launches the MIRVs, is the "bus." The bus has an array of small manoeuvering rockets and its guidance system has been programmed with the appropriate manoeuvers to launch its warheads. As it is coasting through space it manoeuvers so that it is pointing in the right direction and moving at the right speed for the first MIRV to hit the first target. It then releases that MIRV. Next it changes direction and speed slightly and releases the second MIRV to hit another target, and so on until all the MIRVs have been released. The targets have to be within an area about 90 miles by 30 miles, but even so one MIRVed ICBM could destroy every major city in Connecticut or New Jersey.

Each of the individual MIRVs carried on a MIRVed missile is, of course, much smaller and less potent than the single warheads that the missile could carry. In fact, all of them together weigh less and have a smaller explosive yield than the single warhead because of the weight of the bus which carries them. The warhead on the single-warhead version of the SS-18 has an explosive yield of 20 megatons while its MIRVed version has ten MIRVs, each of 500 kilotons (a total of 5 megatons). However, a number of smaller nuclear explosions will do as much damage to a city as one big one of considerably greater yield because the effects of the smaller ones are more evenly distributed. The authors of *The Effects of Nuclear War* calculate that ten 40-kiloton detonations on Leningrad would kill and injure as many people as a single one-megaton explosion. The MIRVed missile can do far more damage than the single warhead missile if each MIRV hits a different city, and its advantages for attacking missile silos are even greater.

Designing an ICBM guidance mechanism is one of the most difficult tasks facing weapons designers. It is difficult not only because of the tremendous range of the weapon, but also because an ICBM is powered, and hence controllable, only for the first 5 minutes (at most) of its 30-minute flight. Once the reentry vehicle has separated from the last of the rocket stages its direction and velocity can no more be corrected than can those of a bullet once it has left the barrel of a rifle. The slightest inaccuracy at this point results in a much greater inaccuracy at the point of impact. (A MIRVed missile's bus continues to manoeuver after the last rocket stage drops off, but its manoeuvers are already programmed into its guidance system and cannot be changed.) A microscopic particle of dust on one of the components of an ICBM's guidance system can make the missile miss its target by more than 100 feet. Stopping the powered portion of its flight one-thousandth of a second too soon or too late will make it miss by nearly half a mile. For an ICBM to hit anywhere near its target requires a degree of precision in manufacturing, installing, and calibrating its guidance system so extraordinary as to be almost inconceivable to the non-quantitative mind.

In the face of these obstacles, American and Soviet weapons designers have during the past quarter century greatly improved the accuracy of ICBMs. The American Atlas D and the Soviet SS-7, built during the late 1950s, had CEPs of about a mile; the latest models of the Minuteman III and the SS-18 have CEPs of about one-eighth and one-fifth of a mile, respectively. (The CEP or "circular error probable" of a missile is a measurement of how close to the target it will land if its guidance system has been programmed with perfect accuracy. A CEP of one mile means that it has a 50–50 chance of landing within a mile of the target.)

This does not mean that if the missiles were ever used in a nuclear war they would attain such accuracies. A missile's CEP is a theoretical maximum, only attainable if its guidance system is programmed with perfect accuracy, which can be done on a test range but would be nearly impossible in a nuclear war. Not knowing exactly where the target is, inaccurate calibration of the guidance system, unexpected strong winds over the target, and minute unmapped variations in the earth's magnetic field can all cause a missile to miss a target which, judging by its CEP, it should be able to destroy. On the other hand, missiles have become far more accurate in recent years and their accuracy is continually being improved.

ICBMs and other ballistic missiles have excellent penetrativity; once they are launched it is extremely difficult to destroy or deflect them in flight. During the 1960s ground-based anti-ballistic missile (ABM) systems were developed to destroy ballistic missile warheads in flight. A 1960s type ABM system is a combination of long-range radars to spot the incoming warheads, short-range radars to track them and guide ABM missiles to intercept them, nuclear-armed ABM missiles, and powerful computers to operate the system.

Destroying a single warhead with an ABM missile is difficult enough, "hitting a bullet with another bullet" although it has been done in tests. Stopping a full-scale missile attack, even with a very large ground-based ABM system, would be many times more difficult. An ICBM bus can launch numerous decoys which cannot be distinguished from its MIRVs until the latter are 30–60 seconds away from their targets. If the ABM system launches its missiles as soon as enemy warheads are detected coming in, it will waste most of them on decoys and not have enough left to deal with the real warheads. If it waits until the decoys and the warheads can be told apart, then it has 30–60 seconds to track and destroy several thousand enemy warheads. It must do so even though when the first ABM missiles detonate their nuclear warheads they temporarily blind the radars guiding the other ABM missiles to their targets. Nevertheless, to defend cities against attack the system must destroy virtually all of the enemy warheads; if even one of the dozens that could be launched against a city were to get through, the city would suffer catastrophic damage.

Ground-based ABM systems were not found to be worthy of deployment, except to a very limited extent, so neither superpower can now defend its cities against a full-scale attack. Revolutionary new technologies may make the defense of cities possible many years from now; the Reagan administration's Strategic Defense Initiative (to be discussed in chapter 19) proposes to do this. However, even ABM enthusiasts do not promise that ABMs will be able to protect an entire country any time in the twentieth century.

The Achilles' heel of the ICBM is its questionable survivability. An ICBM is protected against attack by being housed in a missile silo, a massively-constructed concrete tube whose top end is flush with the surface of the ground (see figure 7.1). The heavy door on top of the silo can be slid back to fire the missile which sits pointing upwards in position to fire and is protected against shock damage by a system of springs and shock absorbers. The command post which controls the missile is several miles away, deep underground and similarly well-protected. Such installations probably could survive a nuclear detonation a thousand feet away, and during the 1960s and the 1970s that was ample protection against any possible attack.

It is not ample protection today. Improved missile accuracy and MIRVs have made ICBM silos much more vulnerable to attack by enemy ICBMs. MIRVs make it possible to send more warheads against each silo and improved accuracy makes each warhead more likely to destroy its target. Together they make it possible for each attacking ICBM to destroy several of the defender's missiles in their silos whereas in the single-warhead days the attacker would have had to expend several of his missiles to destroy one of the defender's. The Soviet Union could today destroy 90 percent of the United States's ICBM force using only a fifth of its own, its 308 SS-18s. Similarly, if the United States were to deploy them, 300 American MXs could destroy over 90 percent of the Soviet Union's ICBM force. The threat that ICBMs pose to the survivability of an opponent's ICBMs may not be as great as that; the figures cited are based on theoretically-attainable CEPs rather than on combat experience. However, the threat has grown in recent years and continues to grow.

Several methods of improving the ICBMs' survivability have been proposed. One option – building stronger missile silos – cannot be exploited much further. The physical limit on silo "hardness" is about 5,000 psi and even a silo able to withstand this enormous overpressure could be overcome by warheads only slightly more accurate than those in service today.

Another proposed option is to adopt a "launch-on-warning" policy, that is to launch one's own ICBMs immediately after the enemy launches his and before his warheads arrive at one's ICBM silos. It is technologically feasible. Early-warning satellites can detect the launching of an enemy missile almost as soon as it happens and modern ICBMs can be launched on only a few minutes' warning. This policy would require that the decision to launch the threatened ICBMs be taken within 10 or 20 minutes – the most momentous decision any political leader could conceivably make, and very likely to be the last. An even more serious difficulty is that even the best early warning system can issue a false alarm and once a missile has been launched it cannot be brought back to its base. There have already been several false warnings of an all-out Soviet attack; a launch-on-warning policy would make it possible for World War III to occur because of a malfunction in a radar or a reconnaissance satellite.

Perhaps a more promising option is to defend ICBM silos with an ABM system. An ABM defense of missiles is much more feasible than an ABM defense of cities because it does not need to work nearly as well. For even a few warheads to get through and hit the defender's cities would be an almost unendurable catastrophe, but a few warheads hitting missile silos would still leave the defender plenty of ICBMs to retaliate with. Under the 1972 ABM Treaty, however, neither superpower can protect more than a few of its ICBMs this way.

It is possible to protect an ICBM by preventing the enemy from knowing where it is, by concealing it, moving it around, or some combination of the two. The Carter administration planned to protect the United States's MX missiles by a multiple protective structure (MPS) system. Each of 200 MX ICBMs would have had 23 protective shelters from which it could be launched with a heavy duty road linking up all the shelters and an 800-ton transporter to carry the missile. This behemoth would grind around and around on the road, stopping at and entering a shelter from time to time and perhaps leaving the missile there, or perhaps not. Playing this gigantic shell game would ensure that the Soviets could never know which shelter contained the missile and would have to hit all 23 to be sure of destroying it. The MPS system would have been, as Herbert Scoville Jr points out, "the largest construction project in history ... surpass[ing] the Panama Canal, the Alaska Pipeline, even the pyramids."[2] It would have cost perhaps 60 billion dollars and occupied 40,000 square miles, an area equal to most of New England. President Reagan cancelled it.

Other proposals would have the MXs based on barges floating around in the Great Lakes, in underwater capsules anchored to the seabed, on very large railroad cars moving around on the common carrier railroads, and in holes 2,000 feet deep covered with quicksand through which the missile would rise to be launched. Over 40 MX basing options have been studied, most relying on mobility or concealment, and most of them more imaginative than feasible.

The Reagan administration proposed to protect MX missiles by a system termed "Dense Pack." Dense Pack exploits "fratricide," the tendency of an exploding nuclear warhead to disable other (unsheltered) warheads in its vicinity. It involves placing a cluster of MX silos very close together, close enough so that the explosion of a warhead over one would disable any warheads descending on the others: only one silo in the cluster would be destroyed. The idea remains untested, which is why Congress refuses to fund Dense Pack. The survivability of ICBMs still seems very dubious.

Ballistic Missile Submarines

The best way to protect a ballistic missile is to send it to sea. Submarines armed with ballistic missiles are by far the most survivable type of strategic nuclear weapon, which makes them the most important element of the United States's strategic nuclear force. A typical missile submarine, a United States "Poseidon" type, is nuclear-powered, 425 feet long, and carries 16 SLBMs (figure 7.2). On patrol it stays underwater continuously for several months, submerging a few miles outside port, moving randomly around its patrol area as quietly as possible, and coming to the surface again only when it returns to port. It remains underwater even to fire its missiles; they are thrust up to the surface and ignite above the water. Each of its Poseidon C3 SLBMs is a MIRVed missile with ten 50-kiloton MIRVs, so one submarine could hit 160 targets in the Soviet Union. The United States has 28 Poseidon boats, and eight much larger Trident submarines. The Tridents carry 24 SLBMs, each with eight 100-kiloton MIRVs and as great a range as an ICBM. The Soviet Union has 62 missile submarines of types roughly comparable to the Poseidon and Trident boats and a few older and much less capable ones.

Locating and destroying a large missile submarine force is an almost impossible task. Submarines on patrol and constantly in motion are much harder to destroy than ICBM silos or bomber bases. They are harder to locate and destroy than other types of submarines because of the great range of their weapons. A torpedo-armed submarine must approach to within several tens of miles of a ship it wishes to sink, bringing it within range of its intended victim's anti-submarine weapons, but a missile-armed submarine can stay thousands of miles away from its target. A Trident submarine can be anywhere within 24 million square miles of ocean and still be close enough to Moscow to hit it. Enemy anti-submarine forces must seek it out rather than wait for it to come to them. In addition, any attack on a force of missile submarines must destroy all of them virtually simultaneously. The destruction of just part of the force would cause the survivors to launch their missiles, and even a few missile submarines could inflict catastrophic damage on any country in the world.

One way to attempt to counter a missile submarine force is to carry out an "open-ocean search," that is search the entire area in which enemy missile submarines might be, determine the location of each one, and then destroy it. Sonar equipped aircraft and ships can do this but not nearly fast enough; even an extremely large anti-submarine force consisting of several thousand ships and aircraft would need several days to search an entire ocean and destroy all the missile submarines beneath it. An array of hydrophones (underwater microphones) on the bed of that ocean might work better. Under ideal conditions a hydrophone can detect a submarine several thousand miles away by the sound it makes moving through the water, so an integrated system of hydrophones can search an entire ocean and continually follow the movements of enemy submarines within it. Once all the submarines have been located each one could be hit with a sudden barrage of missile warheads.

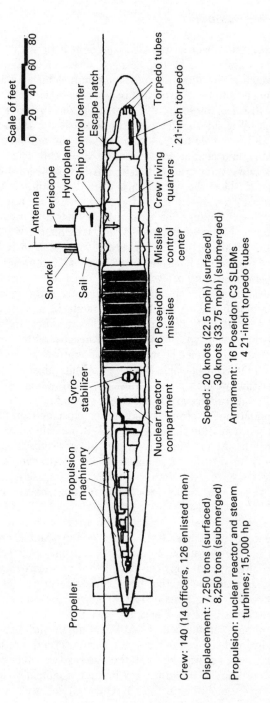

Scale of feet
0 20 40 60 80

Antenna
Periscope
Snorkel
Hydroplane
Ship control center
Escape hatch
Sail
Torpedo tubes
21-inch torpedo
Crew living quarters
Missile control center
16 Poseidon missiles
Gyro-stabilizer
Nuclear reactor compartment
Propulsion machinery
Propeller

Crew: 140 (14 officers, 126 enlisted men)

Displacement: 7,250 tons (surfaced)
8,250 tons (submerged)

Propulsion: nuclear reactor and steam
turbines; 15,000 hp

Speed: 20 knots (22.5 mph) (surfaced)
30 knots (33.75 mph) (submerged)

Armament: 16 Poseidon C3 SLBMs
4 21-inch torpedo tubes

The 19 "Lafayette" class boats were originally fitted with "Polaris" SLBMs but have carried "Poseidon" missiles for most of their service lives. In a strategic nuclear war they would play the role of attack submarines once they had fired off their missiles, which is the primary reason why they are armed with torpedoes as well as missiles. The nuclear reactor generates steam to power the turbines which drive the boat.

Figure 7.2 *A missile submarine – a United States Poseidon submarine of the "Lafayette" class .*

Long range hydrophone detection is not very accurate, however; at best it can tell where the submarine is within several tens of miles. This is because of the time it takes sound to travel underwater. The sound emitted by a submarine takes about an hour to reach a hydrophone 3,000 miles away and in that time the boat, traveling at its normal speed, will have gone 10 or 20 miles. To be sure of destroying the submarine one must saturate with nuclear explosions the entire area of several hundred square miles within which it could be. This requires a large number of warheads, probably more than the submarine's SLBMs carry.

Another great weakness of this method is the inability of a hydrophone array to locate all of the enemy's submarines. There are too many other sources of sound in the ocean – waves, rain, whales, surface ships, oil-drilling rigs, etc. – and the sound made by a slowly-moving submarine is too feeble for long-range hydrophone detection to be as reliable as it needs to be. As Kosta Tsipis puts it, "Trying to detect a nuclear submarine somewhere in the Atlantic Ocean by listening for the sounds it makes is comparable to trying to find a person wearing wooden clogs somewhere in the streets of Los Angeles by listening for the sound of his steps."[3] The hydrophones can locate some enemy submarines but not all of them, and successfully to attack a force of missile submarines one must know where all of them are.

Another possibility is to employ "trailing" tactics against enemy missile submarines. This entails stationing attack submarines outside enemy missile submarine bases. Each missile submarine, when it leaves port, is followed by one or several attack submarines, rather like policemen tailing a suspect. The attack submarines stay several miles behind the missile submarine, using sonar to maintain contact, and are prepared to sink it when given the word. This seems rather a clever alternative to searching the entire ocean, but in practice a resourceful submarine commander can usually shake any pursuer. He can send out a noise-maker that simulates the sound of his submarine, move along the coastline where the "trail" cannot follow him and then make a dash for the open sea, conceal his boat amidst a school of whales, or resort to many other stratagems to foil the pursuit.

"Trailing" tactics or the combination of an ocean-wide search and missile barrages might destroy a few missile submarines very quickly and perhaps many more over an extended period of time. Neither could do what an attack on a missile submarine force would have to do: destroy all of it very quickly. Thus, the missile submarine forces of the superpowers are, and for the foreseeable future will remain, essentially invulnerable to enemy attack.

One defect of missile submarines is that it is impossible to keep them on patrol all the time. They need to spend a lot of time in port for routine maintenance and periodic major overhauls. The United States navy has reduced the "in port" time of its missile submarines to the minimum by having two complete crews for each boat, but even so only slightly more than half of them are on patrol at any given moment: 20 out of a total force of 36. The Soviets keep on patrol only six or seven of their 62 modern missile submarines. Boats in port are easy targets and almost certainly would not survive an enemy missile attack.

Strategic Bombers and Cruise Missiles

Bombers were the first strategic weapons to be developed and were the mainstay of strategic nuclear forces for many years (see figure 7.3). Many people believed that the development of ICBMs and missile submarines would make bombers obsolete, but it has not. Strategic bombers are still an important part of the United States's strategic nuclear force, constituting about two-fifths of its striking power, and the Soviet Union also has a powerful strategic bomber force. They continue to be important strategic weapons because, although they have some obvious defects, they also have important advantages over ICBMs and missile submarines.

The strategic bomber has the most striking power of any nuclear weapon. Even the most powerful ICBMs and SLBMs carry no more than ten warheads, but a B-52 bomber can carry 24 bombs and other nuclear weapons, each with as much striking power as the individual warheads on a missile. The United States's strategic bomber force thus has more striking power than its ICBMs even though there are three times as many ICBMs.

The survivability of strategic bombers is now greater than that of ICBMs but less than that of missile submarines. A bomber on the ground is extremely vulnerable to attack; a one-megaton explosion 12 miles away will disable it. However, if it is on "ground alert" – ready to fly, fully loaded with bombs and fuel and with the crew nearby – it can take off in less than 15 minutes. As there would be almost 30 minutes warning of an enemy ICBM attack, the bomber would have enough time to get into the air and far enough way to survive a nuclear explosion on its airfield. (This is, of course, a "launch-on-warning" policy but "launch-on-warning" is much less dangerous with bombers than with ICBMs because a bomber can return to its base if the warning turns out to be a false alarm.) At present about 30 percent of the United States's strategic bombers are kept on ground alert at all times.

An attack on bomber airfields carried out by missile submarines launching their SLBMs from nearby coastal waters could be a much greater threat. The flight time of a Soviet SLBM from the Atlantic Ocean to some United States bomber bases could be as little as 6 minutes. It may, however, be possible to get a bomber into the air in less than 6 minutes (the exact reaction time of American bombers on ground alert is a very close-guarded secret.) The bombers could also be redeployed to bases distant from the ocean where they would be less vulnerable to SLBM attack. It is even possible to keep a few bombers in the air and bombed-up at all times. From 1961 to 1968 the United States had a few bombers in the air, on "airborne alert," at all times, but this is an extremely expensive practice. It uses up tremendous amounts of fuel and wears out both the aircraft and their crews very rapidly.

The great weakness of the strategic bomber is its questionable penetrativity. There is a good chance that it will be shot down before reaching the target. It is much easier to shoot down than a missile warhead: much bigger, much slower (600 miles an hour rather than 16,000), and requiring much more time to reach the target. It can be destroyed fairly easily by an interceptor or

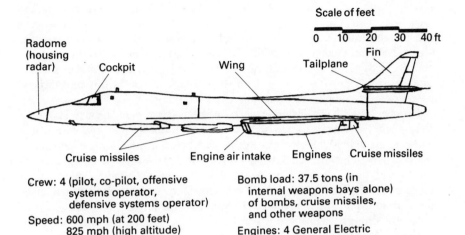

Scale of feet

0 10 20 30 40 ft

Radome (housing radar)

Cockpit

Wing

Tailplane

Fin

Cruise missiles   Engine air intake   Engines   Cruise missiles

Crew: 4 (pilot, co-pilot, offensive systems operator, defensive systems operator)

Speed: 600 mph (at 200 feet) 825 mph (high altitude)

Range: 7,455 miles (unrefueled)

Bomb load: 37.5 tons (in internal weapons bays alone) of bombs, cruise missiles, and other weapons

Engines: 4 General Electric turbofans with afterburners

Wings swept back

Wings extended

All 100 of the B-1Bs to be built for the Strategic Air Command are now in service. The B-1B is designed for low-level attack, coming in as low as 200 feet from the ground to frustrate enemy radars. It can also launch cruise missiles. On a typical nuclear war mission a B-1B would launch several cruise missiles from well outside enemy territory, and then penetrate enemy defenses to attack other targets with bombs and SRAMs (Short Range Attack Missiles). It is a "variable geometry" airplane: the outer wing panels are extended straight out for landings and take-offs, and swept back for high-speed flight. (See top views below). Extending the wings increases the lift and controllability in low-speed flight; folding them back reduces the wind resistance for high-speed flight.

Figure 7.3 *A strategic bomber – the United States's B–1B*

an anti-aircraft missile. There has been considerable debate over how many, if any, of the United States's bombers would reach their targets in a nuclear total war with the Soviet Union. The Soviet Union has devoted very considerable resources to its anti-bomber defenses and they are formidable: about 1,800 interceptors and 9,000 anti-aircraft missiles – six interceptors and 28 missiles for every strategic bomber the United States has in service. Since current plans direct each bomber to attack ten or 12 different targets, its chances of completing a mission seem remote.

On the other hand, when American B-52s were sent against Hanoi in 1972–3 they managed to penetrate what were probably the most formidable anti-aircraft defenses of any city in the world at that time. Only 3 percent of the bombers sent on each mission were shot down, which is comparable to the rate of losses suffered in World War II strategic bomber attacks. In a nuclear war 3 percent, or even 30 percent, losses would not stop a bomber force from wreaking tremendous devastation. Soviet anti-bomber defenses today are stronger than those which defended Hanoi, but the equipment and tactics of American strategic bombers have also been improved.

There are several ways to ensure the penetrativity of a bomber force. If the aircraft come in at a very low altitude, as American strategic bombers would in an attack on the Soviet Union, ground-based radars cannot detect them except at very short range. It is possible to detect low-flying aircraft with ground-based OTH-B (over-the-horizon backscatter) radars – a new technology which still has many limitations – or with "look-down" radars on air defense aircraft. On the other hand, modern "stealth" technology is making aircraft harder and harder for any kind of radar to detect. Also, bombers can attempt to jam or spoof enemy radars and missile-guidance systems: American strategic bombers carry powerful equipment for that purpose. The attacker can also destroy the defender's radar installations, anti-aircraft missiles, and interceptor bases with missile warheads, thus clearing a path for the bombers arriving several hours later.

Probably the best way to ensure a bomber's penetrativity is to equip it with weapon that can be launched beyond the range of the enemy's air defense system. To this end the United States has begun to equip its strategic bombers with air-launched cruise missiles (ALCMs). A cruise missile is a small pilotless bomber which is guided to its target either by radio instructions from its launch control center or by its own internal guidance mechanism. Cruise missiles can also be launched from the ground, from a surface ship or from a submerged submarine, as well as from an airplane.

Cruise missiles have been around for many years – the Germans used them during World War II – but until recently they were too inaccurate to be very effective. Then in the 1970s American weapons designers developed several types of really effective cruise missiles using a new guidance system. This new system, TERCOM (terrain contour matching), uses a computerized map of the terrain the missile is to fly over, showing the elevation of each part of its intended course, and a radar altimeter which shows the elevation of the terrain it actually is flying over. By matching the elevation of each part of the intended course with that of the missile's actual course the guidance system can tell if the missile is going where it should and make the necessary corrections for it to do so. Because a cruise missile, unlike a ballistic missile, is in powered flight and hence controllable all the way to the target, it is inherently the more accurate weapon. Cruise missiles guided by TERCOM are accurate to within about 300 feet.

The result of TERCOM and other sophisticated technologies is a strategic nuclear weapon which is only 21 feet long, comparatively inexpensive (about a million dollars apiece compared with 20 million for an MX ICBM), and able to deliver its nuclear warhead more accurately than any other strategic nuclear weapon. A B-52 can carry 20 ALCMs and launch them from well outside the borders of the Soviet Union. They fly 100 feet above the ground and are almost invisible to radars (an ALCM is about as detectable on a radar screen as a seagull), making them difficult to shoot down. They can be hit, but it is much harder to shoot down 20 small cruise missiles than one very large B-52.

In addition to its questionable penetrativity, another defect of the strategic bomber may be the long time it takes to reach its target. While the flight time of an ICBM is 30 minutes and that of an SLBM may be only 6 minutes, a bomber would take about 10 hours to fly from the United States to the Soviet Union. This is a major defect for any country whose strategy is to attack first and destroy the enemy's strategic nuclear weapons before they can be used. By the time the attacker's bombers reached the enemy's missile silos the ICBMs in them would have been launched. It is not a real defect, however, if the strategy is to deter attack and to launch the strategic nuclear force only after an enemy attack.

Bombers have two important advantages over ICBMs and SLBMs. One is flexibility. A bomber is the only strategic weapon which can seek out a target whose location was not known when it was launched. Unlike a missile, it can be redirected while in flight to attack a different target, to avoid enemy attempts to intercept it, or to take advantage of newly-made gaps in enemy defenses. It is flexible also in the sense that it can employ a variety of different tactics. The B-52 was designed for high-altitude attacks but it can also attack at a very low altitude, as American B-52s would in a nuclear war with the Soviet Union today, and most are now able to attack with ALCMs: three very different tactics. Bombers are also the only strategic weapons which can be used in conventional wars, as was shown by the use of B-52s in Vietnam. No ICBM or SLBM has ever been fitted with a conventional warhead: a missile can only be used once and large missiles are too costly to justify using with conventional warheads. A bomber, however, can be flown many times.

The other great advantage of the strategic bomber is that it is a known quantity; it is the only type of strategic nuclear weapon which has ever been in a war. No country has ever employed an ICBM or SLBM in a war and the United States has never even successfully fired an ICBM from an operational silo. (Four unsuccessful attempts were made, since when all American ICBM firings have been made from special test silos.) Modern strategic bombers have made thousands of flights, both in peace and at war, under all conceivable climatic conditions. On paper ICBMs and SLBMs look very impressive, but there is substantial doubt as to how they would perform in a real war. The bomber may not look as impressive but there is no doubt that it works.

The Command and Control of Strategic Nuclear Forces

No matter how many missiles and bombers it has, a strategic nuclear force is useless unless it also has a well-developed command and control system. The command and control system must:

1 Detect any enemy attack or preparations for it.
2 Convey adequate information about the attack to whoever would make the decision to strike back.
3 Communicate the order to strike back to the bomber bases, the ICBM launch-control posts, and the missile submarines.
4 Survive enemy attempts to destroy it before it has carried out its mission.

To do these things requires a complex and extensive array of sensors, command posts, and communication links.

The United States's strategic command and control system extends all over the world and is manned by some 25,000 servicemen. It has about 30 reconnaissance and communications satellites in orbit at all times. Early-warning satellites, in geosynchronous orbits, hover over Soviet missile fields and areas where Soviet missile submarines might be on patrol; they carry infrared sensors capable of detecting the launching of a ballistic missile within 2 minutes after it happens. Photoreconnaissance satellites fly over the Soviet Union several times a day carrying infrared sensors which work at night as well as in the daytime and cameras sharp-eyed enough to pick out a human figure from the satellite's orbit 160 miles above the earth. Other "ferret" satellites listen for radio transmissions. The satellites are supplemented by radar stations. Eight massive phased-array radar stations, on the borders of the United States and in Greenland and the United Kingdom, search the skies in all directions for enemy missile warheads. DEW Line radars, in the Far North, watch for enemy bombers. There is also SOSUS (sound surveillance system), a network of hydrophones constantly searching the North Atlantic, the North Pacific, the Barents Sea, European waters, and the seas of northeast Asia for Soviet submarines (see map 7.1).

Data from the reconnaissance satellites, the radar stations, and SOSUS goes to North American Air Defense Command (NORAD) headquarters deep inside a mountain in Colorado. There it is continually being collated into a comprehensive picture of what the Soviet strategic nuclear force is doing. From NORAD headquarters warning of a Soviet attack would be sent to the president, who has with him at all times the communications equipment necessary to order a counterstrike. Arrangements have been made to get the president and other key personnel very quickly aboard a specially equipped Boeing 747 from which they could communicate with the entire United States

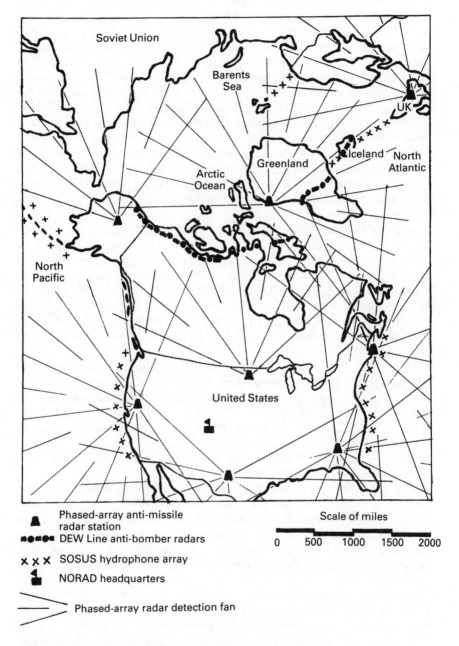

▲	Phased-array anti-missile radar station
●■●■●■	DEW Line anti-bomber radars
✗ ✗ ✗	SOSUS hydrophone array
◢	NORAD headquarters
——	Phased-array radar detection fan

Scale of miles

0   500   1000   1500   2000

Map 7.1 *The United States's strategic command and control system*
Adapted from: Ashton B. Carter, John D. Steinbruner and Charles A. Zraket (eds), *Managing Nuclear Operations* (Washington, DC: Brookings Institution, 1987), p. 312; John M. Collins, *US–Soviet Military Balance: Concepts and Capabilities, 1960–1980* (Oxford: Pergamon Press, 1985), p. 160.

strategic nuclear force. There is another flying command post, with an air force general on board, actually in the air 24 hours a day. (There are several aircraft and several generals assigned to this duty, each working an 8-hour shift.) Even if the president and all the rest of the United States government were destroyed, the general in that airplane could order the launching of air force ICBMs and strategic bombers. The navy keeps two flying command post aircraft in the air at all times, one over the Atlantic and one over the Pacific, to communicate with its missile submarines.

The Soviet Union also has an elaborate command and control system. The Soviets have nothing like SOSUS but they do have reconnaissance satellites and thousands of early-warning and air-defense radar stations. They also have numerous strongly constructed blast shelters for the ruling elite and other essential personnel, some flying command posts, and a secret subway line to carry Soviet leaders from the Kremlin to the airport from which they would be flown to safety.

Strategic command and control systems are vulnerable to attack. "Fewer than 100 judiciously targeted nuclear weapons could so severely damage U.S. communications facilities and command centers that form the military chain of command that the actions of individual weapons commanders could no longer be controlled or coordinated."[4] Reconnaissance satellites might be destroyed in orbit or neutralized by destroying the very vulnerable ground stations to which they transmit the information they have gathered. Large ground-based radars are easy targets, destructible by nuclear explosions several miles away.

Underground command posts are about as vulnerable as missile silos. They are hardened to withstand tremendous overpressures, but also are in fixed, known, easy-to-target locations. Even NORAD's headquarters might be destroyed by very large warheads exploding on its access tunnels. Flying command posts are the most survivable element of a command and control system, because of the difficulty of locating an airplane in the air a continent away. However, even they may soon become vulnerable; satellite-mounted radars able to track aircraft in flight are under development.

The most vulnerable part of a strategic command and control system probably is its apex, the chief executive. The President of the United States or the Secretary-General of the Communist Party of the Soviet Union cannot live in a blast shelter or on a flying command post – they have countries to run. A surprise attack might allow political leaders only a few minutes to get to a place of safety, and the arrangements for doing so are very likely to break down in an emergency.

Destroying its command and control system would not necessarily neutralize a strategic nuclear force. Submarine commanders and ICBM launch-control officers, knowing or suspecting that their country had been attacked, might well launch their missiles without being told to. They might even have been given standing orders to do so. That is the best protection against a "decapitating" attack, the uncertainty of its results. The attacker might hope that destroying his victim's command and control system might prevent retaliation, but he could not be sure that it would not instead lead to every surviving nuclear weapon hitting back at him. The latter is the more likely possibility.

Important as it is to be able to launch the bombers and missiles if the time comes, it is at least as important to ensure that they are not launched until then. A strategic nuclear force must have a safety catch as well as a trigger. In fact, the safety catch must be much more reliable than the trigger. The trigger must ensure only that most of the weapons are launched if the time comes to use them, but the safety catch must ensure that not a single one of several thousand nuclear weapons is ever launched or detonated accidentally. The accidental launching of even a single missile could cause a nuclear total war. In contrast, if there were to be a war despite all attempts to prevent it, failure to use one or two of the thousands of nuclear weapons available would not make much difference.

In principle there are three possible causes of an accidental nuclear war:

1 A nuclear warhead might explode accidentally, thus leading the country on whose territory it exploded to believe it was under attack and launch its weapons.
2 An individual might launch a nuclear weapon on his own initiative.
3 A radar station or a satellite might malfunction or a technician make a mistake and issue a false alarm, indicating that the enemy was attacking when he was not.

All of these are remote possibilities, and much ingenuity and expense have been devoted to making them as remote as possible, but they can never be remote enough as long as nuclear weapons exist.

The first possibility is not worth worrying about; there is no danger that an ICBM warhead will go off while the missile sits in its silo or a hydrogen bomb inside the bomber carrying it. Nuclear weapons have one (and only one) benign characteristic: unlike conventional explosives they cannot explode spontaneously or as the result of a shock. They must be deliberately detonated by somebody. Nuclear warheads and bombs have been accidentally dropped out of airplanes, involved in crashes, and melted down in fires without exploding. Even an explosion a few feet from a nuclear warhead would simply scatter the warhead's fissionable material rather than detonate it. To keep a nuclear warhead from exploding all that is necessary is to ensure that nobody deliberately sets it off.

To make sure that nobody can start a nuclear war on his own initiative the United States armed forces have for several decades followed the "two-man rule" in all handling of nuclear weapons. At least two people must be present whenever anything is done with or to a nuclear warhead or the computer codes necessary to detonate it. At least two carefully-selected officers must act independently to detonate or launch a nuclear weapon. To launch an ICBM both of the officers in the command post which controls it must turn the keys in their consoles within 2 seconds of each other *and* two other officers in another, distant command post must do the same thing. When a bomber is in the air, with nuclear weapons on board, three members of its crew must act independently to arm and launch the bomber's weapons. A missile submarine cannot launch its missiles except by the joint action of at least four of its officers. Modern nuclear warheads are also protected by permissive action links (PALs), locks into which the correct computer code has to be entered to arm the warhead. The Soviet armed forces have taken similar measures to prevent unauthorized use of their nuclear weapons. Not much more could be done without weakening the system's ability to respond to an attack. It is extremely unlikely that anybody could launch a nuclear weapon purely by his own efforts or that all of the people who would have to collaborate to launch one would simultaneously be so psychotic as to do so. One would like to know for certain that it is impossible.

Satellites, radar stations, and other elements of a command and control system certainly can malfunction and issue a false alarm. This has happened several times in the past to the United States command and control system. A flock of geese, radar waves bouncing off the moon, and the failure of a 46 cent computer chip have all been misread as evidence of a Soviet attack coming in. Presumably the mechanical defects that caused those false alarms were corrected, but nothing as complex as the United States strategic command and control system can ever be defect-free.

Fortunately, the size of a superpower's strategic command and control system and the number of indications of an enemy attack the system would generate make it almost impossible for the malfunctioning of a single sensor to cause a nuclear war. An all-out Soviet attack would send a flood of data flowing into NORAD headquarters from all over the world. The SOSUS hydrophones would probably have tracked Soviet missile submarines putting out to sea and moving to their battle stations days or weeks before the attack. Photoreconnaissance and ferret satellites might have caught military units being mobilized and moved into position. The early warning satellites would detect several thousand Soviet missile launchings and the radar stations would pick up the warheads arcing in towards the United States. Knowing this, the commanders of NORAD would not be greatly concerned if a single malfunctioning satellite or radar station were to report an attack. They might order a higher level of alert but they would not decide that an attack was under way unless all the other sensors reported it.

Another protection against a false alarm leading to a nuclear war is the survivabiity of strategic nuclear weapons. If one's strategic nuclear force can survive an all-out enemy attack there is no need to launch it on warning of the attack. One can wait to see whether the attack actually occurs. Having a survivable strategic nuclear force is thus not only the best protection against a deliberate enemy attack, but also very good protection against a nuclear war being triggered accidentally.

Notes

1 A bomber or a missile by itself is a "delivery vehicle;" carrying a nuclear warhead or bomb it is a "nuclear weapon."

2 Herbert Scoville, Jr, *MX: Prescription for Disaster* (Cambridge, Mass.: MIT Press, 1981), p. 169.
3 Kosta Tsipis, *Arsenal: Understanding Weapons in the Nuclear Age* (New York, Simon and Schuster, 1983), p. 225.
4 John D. Steinbruner, "Nuclear Decapitation", *Foreign Policy*, 45 (Winter 1981–2), p. 18.

8
Soviet Nuclear Strategy

Thrice is he armed that hath his quarrel just,
But four times he who gets his blow in fust.

Josh Billings (attributed)

The Soviet World View and Soviet Nuclear Strategy

The Soviet Union's nuclear strategy is determined, as any country's military strategy should be, by the political goals and the world view of its ruling elite. The goal it serves is to protect the Soviet Union against nuclear attack by the United States. The world view it reflects displays a deep fear of the United States, a Soviet conviction that nothing can ensure with absolute certainty that the United States will not attack the Soviet Union.

Soviet leaders feel that the United States's ruling elite is not only inherently aggressive and deeply hostile to the Soviet Union but also defective in its grasp of reality and quite reckless. In their view American leaders do not comprehend reality, partly because, not being Marxist–Leninists, they do not have the intellectual key to it. Even if they had that key, their class interests would prevent them from using it; the future they face is too bleak for them to accept. Capitalism is doomed. The United States, its allies, and the elements which rule it are growing ever weaker as the Soviet Union and the forces allied with it grow ever more powerful. The United States is torn by economic and social conflict, increasingly at odds with its allies, and in retreat all over the Third World. The Soviet Union is united, growing in military and economic strength, and continually increasing its international influence. Soon the Soviet Union will surpass the United States and be the most powerful country in the world. Some day Communism will rule the world.

All this may seem bizarre to a Western reader, aware of the Soviet Union's economic difficulties and the stresses within the Soviet empire, but it is the way Soviet leaders view the world. This view flatters their national pride and conforms to the Marxist–Leninist ideology in which they have been steeped since childhood. Nor is it so implausible that a person strongly motivated to accept it, even an intelligent, well-informed person, could not do so. The United States *does* have its problems, the Soviet Union has indeed grown very powerful, and a fair number of countries have fallen under Soviet control.

The Soviets believe that the combination of American despair of the future and a sudden severe crisis – a major war in Europe, for example – might well lead to a nuclear total war. Launching an all-out attack on the Soviet Union might seem to American leaders the only way to reverse the course of history and avert the doom of capitalism. Although a nuclear attack would destroy capitalism rather than save it, no Soviet leader has ever granted that it might not happen.

The Soviets do concede that this will not *necessarily* happen. Many, perhaps even most, American leaders grasp reality well enough to fear nuclear war. They know that if the United States were to attack the Soviet Union it would be totally destroyed by the Soviet retaliatory blow. As long ago as 1956, Secretary-General Nikita Khrushchev avowed that "war is not fatalistically inevitable." However, he also pointed out that "As long as capitalism survives in the world, the reactionary forces ... will continue their drive towards military gambles and aggression, and may try to unleash war."[1] The Soviets still believe that. They perceive the United States's ruling elite as caught on the horns of a dilemma: fearing nuclear war, but unable to find any other way to prevent the triumph of Communism. They cannot tell which horn the Americans will seize. Because the Americans fear nuclear war, and because of the Soviet Union's military power, war can be prevented. On the other hand, because of the Americans' recklessness and aggressiveness, and the insoluble dilemma they face, they might attack. Soviet strategy for nuclear total war must address both "The necessity of its prevention, and ... the possibility of its being waged."[2]

If there could be a nuclear war, the Soviet Union must be prepared to fight it, to survive it, and perhaps to win it. The tasks set for Soviet nuclear strategy are to prevent a nuclear war if possible, but also to win the war if it cannot be prevented. Soviet leaders, both military and civilian, are very much aware of how destructive nuclear weapons are, and in the past few years many of them have stated that neither superpower could win a nuclear total war. However, they continue to prepare the Soviet Union to fight such a war as well as possible and to survive, if not to win it. They do so because they cannot accept the possibility that the Soviet Union might be destroyed by something which its government cannot control, an American decision to launch an all-out attack. Their deep mistrust of the United States's ruling elite, indeed their entire world view, drives them to search for reassurance that nothing their enemies can do would destroy their country or halt the march of history towards the final triumph of Communism.

In formulating their nuclear strategy the Soviets refuse to choose between deterrence and defense. They seek both to deter a nuclear total war and to fight it as well as possible should deterrence fail. In their eyes there is no conflict between the two objectives and the same strategy can serve both. The more powerful the Soviet Union is, the less likely the United States is to start a war *and* the more likely the Soviet Union is to win any war the Americans do start. Therefore, they see as their best protection against nuclear war not a balance of strategic nuclear capabilities between the superpowers, as United States nuclear strategists tend to do, but the superiority of Soviet power over American power: the greater, the better. They may concede that neither superpower can gain strategic nuclear superiority over the other, but no Soviet leader has ever doubted that it would be very desirable for the Soviet Union to have it.

Soviet Strategy for a Nuclear War

If the Soviet Union is to survive and win a nuclear total war with the United States, the primary task of the Soviet armed forces must be to protect the Soviet Union against the United States's strategic nuclear weapons. It is not enough to save lives; Soviet defenses must also ensure that the Communist Party continues to rule the Soviet Union, and preserve the country's industrial base, its ability to wage war. This is a formidable task. Just two-thirds of the United States's SLBMs, without the aid of a single bomber or ICBM, could kill more than a quarter of the Soviet Union's population and destroy three-quarters of its industry.

There are three ways to defend a country against strategic nuclear weapons: counterforce attack, active defense, and civil defense. The Soviets would have to employ all three, in concert and to the greatest extent possible, to give the Soviet Union any chance of surviving a nuclear total war with the United States. They would have to begin the war with a counterforce attack: an all-out attack against the United States's strategic nuclear force. Then they would have to employ very strong active defenses – ABMs, interceptors, and anti-aircraft missiles – to shoot down as many as possible of the missiles and bombers which had survived the counterforce attack. Finally, they would need to have an extensive and effective civil defense system to deal with the United States's strategic nuclear weapons which had got past the active defenses.

The Soviets grant that the United States might conceivably launch a "bolt from the blue" surprise attack at any time. If that happens, they will hit back. They anticipate launching their own weapons the moment they are certain the attack is in progress, very likely even before enemy warheads explode on Soviet territory. However, they believe the "bolt from the blue" to be only a remote possibility.

In their eyes, a nuclear total war is much more likely to result from a severe political crisis, a conventional war, or a tactical nuclear war. Only then would American leaders, reckless and despairing as they may be, bring on a nuclear holocaust. Therefore, Soviet leaders would have at least several days to anticipate the attack, observe preparations being made for it, and make their own preparations.

This is a crucial assumption, because Soviet plans for fighting a nuclear war are heavily dependent on adequate warning. Ordinarily, much of the Soviet Union's strategic nuclear force is not in instant readiness to fight. Almost all the ICBMs are on alert, but only a fifth of the missile submarines are on station, and none of the bombers is on ground alert. Major elements of the Soviet command and control system would also have to be made ready for war. The Soviet civil defense system is particularly dependent on warning time; it relies primarily on moving most of the country's urban population into the rural areas before an attack, which would require at least several days.

Above all, the Soviets would need warning that they were about to be attacked so they could carry out a preemptive attack of their own first. This Soviet first strike would be directed against the United States's ICBMs in their silos, missile submarines in port and at sea, bombers on their airbases, and the various components of the American command and control system. It might start with a first salvo of SLBM warheads aimed at United States command posts and bomber bases. (SLBMs launched from missile submarines just off the coast might afford only six minutes or so warning of their attack.) The second, much heavier, salvo would be ICBM warheads directed against United States ICBM silos and all other elements of the United States strategic nuclear force.

No American can know for certain that Soviet leaders have planned a preemptive attack: only the Soviets know what their plans are. The evidence that they have is, however, fairly strong. This is what they would have to do in order to survive and win a nuclear war. It is what Soviet writings on military strategy suggest they would do: for the past 30 years Soviet military strategists have hinted at, or even openly avowed, the desirability of starting a nuclear war with a preemptive attack. And, it is what the Soviet Union's strategic nuclear force is designed to do.

The best weapon to carry out a preemptive attack is an ICBM; the next best is a missile submarine. ICBMs are best because of their great accuracy, their short flight time, and the relative ease of communicating with missile command posts and thus of coordinating a massive attack. The ICBM's greatest weakness, its poor survivability, is not important if it is used to strike first. Missile submarines are less useful because it is hard to communicate with submarines at sea well enough to coordinate an attack. Despite this they can participate effectively in a preemptive attack and the very short flight time of their missiles may facilitate it. Bombers are not useful at all. They are so slow that by the time one reaches an enemy missile silo the ICBM in it will probably have been launched. One would expect, then, that a country that plans to strike the first blow in a nuclear war would devote its primary effort to building ICBMs, a strong secondary effort to missile submarines, and little effort to bombers. This is what the Soviet Union has done. Seventy-five percent of the destructive power of the Soviet strategic nuclear force is carried on its ICBMs, 25 percent on its missile submarines, and only 5 percent on its bombers.[3] The clear Soviet preference for first-strike weapons, Soviet writings on nuclear strategy, and Soviet determination to survive and win a nuclear war convey an ominous message. They suggest that, in a desperate crisis, the Soviets might start a nuclear total war.

A massive, well-coordinated preemptive attack taking the United States by surprise probably could destroy about three-quarters of the United States's strategic nuclear weapons. Perhaps 90 percent of the ICBMs would be knocked out in their silos. At least 70 percent of the bombers, those not on ground alert, would be destroyed on their airfields. The missile submarines in port (45 percent) would be sunk. So would some of those at sea if the

Soviets had attack submarines trailing them or were able to direct missile barrages against them. The Soviet armed forces certainly have enough missiles to hit all these targets.

Taking the United States by surprise would be very difficult. Any crisis severe enough to induce the Soviet government to make a preemptive attack would probably also make the United States government put its forces on alert. United States ICBMs could be put in a "launch-on-warning" posture, the bombers put on ground alert and a few of them even on airborne alert, and missile submarines sent to sea. A preemptive attack against a fully-alerted United States strategic nuclear force almost certainly would fail.

Even the most successful Soviet attack conceivable would leave about 400–500 American missiles and bombers able to retaliate. These would have to be dealt with by the active defenses. Soviet anti-bomber defenses are modern and very extensive and might be able to destroy most of the attacking bombers, but their effectiveness cannot really be tested without fighting a nuclear war. Soviet anti-missile defenses are very weak – fewer than 100 ABMs, many of them obsolescent, defending Moscow. Active defense is the weakest of the Soviet Union's three lines of defense against American strategic nuclear weapons, but it would take a certain toll of an attacking force.

Civil defense is the last line of defense against nuclear retaliation from the United States. It would have to protect the Soviet Union's administrative structure, industrial base, and people against the effects of more than 1,000 bombs and warheads exploding on Russian cities. The Soviet civil defense system is extensive but it serves some Russians better than others. The highest-ranking leaders, deep inside their blast shelters, would be as safe as anybody can be in a nuclear war. Several hundred thousand lesser but still important people – Party and government officials below the very highest rank – would be accommodated in strongly-built blast shelters near their places of work. Another 10 or 20 million city dwellers have less strongly constructed shelters, often in the basements of factories or apartment buildings. The other 100 million or so would have to get out of the cities before the bombs arrive. Plans have been made to do this, but the planners estimate that it would take 72 hours.

The Soviet government invests considerable resources in civil defense. It has over 100,000 full-time civil defense workers and has constructed tens of thousands of blast shelters in the urban areas. Almost all Russians have received at least some civil defense training. It is taught to children in the schools, at the ages of eight, 11 and 15, and at places of work. More than 60 million copies of the basic civil defense manual have been issued, and gas masks, protective clothing, and other essential items have been supplied in quantity. Every Russian factory has its civil defense director who instructs its employees in civil defense and conducts civil defense exercises. The Soviets would not have made this very considerable investment of their limited resources if they did not think the Soviet system could survive a nuclear war.

How effective it all would be is another question. Ordinary Russians have a rather cynical joke about the subject:

"What do you do in the case of nuclear attack?"
"Wrap yourself in a sheet and crawl slowly to the cemetery."
"Why slowly?"
"So as to avoid causing a panic."[4]

There is some justification for this cynicism. Moving 100 million people out of the cities, many of them on foot, would be more likely to take several weeks than 72 hours. Feeding this horde in the impoverished Russian countryside and constructing fallout shelters for them would be a stupendous task. The necessary warning time is very likely not to be available. Also, even if this exodus were completed successfully, it would almost certainly put the American government on alert – and conceivably even trigger an American preemptive attack.

The implementation of Soviet civil defense plans *might* save many Russian lives. A Central Intelligence Agency report estimates that, given a week or more to prepare, the Soviet civil defense system could hold down the number of Russian casualties to "the low tens of millions."[5] On the other hand, little of the Soviet Union's industrial base would survive a nuclear total war. Soviet industry is almost entirely located in the large urban areas and factories are very hard to protect against nuclear strikes. The Soviet ruling elite is well protected and probably most of its members would survive. Whatever was left of the Soviet Union after a nuclear total war, the Communist Party would continue to rule it.

Neutralizing the United States's strategic nuclear force is only part, although the most important part, of the Soviet strategy for a global nuclear war. The Soviet Union would attempt not only to survive the war but also totally to destroy its enemies. The Soviet armed forces would be called upon decisively to defeat the United States and its allies; annihilate their armed forces, destroy their economic assets, occupy their territories, and impose Communist governments on them. Soviet strategic nuclear forces would carry out "the destruction of the largest industrial and administrative-political centers, power systems, and stocks of strategic raw materials and materials; disorganization of the system of state and military control; destruction of the main transport centers; and destruction of the main groupings of troops, especially of the means of nuclear attack."[6] In short, they would hit every target of military, political, or economic value. The initial preemptive attack would hit the United States's strategic nuclear force. Then there would be a series of follow-up attacks, one every several hours, as long as the Soviet Union had missiles to fire off, directed against all the other targets.

That would just be the beginning. "A future world war . . . will be a global opposition of multimillion coalitional armed forces unprecedented in scale and violence and will be waged without compromise, for the most decisive political and strategic goals. In its course all the military, economic, and spiritual forces of the combatant states, coalitions and social systems will be fully used."[7] Soviet planners envisage fighting all over the world with conventional and tactical nuclear weapons, huge land armies sweeping across

Eurasia and navies battling in all of the world's oceans. Even the millions of casualties caused by the mass use of nuclear weapons would not stop the fighting; whole armies of reserves would be mobilized in their nuclear-devastated homelands and sent to fill the gaps. Months after the initial nuclear strikes remnant United States and Soviet forces might still be grappling in Lapland or Korea, the homelands behind them in ruins. In the end, Soviet planners hope, superior Soviet numbers, morale, and preparation for nuclear war would bring victory.

This would be the most decisive war in history and the last. By the time it ended capitalism would be no more. The Soviet Union, despite having suffered catastrophic losses, would survive and dominate the post-war world. All of Eurasia, at least, would be under Soviet control. Soviet military leaders have not made it clear how they plan to occupy the United States's territory in a global nuclear war, but they can undoubtedly take Eurasia.

Soviet Nuclear Strategy and the United States

Soviet discussions of nuclear strategy make the American reader very nervous. It seems incredible that sane, well-informed, responsible men should make plans to fight and win a nuclear total war. Surely these plans are not serious? Surely they are just propaganda, designed to mislead Americans, reassure the Russian people about nuclear war, or advance the interests of the Soviet military-industrial complex? Unfortunately, they are not. Soviet nuclear war strategy has been worked out in considerable detail over the past 30 years, widely discussed, and taught to all ranks of the Soviet armed forces as established doctrine. This has not been done just to mislead Americans or to reassure the Russian people about nuclear war. As Raymond Garthoff once pointed out, "in the case of doctrine . . . the Soviets simply cannot afford to mislead their own officer corps merely in order to try to mislead us."[8] Soviet plans to fight and win a nuclear war are also reflected in the kinds of weapons and forces the Soviet Union has: the preference for ICBMs over bombers and missile submarines, the powerful anti-bomber defenses, the civil defense system, and the ability to mobilize much larger ground forces than would ever be needed in a conventional war. None of these are necessary to a purely deterrent nuclear strategy – the United States does not have them – but they serve a nuclear war-fighting strategy well.

The Soviet strategy for nuclear war must, then, be taken seriously, but it is not as menacing as it is sometimes portrayed. It is a preemptive attack, not a preventive war, strategy. The Soviets have adopted this strategy because they fear American aggression, not because they wish to conquer the world. They see it as the best means they can devise to defend the Soviet Union, not as a "blueprint for world conquest." It is not going to lead them to launch a "bolt from the blue" surprise attack on the United States without immediate provocation. If they ever attack, it will be because they think they are about to be attacked.

Soviet nuclear strategy does increase the danger of a nuclear total war. If things get bad enough, if there is another Cuban missile crisis, this one badly mishandled by both sides, or a war in Europe, the Soviets might well believe a nuclear war to be unavoidable. If they conclude that it is, they might well try to get in the first blow, as their strategy dictates. Reliance on preemption places the Soviet government under heavy pressure to escalate a lesser conflict into a nuclear total war. So far, the actions of Soviet political leaders have been wiser and more cautious than their military plans. There has not been a nuclear war, or even a serious danger of one. However, as a crisis deepens and war comes more and more to seem inevitable, the plans military men have made to fight the war tend to get implemented even if that means throwing away any chance to avert it. That is what happened in August, 1914: the plans made for mobilizing the armies were implemented once war seemed likely, and any chance of averting World War I was lost. It could happen again.

That does not mean that it will happen again, or that Soviet leaders are totally committed to the preemptive attack strategy. The Soviets are very much aware of the devastating power of nuclear weapons and of the catastrophic damage *any* nuclear total war would inflict on the Soviet Union. They should also be aware of the many obstacles to a successful preemptive attack. They have other options and, at the moment of truth, may decide to exercise one of them. Even if a nuclear total war seems virtually certain, they might refrain from striking the first blow if they see some small chance that the war could be averted and are not sure that the Soviet Union would ensure its survival by striking first. On the other hand, they have revealed their intent to strike first and made extensive preparations to do so. Prudent United States military planners must frame their nuclear strategy on the assumption that the Soviets think a nuclear total war can be won and might be tempted to launch a preemptive attack.

This is not to say that the United States should adopt a nuclear warfighting strategy of its own. Soviet nuclear strategy makes it harder to deter an attack, but certainly does not make it impossible. If Soviet nuclear strategy is reckless and unrealistic that does not justify the adoption of a similarly flawed United States nuclear strategy. Instead, it demands the conduct of United States national security policy with an extra measure of strength, wisdom, and restraint.

Notes

1 Report to the 20th Party Congress, February 1956, quoted in David Holloway, *The Soviet Union and the Arms Race* (New Haven, Conn.: Yale University press, 1983), p. 32.

2 "On the Question of the Role of Economics in Nuclear Warfighting", *Voyennaya mysl'* [Military Thought], 11 (November 1965), quoted in Joseph D. Douglass, Jr. and Amoretta M. Hoeber, *Soviet Strategy for Nuclear War* (Stanford, Ca.: Hoover Institution Press, 1979), p. 9.

3 In contrast, for a second-strike deterrence strategy the missile submarine is the most useful weapon, then the bomber, and thirdly the ICBM.

4 Andrew Cockburn, *The Threat: Inside the Soviet Military Machine* (New York: Random House, 1983), p. 235.
5 US Central Intelligence Agency, "Soviet Civil Defense", NI 78–10003 (July 1978), p. 12.
6 Major General Vasily I. Zemskov, "Characteristic Features of Modern War and Possible Methods of Conducting Them", in *The Soviet Art of War*, ed. Harriet Fast Scott and William F. Scott (Boulder, Colo: Westview Press, 1982), p. 213.
7 Marshal of the Soviet Union Nikolay V. Ogarkov, "Military Strategy", in *The Soviet Art of War*, ed. Scott and Scott, pp. 246–7.
8 Raymond L. Garthoff, *Soviet Strategy in the Nuclear Age* (New York: Frederick A. Praeger, 1958), p. 272.

9
United States Nuclear Strategy

Thus far the chief purpose of our military establishment has been to win wars. From now on its chief purpose must be to avert them. It can have almost no other useful purpose.

Bernard Brodie, *The Absolute Weapon*

There are two salient realities of the nuclear predicament. One is that no matter what might be done to limit its destructiveness, a nuclear total war would be the worst catastrophe ever suffered by the human race. The other is that no matter what is done to prevent it, a nuclear total war might still occur. There is no policy a country can follow – preventative war, preemptive attack, deterrence, arms control, disarmament, or anything else – which guarantees that it will never suffer a nuclear attack. There is no way it can adequately protect itself if it is attacked.

These realities place before nuclear strategists a most difficult choice: to rely exclusively on trying to prevent a nuclear war, or to try to limit the damage it would do, perhaps at the price of making it more likely. Soviet military planners have chosen the second option. Their American counterparts, generally more hopeful about preventing a nuclear war and less so about limiting the damage, have been of two minds. Most wish to put all the country's eggs in the one basket of nuclear war prevention and then watch that basket very carefully. These are the advocates of a MAD (as it has been termed by its opponents) or "Mutual Assured Destruction" strategy. Others, advocates of various "Damage Limitation" strategies, would put the country's eggs in two less closely-watched baskets. They feel that the United States, while attempting to deter a nuclear war, should also be prepared to fight it if deterrence should fail. (This approach is sometimes referred to as NUTS ["Nuclear Utilization Target Selection"] by the advocates of MAD. A nuclear strategist can be either MAD or NUTS.)

No clear and definitive choice between Mutual Assured Destruction and Damage Limitation has been made by the United States government. The requirements of Mutual Assured Destruction have dictated most aspects of the United States strategic nuclear force: the rationale for its existence, the types of weapons the force employs, the great effort to make them survivable, and the virtual absence of active defense and civil defense forces. On the other hand, Damage Limitation has dictated the selection of the targets to be hit in a nuclear war and the development of missiles that are much more accurate than MAD requires or even allows. The result has been a strategic nuclear force which serves both strategies to some degree but neither one as well as it might.

Mutual Assured Destruction

The Mutual Assured Destruction strategy is designed to do just one thing: prevent a nuclear total war. It works on the premise that a nuclear total war would be so great a catastrophe for the entire human community that it must never be allowed to happen. The United States must never, for any reason whatever, start a nuclear total war, and must do everything it can to prevent the Soviet Union starting one. MAD holds that that is the only function the United States strategic nuclear force can possibly serve. *"Nuclear weapons serve no military purpose whatsoever. They are totally useless – except to deter one's opponent from using them."*[1] Whatever the bombers and the missiles do after a war breaks out, or whether they do anything at all, is of little consequence; no exercise of military power could help the United States much at that point. The only possible war aim, once the missiles fly, is to stop the war before both sides have been obliterated, not that MAD offers much hope that this could be done or much guidance as to how to do it.

Mutual Assured Destruction is, then, a deterrence strategy. Its purpose is to prevent a Soviet nuclear attack on the United States. If the Soviets ever were to mount such an attack it would be for two reasons:

1   They believe they can get away with it, that the United States will not retaliate or its retaliation would be so weak that the Soviet system would survive.
2   They believe that the United States is about to attack the Soviet Union, that a nuclear total war is inevitable, and that the Soviet Union can avert being destroyed only by hitting first.

They probably will not attack unless they are convinced of both these things. No matter how vulnerable the United States is, there are enough uncertainties in any preemptive attack that the Soviets will not launch one unless they are almost certain that a nuclear war is inevitable whatever they do. On the other hand, no matter how imminent a nuclear war seems to be, they will not strike the first blow unless they feel that doing so will enable the Soviet Union to survive. Only if the war cannot be averted *and* if striking first will save their country will the Soviets attack.

Deterrence depends, therefore, on convincing the Soviets that:

1   They cannot get away with it; no matter how heavy, lucky, or carefully-planned their preemptive attack is, the United States can and will hit back and destroy them.
2   The United States will not attack the Soviet Union, no matter what the provocation, unless it has first been attacked.

These messages are best conveyed by the kind of strategic nuclear force the

United States builds, and by whatever else it does or refrains from doing to deal with the threat of nuclear war. Deterrence depends on demonstrating what one intends to do as well as what one is able to do, but the best way to demonstrate intentions, at least in this case, is through capabilities. The Soviets judge American intentions much more by what the United States has prepared itself to do than by what its government says it will do. The best way to convince them that the United States would retaliate for a Soviet attack is to be able to retaliate, under any circumstances and with crushing power. The best way to convince them that the United States will not launch a preemptive attack is for it to be unable to do so successfully.

The United States must have a powerful and well-protected strategic nuclear force. That force is to be used only in retaliation for an enemy attack and would be launched only after the United States had been hit. Therefore, the test of its adequacy is how much damage it could do after suffering a full-scale attack designed to destroy it. Nuclear strategists draw an important distinction between a strategic nuclear force's "first-strike" and "second-strike" capabilities. The first-strike capability is what it could do in a preemptive attack, using all the weapons that could be made ready for action in peacetime. The second-strike capability of a nuclear force is what it could do with the weapons it would have left after the enemy had done his best to obliterate it. It is the second-strike capability which achieves Mutual Assured Destruction.

This United States retaliatory force must be a "countervalue" rather than a "counterforce" instrument; its weapons must be designed to destroy cities and industrial plant and kill civilians rather than to hit military targets. Its retaliatory attack could do little damage to the Soviet Union's most important military targets, the strategic nuclear weapons, because they would already have been launched: the missile silos would be empty, the bombers flown, and the missile submarines would have fired off their missiles. Instead, the United States force deters the Soviets by threatening them with the destruction of what they value most, the Soviet Union. By its existence it says to them, "If you attack the United States, the United States will do so much damage to your country, its people and its economy, that nothing you could hope to gain would outweigh your loss." Robert McNamara held that the loss of 25 percent of the Soviet Union's population and 70 percent of its industry would be "unacceptable damage" in Soviet eyes. If there were no doubt that a United States retaliatory attack could inflict such losses then deterrence would be secure. Of course, MAD like any other deterrence strategy, is based on unverifiable assumptions about the opponent's state of mind. However, devastation of the magnitude McNamara envisaged should be enough to appall even the toughest Soviet leader.

That is what the United States strategic nuclear force must be able to do but what it must *not* be able to do is equally important. It must not be capable of a successful counterforce attack. Because a counterforce attack can only succeed if it is a first strike, having the ability to execute one inevitably suggests to the opponent that one might launch a first strike. Fearing that, he might decide to hit first himself, launch his missiles before they could be destroyed, on the "use them or lose them" principle.

Mutual Assured Destruction is best secured by weapons that are ideal for a second-strike countervalue attack and of no use at all in a first-strike counterforce attack. They must be very survivable, slow to reach their targets, and not very accurate. Survivability, not important at all in a first-strike weapon, is the most important characteristic of a second-strike weapon. The missiles must be placed in silos, on mobile land-based launchers, or on missile submarines, and the bombers kept on ground or airborne alert. Weapons which are slow to reach their targets cannot be used in a counterforce attack because the weapons they might be aimed at could be launched before they arrive. They can, though, do tremendous damage in a countervalue attack; a city will still be there when the bomber arrives. Only a counterforce weapon needs to be accurate. A warhead aimed at an ICBM silo must land within a few hundred feet of its target to destroy it but one aimed at a city can miss by a mile and still kill a great many people.

The missile submarine serves the Mutual Assured Destruction strategy better than any other strategic nuclear weapon. It is by far the most survivable of strategic nuclear weapons and its missiles are less accurate than an ICBM. If both superpowers had nothing but missile submarines there would be little advantage to striking first and thus little danger that even the most ominous crisis would lead to a nuclear war. The ICBM does not serve MAD's purposes very well at all. It is too vulnerable to an enemy counterforce attack and, with its great accuracy, is itself too good a counterforce weapon. The strategic bomber is better than the ICBM, but less suitable than the missile submarine. Its vulnerability while on the ground tempts an enemy counterforce attack but it is more survivable than the ICBM. The long time it takes to reach a target disqualifies it for counterforce purposes, while it is still very useful in a countervalue attack.

The Mutual Assured Destruction strategy prohibits any kinds of protection against nuclear retaliation; it prohibits active defense and civil defense as well as preemption. MAD's proponents argue that neither active defense nor civil defense can possibly serve any purpose, that there is no way a country can survive an all-out nuclear attack. Even the strongest possible active defense would let so many enemy warheads through that even the most extensive civil defense measures could not prevent the death and injury of many tens of millions. Any defense against nuclear attack is a waste of money. Even if it were to offer some degree of protection, the opponent could and would negate its effects by strengthening his strategic nuclear offensive forces. Both sides would soon be back where they were after spending vast sums of money on defensive measures and offensive weapons to overcome those defensive measures.

The advocates of MAD hold that for the United States to spend the tens of billions of dollars necessary to build a nation-wide ABM system, effective anti-bomber defenses, and an extensive civil defense system would not be just a waste of money. It would also significantly increase the danger of a nuclear total war. That investment would be made only if the American people and their government were convinced that it would enable the United States to survive a nuclear total war. So deluded, they might act more aggressively in a war-threatening crisis, perhaps so aggressively as to bring on a nuclear war. They would find – too late – that there *is* no defense against nuclear weapons. American defensive measures might also, oddly enough, make the Soviets more reckless. These measures would inevitably be, to the Soviet mind, evidence of American willingness to fight a nuclear war. Fearing attack, the Soviets might well resolve to avert it by a preemptive attack of their own. The United States government can prove that its intentions are peaceful only by leaving the American people totally vulnerable to Soviet retaliation.

The MAD strategy does permit defending certain targets: ICBM silos, command posts, bomber bases, and missile submarine bases. Defending these targets strengthens deterrence, as long as the cities are left undefended, by making a counterforce attack less likely to succeed. MAD implies a rather repellent paradox: missiles may be protected against nuclear attack but people may not be. Killing weapons is bad, killing people is good.

The heart, then, of Mutual Assured Destruction is what Wolfgang Panofsky calls a "mutual-hostage relationship" between the superpowers.[2] The people of each country are being held hostage for the good behavior of their government. If a government starts a nuclear war its people will be annihilated. This, as are other aspects of *Mutual* Assured Destruction, is a shared condition. The peoples of both superpowers must be equally and totally vulnerable and the governments of both must realize and accept this vulnerability. Mutual Assured Destruction cannot be achieved by the efforts of just one superpower, and if it is achieved it protects both of them. It is not enough for the United States to refrain from building counterforce weapons; the Soviet Union must protect its retaliatory force. It is not enough for the United States to renounce active defense and civil defense; the Soviet Union must do so also.

American MAD strategists do not want the United States to be militarily superior to the Soviet Union. They see the security of their country to lie in an equal balance between its power and that of its opponent, in both having the same capabilities and vulnerabilities. This may require that the United States deliberately limits certain of its capabilities – for example, the accuracy of its missiles. It also means that an increase in certain Soviet capabilities may make the United States more secure. For the Soviets to improve the survivability of their strategic nuclear weapons makes a nuclear war less likely by reducing both Soviet and United States incentives to preempt.

The Mutual Assured Destruction strategy was formulated by civilians, experts who had never learned or had totally rejected the traditional military approach to war. Perhaps only civilians could have formulated such a strategy because many of its aspects are fundamentally alien to traditional military thought. Military men throughout the ages have sought to make their armies as strong as possible and have hoped their opponents would be weak. MAD teaches that it is better to be equal to an opponent than to be stronger than him, that it may be desirable deliberately to limit one's strength, and that sometimes one should welcome the opponent's becoming stronger. Armies have always been used to fight wars, but MAD teaches that the only function of a strategic nuclear force is to avert a nuclear total war. Military strategy has always guided the conduct of wars but MAD teaches only how to avoid a nuclear total war. Once that war starts the MAD strategy has little guidance to offer. In that sense it is, perhaps, not a military strategy at all.

Damage Limitation

The fundamental defect of the Mutual Assured Destruction strategy is that it stakes everything on preventing a Soviet first strike. It offers no useful guidance as to what to do if deterrence fails and little hope of averting a nuclear total war. In that event, MAD does, of course, offer a plan of action – a massive American retaliatory attack on the cities and people of the Soviet Union – but it is a very unwise plan. To threaten a massive retaliatory attack may be the best way to prevent a nuclear war, but actually to carry it out would be totally irrational and immoral.

It would be irrational because destroying several hundred Russian cities and upwards of 100 million Russians would serve no purpose whatever. It might not save a single American life and it certainly would not save democracy in the United States or anywhere else. The damage would already have been done, irreversibly, before the first United States warhead exploded on Soviet territory. The American retaliatory blow would not even be a valid act of revenge. The Soviet leaders would be safe in their shelters while their innocent subjects died. All it would accomplish would be to ensure that the holocaust which had consumed the United States would destroy the rest of the world as well.

It would be immoral because it would entail the killing of hundreds of millions of innocent people. As one of MAD's most able critics puts it, "our method for preventing nuclear war rests on a form of warfare universally condemned since the Dark Ages – the mass killing of hostages."[3] The political purposes which conceivably might have justified so many deaths would have evaporated before the retaliatory blow was struck. There would no longer be a democratic political system to defend in the United States, and perhaps not even an American nation.

The MAD strategy makes sense only if it is never implemented, only as long as there is not a nuclear war. Unfortunately, neither MAD nor any other policy the United States could follow can guarantee that there is never a nuclear war. Not even the combined efforts of the United States, the Soviet Union, and all the other countries in the world can guarantee that. Herman Kahn wisely points out that "*history has a habit of being richer and more ingenious than the limited imaginations of most scholars or laymen.*"[4] Nobody can anticipate and thus forestall all of the combinations of circumstance that might lead to a nuclear war. As long as nuclear weapons exist and sovereign states exist, it might still happen. If so, then perhaps the United States should have a nuclear strategy that offers some hope of national survival even in a nuclear war, a strategy for conducting such a war and attaining the best outcome possible.

Another significant flaw of MAD is that the Soviets do not accept it. They still have not given up hope that the Soviet Union could survive a nuclear total war and they continue preparations to fight it. This suggests that deterrence can never be as secure as the MAD strategy requires. Deterrence is made even more doubtful by Soviet reliance on a preemptive attack as the primary means to survive a nuclear war. An opponent who plans to get in the first blow *can* be deterred, but not nearly as easily as one who does not. MAD is, after all, *Mutual* Assured Destruction. One superpower can, by its own efforts, do much to achieve it, but it is only fully assured if both accept and work to strengthen it. This the Soviet Union apparently is not prepared to do.

One more defect of MAD is that it serves only one of the deterrent objectives of United States national security policy – deterrence of a massive Soviet nuclear attack directly on the United States. It does not even deter a Soviet attack on Western Europe, let alone threats to other vital American interests. MAD leaves the people of the United States so vulnerable that for the United States to hit the Soviet Union in retaliation for a Soviet invasion of Western Europe would lead to the swift and total destruction of the United States. No United States president would launch a nuclear attack under those circumstances, and so no threat to do so can be credible enough to protect Western Europe. Much as Americans value their European allies, they are not prepared to commit suicide on the Europeans' behalf. If the threat of nuclear retaliation is to protect Western Europe at all, it must be as part of a strategy that would enable the United States to survive a nuclear war.

Mutual Assured Destruction has been generally accepted as the United States's nuclear strategy for more than 30 years, but it is not an easy strategy to like. No administration has been entirely happy with it and all, beginning with John F. Kennedy's, have sought alternatives that would be more flexible and humane and would not doom the entire world to a nuclear holocaust should deterrence fail. McNamara's "city-avoidance" proposals, the "Schlesinger Doctrine" advanced by Secretary of Defense James Schlesinger, the Carter administration's "countervailing strategy," and the Reagan administration's "Defense Guidance" plan have all been attempts to escape the iron logic of MAD. Unfortunately, it has proven easier to see the faults in MAD than to devise a good alternative.

All these alternatives to MAD are variants of the Damage Limitation strategy which attempts to enable the United States to survive a nuclear war if it is attacked. Few will quarrel with that objective, although many consider it unattainable. Another and much more controversial objective of Damage Limitation is to make it credible that, under some circumstances, the United States might start the nuclear war. As long as any exchange of nuclear blows between the two superpowers is sure to kill 100 or 200 million Americans, no United States government will ever begin it and no American threat to do so can ever be credible. However, if the number of American dead can be limited to 10 or 20 million, perhaps it is conceivable that a United States president might launch an attack. It would be the most agonizing decision he could possibly make and would require some extreme provocation – a Soviet invasion of Western Europe, or a limited attack on United States military targets – but he could do it. Because he could, the threat to launch that attack would be credible enough to protect Western Europe or to deter a Soviet counterforce strike on the United States. This does not make Damage Limitation a plan to start a nuclear war. Both Mutual Assured Destrucition and Damage Limitation are essentially deterrence strategies. The difference between them is that MAD undertakes only to deter an all-out Soviet attack on the United States while Damage Limitation undertakes to deter a number of less extreme Soviet actions, and to achieve the best outcome possible if deterrence should fail.

Damage Limitation relies heavily on counterforce strikes to destroy as many Soviet strategic nuclear weapons as possible before they could be used to hit American cities. It is inherently a counterforce rather than a countervalue strategy. American strategic nuclear weapons are to be used exclusively against Soviet military targets and primarily against the Soviet strategic nuclear force. This dictates the use of small, extremely-accurate warheads: accurate enough to destroy command posts and missile silos, and small enough so they would not destroy nearby cities. It also implies, at least to Soviet observers, that the United States might strike first.

The advocates of Damage Limitation deny that theirs is necessarily a first-strike strategy while recognizing that it can serve as one. They hold that in some circumstances the United States would strike first and in other cases be prepared to make a counterforce second strike. If the Soviet Union invades Western Europe then the United States should make a first strike, hit Soviet strategic nuclear weapons and other military targets even though the United States itself has not yet been hit. On the other hand, if the Soviet Union mounts a first strike on the United States, particularly if it hits only American military targets, the United States retaliatory strike should hit only Soviet military targets. There would still be Soviet military targets worth hitting: missiles which had malfunctioned or been held back for a follow-up salvo, bombers which had returned to their bases, and conventional forces.

The Damage Limitation strategists have shown some interest in active defense and civil defense as supplements to counterforce strikes. They recognize, however, that if the United States and the Soviet Union use all the nuclear weapons they have and hit all the targets they can hit, it will not matter which strikes first or what active defense and civil defense measures each has taken. Both countries will be destroyed. Only if both countries deliberately limit the amount of damage they do can either survive. The Soviets must be induced to limit their nuclear strikes to military targets and not hit American cities. They are much more likely to do so if Russian cities have not been hit. The United States has two reasons for making a counterforce rather than a countervalue attack. One is to destroy the Soviet weapons which threaten American cities; the other is to spare the Russian cities whose destruction would inevitably lead to a Soviet retaliatory attack on American cities. At the same time, the United States should hold back from its counterforce attack enough weapons to destroy the Soviet Union's cities should that become necessary. This would deter the Soviets from making a countervalue attack and force them to conduct the war according to "rules of the game" dictated by the Damage Limitation strategy – and by the United States government. If both sides follow these rules, even a strategic nuclear war would be like most wars fought prior to the twentieth century: a contest between the military forces of the belligerents, involving their civilian populations as little as possible. It would still be a horribly brutal, costly business, but less so than the global holocaust which MAD promises if deterrence should fail.

The reader will have noticed a certain resemblance between Damage Limitation and the Soviet Union's nuclear strategy. Both are, in large part, "war-fighting strategies", not in the sense of being intended to lead to a nuclear war but in that they are intended as guides to the conduct of a war if deterrence should fail. Both are also, in part or *in toto*, counterforce rather than countervalue strategies. Because of their reliance on counterforce strikes, both are dangerously dependent on getting in the first blow.

The most important difference between them is that the Soviet Union's is a strategy for a nuclear total war while Damage Limitation attempts to keep even a strategic nuclear war limited. The Soviet strategy assumes that both parties will use all the weapons they have, that they will attack every target they can, both military and civilian, and that the war will end in the total obliteration of the loser. Damage Limitation attempts to prevent the obliteration of either party, to limit the destruction to military targets alone, and to bring the war to an end before all the nuclear weapons have been used. Most Damage Limitation advocates accept that nobody can win a strategic nuclear war; it must be terminated by a compromise peace if it is not to lead to a global holocaust. For this strategy to be successful requires, therefore, the tact or explicit cooperation of both sides. One of the strengths of the Soviet Union's nuclear strategy is that it requires no cooperation whatever between the adversaries. It assumes that both sides will do their worst and is designed to achieve victory for the Soviet Union regardless of what the United States does. Damage Limitation, on the other hand, depends even more than MAD on the opponent following the same strategy. It requires that the adversaries reach a whole series of tacit bargains as to how many nuclear weapons may be used and what targets may be attacked, and that they do so in the middle of a nuclear war. But if they could not cooperate well enough before the war started to prevent it, why should they be able to cooperate well enough to limit it while it is going on?

A Nuclear Strategy for the United States

Winston Churchill is said to have described democracy as "the worst system devised by the wit of man, except for all the others." This is also a just verdict on Mutual Assured Destruction. MAD is not a good nuclear strategy but it is better than any alternative. Damage Limitation suffers from two great defects:

1 It makes a nuclear war more likely.

2   It cannot prevent catastrophic damage to the United States if there is a
   nuclear war.

"One man's damage limitation, at a certain point, becomes another man's
feared first strike."[5] For the United States to adopt a Damage Limitation
strategy and acquire the means to implement it inevitably suggests to Soviet
leaders that the United States is preparing a preemptive attack on their
country. Damage Limitation advocates insist, no doubt sincerely, that they do
not intend that, that neither side could mount a successful first strike, and
that they do not want either side to be able to do so. Unfortunately, that is
not how the Soviets see it. They have described every statement of a Damage
Limitation strategy, from "city-avoidance" on, as evidence that the United
States is planning a nuclear war. To some degree these cries of alarm are
propaganda, designed to discredit the United States and justify to the Russian
people continued heavy sacrifices for national defense. They also reflect very
sincere Soviet anxieties. The Soviets expect the United States leaders to be
plotting nuclear war against the Soviet Union. It is what their world view
strongly inclines them to believe, and probably what they would be doing if
they were the leaders of the United States. Their own interest in preemptive
nuclear strikes may make it seem more plausible to them that American
leaders have something similar in mind. Also, there are aspects of the Damage
Limitation strategy which would disquiet even a more judicious ruling elite,
and thus must doubly disquiet the leaders of the Soviet Union. There are
cases in which it openly calls for the United States to hit the Soviet Union
before itself being hit – a Soviet invasion of Western Europe for example.
Even if there were not, Damage Limitation is a counterforce strategy and as
such inevitably suggests first strike intentions. Any counterforce attack works
much better as a first strike than as a second strike. If, after the enemy has
attacked, one can destroy the weapons he has held in reserve, probably one
could have destroyed his entire force before he attacked. Why not do so
rather than let him hit first and inflict catastrophic damage on one's country?
Americans who consider Soviet nuclear strategy to be aggressive and reckless
should not be surprised to see a somewhat similar United States nuclear
strategy regarded with suspicion in the Soviet Union.

   The Soviet Union's leaders do not plan to wait passively for an American
attack. If they see it coming, they firmly intend to hit first themselves and
crush the attack before it can be delivered. This would probably be their plan
even if American military strategists had never even mentioned Damage
Limitation, but statements of the Damage Limitation strategy make an already
bad situation worse; they make it more likely that Soviet plans for a preemptive
attack will lead to a nuclear war. It is dangerous enough for the Soviet
government to plan on getting in the first blow, but for the United States
government even to hint that it might try to do the same increases the pressure
on the Soviets to attack before they are attacked. And, if the United States
really does plan to hit first then any violent confrontation, a rapidly escalating
local war for example, can become a race between the superpowers to be the
first to start a nuclear war. It will not matter very much who wins that race.

The Damage Limitation strategy, while making a nuclear war more likely, probably would not make it significantly less costly. There is no way to fight a strategic nuclear war without killing millions of civilians. No matter how carefully the nuclear strikes are directed at military targets alone, so many of those targets are near cities that a great many city-dwellers would die from the blast, thermal flash, and fallout from nuclear explosions on those targets. A strategic nuclear war in which both sides observed the Damage Limitation rules would probably kill about 20 million Americans and an equal number of Russians. This is, of course, far fewer than the hundreds of millions who would die in a war in which cities were deliberately destroyed, but it is still a horrendous toll:

> The idea that attacks of this sort could be considered a limited 'surgical' attack had no reference to anything within human experience, only to a knowledge that it could be worse. Up to 1975, all the wars of America's two centuries had resulted in deaths of less than 1,200,000.[6]

There is no assurance that the war would be waged according to the Damage Limitation rules, even if the United States went into it planning to do so. It takes both sides to limit a war. The Soviets have clearly and repeatedly disavowed any intention of abiding by the Damage Limitation rules in a nuclear war. They describe Damage Limitation, with some justice, as a means of making it easier for the United States to threaten or to fight a nuclear war. Soviet nuclear strategy states quite clearly that in a strategic nuclear war the Soviets would use every weapon they have and hit every target they can reach. They might change their minds once the war starts – but nobody should count on it.

Even if both sides try to keep the war limited they may not succeed. Wars tend to get out of hand, even wars much less devastating than a strategic nuclear war would be. To limit such a war to any meaningful degree both sides would not only have to avoid hitting cities in the initial nuclear strikes but also have to stop fighting before all the nuclear weapons they had were used. To do this would be, as Bernard Brodie puts it, "tantamount to negotiating complete disarmament with a 24-hour deadline, or less."[7] It is just barely conceivable that after an exchange of counterforce strikes, with 20 million dead on each side and several thousand nuclear weapons still unused, the warring governments would take a calm, rational decision to stop the war. Much more likely, the leaders of one or both countries, enraged if not half-crazed by the losses their people had already suffered, would continue to fire off all the missiles they had, at civilian and military targets alike, until there was very little left of the human race.

As there is so little hope of limiting a strategic nuclear war once it starts, United States nuclear strategy must do everything possible to prevent it. This means that Mutual Assured Destruction must be the United States's nuclear strategy. MAD has served the United States well. It has prevented a nuclear war for many years and seems the best bet to prevent it for the foreseeable future. However, it is an expedient, one may hope a temporary expedient, and probably not a permanent solution to the United States's national security needs.

Mutual Assured Destruction is a sound strategy only because strategic nuclear weapons can be protected against attack and cities cannot be. The ability to protect weapons makes MAD possible by ensuring that a well-protected strategic nuclear force can survive and hit back after even the heaviest preemptive attack. The inability to protect cities makes MAD necessary by depriving the superpowers of any other means to guarantee their national survival. Both of these circumstances, based as they are on the present level of development of weapons technology, may change. There has never before in history been a weapon that could not be destroyed and against which there was no defense. It may be that the strategic nuclear weapon is such a weapon, but 40 years into the Nuclear Age is still too soon to be sure. All three types of strategic nuclear weapon, even the missile submarine, may eventually become vulnerable to attack, in which case Mutual Assured Destruction will cease to prevail and a new strategy will have to be found. Perhaps, also, an effective defense of cities against missile attack will be developed, offering the nuclear powers a less precarious guarantee of national survival than the threat of mutual annihilation.

The United States needs to have approximately as many strategic nuclear weapons as the Soviet Union. This is about as many as it does have: 1,957 SLBMs, strategic bombers, and ICBMs. To some people such an arsenal seems much too large, an example of unnecessary and menacing "overkill." They point out that one-tenth of these weapons could destroy the Soviet Union and ask why it is necessary to be able to destroy the Soviet Union ten times over. This is not to comprehend the purpose of the United States's strategic nuclear force. No doubt a small fraction of the American force could destroy the Soviet Union *in a first-strike, countervalue attack.* However, its purpose is, or should be, to deter the Soviets by being able to strike back after a Soviet attack. The measure of its adequacy is its second-strike, not its first-strike, capability. So, having 1,957 strategic nuclear weapons in service, theoretically available for a first strike, is not "overkill;" it is necessary to ensure that there would be enough weapons – 200–400 – available for a second strike.

Another justification for a large American strategic nuclear force is the strong inclination of Soviet leaders to believe that the Soviet Union could survive a nuclear total war. This makes deterrence much more difficult than it would be if the Soviets were wholehearted believers in MAD. Thus the American strategic nuclear force must be so powerful and well protected that even Soviet leaders who firmly believe their country could survive a nuclear total war will realize, while they are under the strongest possible pressure to make a preemptive attack, that they were wrong. This requires a much stronger force than seems reasonable to those who judge a deterrent force's adequacy by whether or not it would be strong enough to deter *them.*

Strategic nuclear weapons are symbols of national power and prestige as well as instruments of war. That is one more reason why the United States should have as many of them as the Soviet Union; the United States should not be perceived as militarily inferior to the Soviet Union. This is an argument for the United States's keeping approximately equal with the Soviet Union in force levels, but not for matching the Soviets weapon for weapon in every category or by every conceivable index of strategic military power. For the United States to fall behind in missile "throw weight," or "equivalent megatonnage" does not have much political significance. Few people know or care what throw weight or equivalent megatonnage are.[8] On the other hand, most political leaders and concerned citizens *can* compare total numbers of ICBMs, bombers, and SLBMs, which is why those numbers influence perceptions of national power.

The United States should have the types of weapons that best maintain Mutual Assured Destruction: a strong missile submarine force, and strategic bombers as insurance that the United States would still have a deterrent even if missile submarines became vulnerable to attack. ICBMs soon will not have much to contribute to the American deterrent unless and until they can be made more survivable.

Simply to keep the American strategic nuclear force at its present level will require building many more bombers and missile submarines during the next decade. Such weapons, like other machines, wear out as time goes by. All of the United States's B-52 bombers were built from 1958 to 1962, so they are all at least a quarter-century old, older in many cases than the pilots who fly them. The Poseidon boats which still comprise three-quarters of the United States missile submarine fleet were built from 1962 to 1967 and, like the B-52s, they are nearing the end of their useful lives. There are plans to keep both types operational into the 1990s, by dint of extensive rebuilding, but there is a limit to how long any weapon can be kept operational by patching it up. Soon the B-52s and the Poseidons will be comparable to the legendary Irishman's hammer: "a hundred years old, only two new heads and three new handles."

Trident submarines are slowly replacing the Poseidons. Eight Tridents had put to sea by the end of 1987 and more are being constructed at a rate of one a year. These are very large, expensive submarines, with more than twice the displacement of the boats they are replacing, and costing about $1\frac{1}{2}$ billion dollars apiece. Despite their size they are quieter and harder to find than their predecessors. An even more valuable asset is the tremendous range of the missiles they carry: 4,600 miles for the Trident C4 on the boats now in service and 6,000 miles for the D5 missile to be fitted on the later models. These ranges allow a Trident submarine to hit targets in the Soviet Union from almost anywhere in the ocean, thus making it virtually impossible for the Soviet navy to search the entire area in which it might be. Submarines with the D5 will be able to hit the Soviet Union from well south of Australia or from the coast of Peru.

The Trident's one real defect is its tremendous cost which ensures that no more than 20 will be built to replace the 41 Polaris and Poseidon boats the United States had during the 1970s. Of these only 13 will be on patrol at any given moment, not many for a force that some day may be the United States's only reliable deterrent. Although they are virtually immune from open-ocean search tactics and will be for a long time to come, they may become vulnerable to trailing tactics. The Soviet Union has nearly 80 nuclear-powered attack submarines suitable for trailing missile submarines and is building more. If the number of United States missile submarines continues to drop as the Poseidon boats are taken out of service, the Soviets may eventually be able to assign four, five, or six attack submarines to trailing each Trident. This, combined with a dramatic breakthrough in Soviet anti-submarine warfare (ASW) technology, could make the Tridents more vulnerable than is acceptable.

The B-1 bomber has been under development for more than 15 years as a replacement for the B-52 (see figure 7.3, p. 97). The Strategic Air Command now has 100 B-1Bs. Without these aircraft the United States would not have an effective strategic bomber force beyond the next few years with the B-52s nearing the end of their useful lives. The B-1 is a much more capable airplane than the B-52. It is faster, has about twice the payload, is designed to fly as low as 200 feet above the ground (B-52s now practice low-level attacks but they were not designed to do so), and is 99 percent less detectable by radar. These qualities should enable it to penetrate Soviet airspace throughout the 1990s and to be an effective cruise missile carrier into the twenty-first century.

There is an even more advanced strategic bomber under development, the B-2 "Stealth" aircraft, which is designed to be virtually undetectable by enemy radars. The B-2 is a flying-wing design, with an airframe made out of radar-deadening materials. Very few people know much more about it than that; perhaps it is appropriate that an airplane designed to be invisible to Soviet air defenses should also be invisible to the American public. Some people feel that the air force, rather than building B-1s, should have the B-52s soldier on until the B-2 is in service. However, developing this new bomber will clearly require radical breakthroughs in aircraft technology, and technological breakthroughs are seldom carried to fruition overnight. If the B-2's development history is at all comparable to that of most other state-of-the-art weapons, it will not be ready for service much before the twenty-first century. The United States air force will need a new strategic bomber before then.

The new and very controversial American ICBM is the MX. It should never have been developed and should not now be in production. The MX is undoubtedly a very potent weapon, nearly three times as large as the Minuteman ICBM it is to replace, carrying ten warheads to Minuteman's three, and able to deliver them with point-blank accuracy. Its CEP is said to be less than 400 feet. All these qualities make it probably the most potent first-strike weapon in existence. As a second-strike deterrent weapon the MX has the same defect as Minuteman or any other ICBM, poor survivability. Whether it is in a silo, an MPS complex, a Dense Pack silo cluster, or any other basing mode, an MX is just as vulnerable as a Minuteman. If anything it is even harder to protect because the MX is a much larger missile and thus requires a larger shelter or transporter.

If the United States keeps ICBMs in its arsenal they should not be MXs. The existing Minuteman missiles are powerful enough for their second-strike countervalue role and more easily concealed or sheltered than MXs. Should a new ICBM be deployed it should be the proposed "Midgetman:" a small, single-warhead missile which could be built in much greater numbers than the MX and would be even easier to shelter or conceal than Minuteman.

Notes

1 Robert S. McNamara, "The Military Role of Nuclear Weapons: Perceptions and Misperceptions", *Foreign Affairs*, 62 (Fall 1983), p. 79. (emphasis in original.)
2 Wolfgang K.H. Panofsky, "The Mutual-hostage Relationship between America and Russia", *Foreign Affairs*, 52 (October 1973).
3 Fred Charles Ikle, "Can Nuclear Deterrence Last Out the Century?", *Foreign Affairs*, 51 (January 1973), p. 281.
4 Herman Kahn, *On Thermonuclear War* (Princeton, NJ: Princeton University Press, 1961), p. 137 (emphasis in original).
5 Harvard Nuclear Study Group, *Living with Nuclear Weapons* (Cambridge, Mass.: Harvard University Press, 1983), p. 146.
6 Lawrence Freedman, *The Evolution of Nuclear Strategy* (New York: St Martin's Press, 1981), p. 390.
7 Bernard Brodie, *Strategy in the Missile Age* (Princeton, NJ: Princeton University Press, 1959), p. 404.
8 "Throw weight" is the total weight of the payloads of a force's missiles; "equivalent megatonnage" is a measure of the area destroyed by a warhead or a number of warheads.

10
The Nuclear Deterrents of France, China, and the United Kingdom

The United States and the Soviet Union are not the only nuclear powers. France, China, and the United Kingdom also have strategic nuclear weapons: bombers and missiles with enough range to hit a superpower. India has set off a nuclear device, although it does not have any combat-ready nuclear weapons, and Israel is believed to have a small number of atomic bombs hidden away. Pakistan, Argentina, Iraq, and South Africa have made considerable progress towards being able to build nuclear weapons, and there are 20 or 30 other countries which probably could "go nuclear" within a decade if they chose to do so. Thus far, though, only France, China, and the United Kingdom have nuclear forces that the superpowers must reckon with.

Each of the three medium-rank nuclear powers has a strategic nuclear force about one-fifteenth as powerful as that of a superpower. Like the superpowers, they have a variety of nuclear weapons: ballistic missiles, missile submarines, and bombers. Their missiles and bombers are shorter range weapons than those of the superpowers and would be considered tactical rather than strategic weapons in the United States or Soviet arsenals. These are shorter range weapons because they are designed almost exclusively for use against the Soviet Union and all of these countries are much closer to the Soviet Union than is the United States. An IRBM or medium range bomber can hit most parts of the Soviet Union from bases in France, China, or the United Kingdom. Close proximity to the Soviet Union also makes their bombers and missiles much more vulnerable to Soviet attack than are those of the United States. This is an inherent defect of land-based tactical missiles and bombers, as will be explained in chapter 11 in further discussion of the weapons themselves (see pp. xxx–xxx). United States tactical nuclear weapons in Europe are equally vulnerable, as are their Soviet counterparts.

Because bombers and missiles on their territories are so vulnerable to Soviet attack, the French and the British now place almost entire reliance on their missile submarines. France has six missile submarines and the United Kingdom four, each about as capable at the United States's Poseidon boats. France can now keep three missile submarines on patrol at all times and the United Kingdom one or two. Before they had missile submarines the two countries relied on medium range bombers much more than they do now. The British V-bombers ("Valiant," "Victor," and "Vulcan" types) were the entire British strategic nuclear force until 1968, when the first British missile submarine became operational. At one time there were 180 of them. Similarly, the Mirage-IV was the first French nuclear weapon. In recent years the V-bomber force has been retired and the Mirage-IVs, although still in service, are no longer France's primary nuclear deterrent. The United Kingdom has never had a ballistic missile in service. France has 18 intermediate range ballistic missiles (IRBMs) but few, if any, of them would survive a Soviet attack.

In 1988 China has two modern missile submarines, nuclear-powered boats comparable to the Poseidon submarines except that each carries 12 SLBMs instead of 16. Several more are reported to be under construction so by the 1990s China should have a missile submarine force comparable to that of France or the United Kingdom. Until very recently, though, she had to rely almost entirely on land-based missiles despite the very considerable disadvantages of such weapons. China has slightly more than 100 IRBMs and medium range ballistic missiles (MRBMs), able to hit targets in Siberia and Soviet Central Asia, and half a dozen ICBMs. The longest ranged Chinese ICBM type can reach any part of the Soviet Union and, for that matter, most of the United States, but only two of this type have been built. None of these weapons is any more survivable than other land-based missiles. The Chinese endeavor to protect their missiles by hiding them in caves, camouflaging them, and moving them around, but they can never be sure that the Soviets have not located most or all of them. China also has a few obsolescent bombers but few of them would survive a Soviet pre-emptive attack and very few of the survivors would be able to get past Soviet air defenses. Until she gets to sea a substantial force of missile submarines China will be more vulnerable to a Soviet preemptive attack than either France or the United Kingdom.

France, China, and the United Kingdom have nuclear weapons partly to enhance their national prestige, to show that they are forces to be reckoned with. For the United Kingdom, having nuclear weapons increases her influence within NATO. In the words of Prime Minister Harold Macmillan, "It makes the United States pay a greater regard to our point of view." Nuclear weapons give France the freedom either to stand apart from NATO or to cooperate with the alliance on her own terms. They mean, as President Charles De Gaulle put it, that France can "command its own destiny."[1] Chinese nuclear weapons strongly reinforce China's claim to equality with the United States and the Soviet Union. For all of the medium-rank nuclear powers possessing nuclear weapons means that they cannot be ignored when war and peace, arms control and disarmament are being discussed. If they are not superpowers, they are in the next rank down from the superpowers.

While prestige is an important consideration, France, China, and the United Kingdom have nuclear weapons primarily for deterrence. Each fears being attacked by the Soviet Union and seeks to guard against it. That is the basic similarity in their nuclear strategies. There are also important differences between them, differences that arise mostly from the different relationship each has with the United States. The United Kingdom is a close ally of the United States, France a rather distant ally, and China no ally at all.

China's relations with the United States are somewhat less hostile than her relations with the Soviet Union. China broke with the Soviet Union in 1960 without ceasing to be a fierce enemy of the United States, and for the next 9 years contended against both superpowers impartially. Since 1969 relations between China and the United States have improved, mostly because the Chinese government has come to see the Soviet Union as the greater threat. China has a long border with the Soviet Union, with a large part of the Soviet army deployed along it; it has no border with the United States. This does not make the United States and China allies, least of all military allies. China is still a Communist-ruled state with the same dogmatic hostility towards liberal democracy as the Soviet Union or any other Communist-ruled state. The United States has never had a military alliance with the Chinese Communist regime.

Chinese military planners must assume that in any war with the Soviet Union China will stand alone; the United States will not come to China's aid. The Soviet Union is more likely to attack China than to attack France or the United Kingdom because of the opportunities offered by the long Sino–Soviet border, and also because an attack on China is less likely to lead to a nuclear total war. China, therefore, needs a strong nuclear deterrent more than the other two. At the same time, China has not yet managed to build a strategic nuclear force as secure from attack as are those of France and the United Kingdom – yet another reason why China is more likely to suffer a Soviet first strike. Perhaps that is why the Chinese government has invested a considerable part of its limited resources in civil defense.

If China has a clear need for a strong nuclear deterrent but has had difficulty building one, the United Kingdom has been able to build an adequate force but has had difficulty showing a need for it. The United Kingdom has long been a close ally of the United States and a loyal member of NATO. The NATO treaty obliges the United States to treat any attack on another NATO member as if it were an attack on the United States, and to render all the aid necessary to defeat the attack. British leaders have consistently read this as requiring the United States to retaliate with all its strength should the Soviet Union hit any country in Western Europe with nuclear weapons. They profess to have no doubts whatever that the United States would fulfill this obligation. If the United Kingdom is, then, fully protected by the American nuclear shield why does it need its own nuclear weapons?

The British strategists' answer to this question is that, although the United States would of course come to the aid of its allies even at the risk of its own existence, the Soviets might not expect it to. They might think they could attack Western Europe without triggering an American response. The British deterrent is necessary, therefore, to supplement the American deterrent. As a British Secretary of State for Defense, Francis Pym, put it:

The United States, by their words and deeds, has constantly made clear its

total commitment to come to the aid of Europe, and to help defend Europe by whatever means are necessary, without exception. . . .
[However] in a crisis, Soviet leaders . . . might conceivably misread American resolution. They might be tempted to gamble on the United States's hesitation.

It is best to have "a second centre of nuclear decision-making within the Alliance."[2] This makes it crystal-clear to Soviet leaders that if they attack Western Europe their country will suffer nuclear retaliation – from the United Kingdom if not from the United States. This is a good argument but it does make the British strategic nuclear force a luxury – an insurance factor rather than a necessity.

Even though France is an ally of the United States, her reasons for having nuclear weapons are much more like China's than the United Kingdom's. The basic premise of French nuclear strategy is this: the United States would not launch a nuclear attack on the Soviet Union in retaliation for a Soviet attack on France. An attack on the Soviet Union would trigger a Soviet counterstroke that would destroy the United States; no country will commit suicide in defense of an ally. The United States will resort to nuclear weapons only if its own territory has been hit by an enemy. An American pledge to go to nuclear war on behalf of an ally is bound to be disregarded at the moment of truth. The Soviets know this as well as anybody; they cannot be deterred by a threat the United States would never carry out.

From this it follows that France's only real protection against Soviet aggression is a French strategic nuclear force able to hit back if the Soviet Union attacks France. France withdrew its troops from the NATO Military Command in 1966 basically because President De Gaulle felt that the United States pledge to fight a nuclear war if France were attacked could not be counted on in a crisis; it was not worth the sacrifice of France's freedom of action to obtain it. This does not mean that France has ceased to be an ally of the United States or of the other members of NATO. She has troops stationed in West Germany and promises to fight with conventional weapons if West Germany were invaded. But the French do believe that when national survival is at stake, as it inevitably is when nuclear weapons are used, no alliance can be greatly relied on. As De Gaulle once said, "Alliances are like roses and young girls. They last while they last."

The strategic nuclear forces of France, China, and the United Kingdom are so much smaller than those of the superpowers that perhaps they are not powerful enough reliably to deter the Soviets. This is probably the most cogent of the several arguments against those forces. Certainly they are far too small to mount a serious counterforce attack on the Soviet Union. Even if every one of the 200 or so nuclear warheads a medium-rank power might have were to destroy a Soviet ICBM silo, almost all of the Soviet Union's strategic nuclear force would be left intact for retaliation. These forces can only be used for second-strike, countervalue purposes, only to retaliate against Soviet cities. Are they adequate for that purpose, i.e., would the remnant left after a Soviet attack on them be adequate?

The French, British, and Chinese governments clearly feel that their strategic nuclear forces are adquate. French nuclear strategists point out that a medium-rank nuclear power does not need to be able to totally destroy the Soviet Union. It needs only to be able to inflict enough damage to outweight whatever the Soviets might hope to gain by attacking it. This theory of "proportional deterrence" asserts that it is easier to deter a Soviet attack on France or another medium-rank power than one on the United States because the Soviet Union has far more to gain from destroying the United States than from destroying France. To destroy the United States would put the Soviet Union within grasping distance of world domination and absolute security while to destroy France would probably leave the Soviet Union essentially no more secure than before. So, whereas the United States needs to be able to destroy the Soviet Union, France needs only to be able to damage it, "tearing off an arm," as De Gaulle put it. This a medium-rank nuclear power can do. The two missile submarines which France or the United Kingdom usually have on patrol can destroy ten of the largest cities in the Soviet Union. These ten cities contain 6 percent of the country's population and 15 percent of its industrial capacity, the loss of which would more than outweigh whatever the Soviet Union might gain by destroying France or the United Kingdom.

The "proportional deterrence" theory holds up if it be granted that the medium-ranked power would have those two missile submarines still intact after a Soviet preemptive attack. However, it might well be possible for the Soviet armed forces to trail and destroy both of them and, by a simultaneous IRBM attack, destroy the submarines in port and all of the victim's other nuclear weapons. Eighty Soviet nuclear-powered attack submarines should be able to trail and destroy two missile submarines if they are assigned no other tasks. That a medium-ranked power's first-strike capability is one-fifteenth that of a superpower does not mean that its second-strike capability would be. It might be left with no second-strike capability at all.

The problem with even a successful counterforce attack on a medium-rank nuclear power is that the Soviet Union would have to employ nuclear weapons in that attack. This would make an American retaliatory strike much more likely than if the victim had been conquered with conventional weapons alone. The "nuclear barrier" would already have been broken by the Soviet Union, probably causing heavy civilian casualties. It would be psychologically much easier for the United States to use nuclear weapons against the Soviet Union if the Soviet Union had just used them on another country. This may be the best argument for the nuclear deterrents of the medium-rank powers. Perhaps they serve their possessors best not by making France, China, and the United Kingdom independent of the United States's nuclear shield, but by making the shield more effective. Because these countries have nuclear weapons, it is much more likely that the United States would go to war if they were attacked.

Notes

1 Quoted in Lawrence Freedman, *The Evolution of Nuclear Strategy* (New York: St Martin's Press, 1981), pp. 311, 313.
2 Quoted in Lawrence Freeman, *Britain and Nuclear Weapons* (London: Macmillan, 1980), pp. 128–9. (The second quote is from the words of Secretary of State for Defense Fred Mulley.)

PART IV
Land War and the Defense of Europe

Who controls the Rimland rules Eurasia; who rules Eurasia controls the destinies of the world.
Nicholas Spykman, *The Geography of the Peace*

11
The Weapons and Conduct of Land War

And when he gets to Heaven,
To St Peter he will tell:
"One more Marine reporting, sir –
I've served my time in Hell."
On the grave of a United States Marine,
Guadalcanal 1942

Nuclear Weapons and Conventional Weapons

Nuclear weapons are so much more devastating than any others that it is rather surprising to see so-called "conventional" weapons used at all in the Nuclear Age. It is as if spears and swords had continued to be important weapons after the invention of the machine gun, even for countries which possessed machine guns. And yet conventional weapons do continue to be used in the Nuclear Age, in tremendous quantities. The nations of the world have tens of thousands of military aircraft, hundreds of thousands of tanks and other armored vehicles, and tens of millions of rifles and machine guns. Even the nuclear powers spend several times as much on conventional as on nuclear forces.

It is obvious why countries that do not have nuclear weapons spend so much on conventional weapons: they must fight with what they can get, even if it is tragically inferior to the weapons others have. Nineteenth-century African warriors fought with spears and swords against British machine guns because those were the only weapons they had. It is not so obvious why even the nuclear powers rely so heavily on conventional weapons, but their doing so shows the limitations of nuclear weapons.

Never before has a weapon built in such large numbers seen such little use. Only two nuclear weapons have ever been used in war and that was more than 40 years ago. Since then the nuclear powers have constructed tens of thousands of nuclear weapons and not used any part of this massive arsenal despite considerable temptation. The United States accepted defeat in Vietnam and came close to suffering a military disaster in Korea without resorting to nuclear weapons. The Soviet Union has apparently accepted defeat in Afghanistan, having tried almost everything but nuclear weapons. In the past decade China has been at war with Vietnam, the United Kingdom at war with Argentina, and France engaged in several small-scale military adventures in Africa, all without using nuclear weapons. It seems that it is not easy to use nuclear weapons in a war.

The fundamental reason why no country has used nuclear weapons since 1945 is the fear that doing so might lead to retaliation in kind. This fear is very clearly the reason why no nuclear power has ever attacked another. It has also inhibited attacks by nuclear powers on countries which do not have nuclear weapons. The victim of the attack is likely to have nuclear-armed allies who could retaliate in kind upon the aggressor. A Soviet nuclear attack on Israel or West Germany would very likely make the United States hit back with nuclear weapons, and the Soviet Union might hit back if China were to use nuclear weapons on Vietnam. A nuclear attack even on a country as diplomatically-isolated as South Africa or Iran would break down the inhibitions on using nuclear weapons which have built up since 1945. It would make the future use of nuclear weapons more likely, and thus threaten everybody. Therefore, all the nuclear powers have thought that to use nuclear weapons against any country is too likely to lead to the use of these weapons against other countries – including themselves. One day perhaps some government will change its mind, but only if it faces a crisis far more acute than any which has arisen since 1945 or is far more reckless than any regime currently in power.

The nuclear powers have strong conventional forces to fight all the wars they dare not fight with nuclear weapons. They also maintain conventional forces for other purposes. Even in a war in which nuclear weapons were used in great numbers there would still be a role for conventional forces. Nuclear weapons can destroy the enemy's army, but only the ground forces can occupy his territory and subjugate his people. If the Soviet Union were to invade Western Europe, it might use tactical nuclear weapons to open gaps in the NATO defenses, but it would also need ground forces to pour through the gaps, destroy the remnants of the NATO armies, and occupy Western Europe. Similarly, if NATO were to use tactical nuclear weapons to defend itself, it would also require conventional forces to exploit the effects of its nuclear strikes. Conventional forces may also strengthen nuclear deterrence. If a strong conventional defense of Western Europe would force the Soviets to use nuclear weapons to break it, they might shrink from doing so, for fear that their use of nuclear weapons in Europe might provoke an American nuclear attack on the Soviet Union. Their need to use nuclear weapons to win a war helps keep them from starting it. In these various ways conventional forces serve as powerful auxiliaries to nuclear forces, making nuclear weapons both more effective in war and more credible as deterrents.

Nuclear weapons clearly have not made conventional weapons obsolete, not even in the armories of the nuclear powers. On the other hand no country which has nuclear weapons can fight a conventional war as if nuclear weapons did not exist. Non-nuclear powers can, but any war in which a nuclear power is engaged, and particularly any war in which there are nuclear powers on both sides, is very likely to become a nuclear war. Nuclear powers can fight conventional wars, but they fight under the shadow of a possible nuclear war.

The Tank–Infantry Team

The most important weapon of land warfare continues to be the tank. Contemporary main battle tanks (there are also light tanks used for reconnaissance) weigh 35–60 tons and have three- or four-man crews (figure 11.1). They can move 30 or 40 miles an hour on roads but much more slowly cross country because a higher speed would shake up the crewmen so much they would not be able to fight. The big gun on a typical main battle tank, the Soviet T-62, can penetrate 11 inches of steel armor plate at a range of more than a mile. The T-62's own armor is 7 inches thick on the turret front.

Tanks are large, expensive machines (the latest models cost one or two million dollars apiece), difficult to keep in operating condition, and much more vulnerable than they look. They are also indispensable. No other weapon offers the combination of firepower, protection, and mobility that a modern main battle tank has. Guns and missiles have greater firepower, armored cars are faster, and well-dug-in infantry have greater staying power, but tanks have the best overall balance of desirable qualities. These qualities are as valuable on the defensive as on the attack. An effective defense requires mobility as well as firepower and protection because the defender must be able to move his forces to counter enemy breakthroughs or to take advantage of enemy vulnerabilities. The main battle tank leads the attack and it is also the heart of the defense.

Because tanks are the dominant weapon on the battlefield the first task of any tank unit is to eliminate all enemy tanks within range. Only after that should the tanks go after enemy infantry, artillery, or other foes. Tanks mostly fight other tanks, which is why their guns are designed primarily to fire anti-tank shells. The best anti-tank weapon is another tank.

In recent years the main battle tank has encountered a formidable rival: the Anti-Tank Guided Missile (ATGM). This is a small rocket which can be carried and launched by one or two men. It has a range of a mile or two, is extremely accurate even at long range, and has a warhead capable of penetrating 15 or 20 inches of steel armor plate. On a target range, if not perhaps in combat, an American TOW ATGM has a four-out-of-five chance of destroying a typical Soviet tank at a range of 2 miles. The potency of the ATGM was demonstrated in 1973, in the Yom Kippur War, when Egyptian troops used Russian-built "Sagger" ATGMs with deadly effect against Israeli tanks. Since then some authorities have maintained that ATGMs have made tanks obsolete.

A 10,000 dollar missile which can destroy a million dollar tank does seem a formidable weapon. However, there are ways to counter the ATGM and already it looks much less formidable than a decade ago. The most important counter to the ATGM is the development of compound armor plate for tanks. Compound armor is made up of multiple layers of steel, ceramics, plastics and other substances – its exact composition is a closely-guarded secret. One inch of compound armor is as effective against the shaped-charge warheads

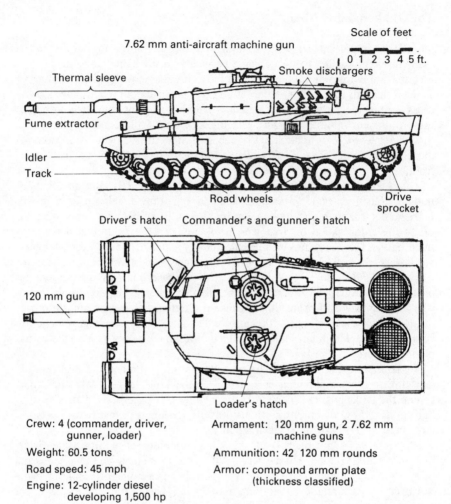

Scale of feet

0 1 2 3 4 5 ft.

7.62 mm anti-aircraft machine gun

Thermal sleeve

Smoke dischargers

Fume extractor

Idler

Track

Road wheels

Drive sprocket

Driver's hatch Commander's and gunner's hatch

120 mm gun

Loader's hatch

Crew: 4 (commander, driver,
    gunner, loader)

Weight: 60.5 tons

Road speed: 45 mph

Engine: 12-cylinder diesel
    developing 1,500 hp

Armament: 120 mm gun, 2 7.62 mm
    machine guns

Ammunition: 42 120 mm rounds

Armor: compound armor plate
    (thickness classified)

The Leopard II, which entered service with the West German Army in 1979, is typical of the main battle tanks entering service in the 1980s. Its 120 mm smoothbore gun is as accurate as a rifled gun and has a very high muzzle velocity. The thermal sleeve improves the gun's accuracy by keeping the barrel at an even temperature; the fume extractor prevents powder fumes from drifting back into the turret. The tank's angular profile betrays the use of compound armor in its construction.

Figure 11.1 *A main battle tank – West Germany's Leopard II*

of ATGMs as several inches of conventional armor. Contemporary ATGMs cannot penetrate the frontal armor of tanks fitted with compound armor plate. The American M-1, the British Challenger, the West German Leopard II, and probably the Soviet T-80, all have compound armor, and the United States, British, West German, and Soviet armies are being reequipped with these types as fast as possible. ATGMs can also be defeated by artillery firing over the heads of the advancing tanks, obliterating woods, buildings, or anything else that might conceal the missile crews. ATGM gunners on foot are extremely vulnerable to artillery fire. If all else fails, a tank force facing ATGM gunners on foot can simply mass to attack a weak point in the defenses, break through at whatever cost, and then overrun whatever the ATGMs were defending. That is how the Israeli tank forces eventually defeated the Egyptian ATGMs in 1973. The mobility of tanks allows them to concentrate and attack a small sector of front much more rapidly than man-carried ATGMs can be concentrated to defend it.

The weaknesses of the ATGM can be remedied by making larger missiles and mounting them on tracked, armored vehicles. ATGMs powerful enough to penetrate compound armor would have to be mounted on some kind of vehicle, they would be too heavy to be man-portable, and only a tracked vehicle can go everywhere that tanks go. The ATGM-carrier should be armored to stand up under artillery fire. But, an armored, tracked vehicle carrying ATGMs, far from making the tank obsolete, would itself be a kind of tank. It could be lighter and less expensive than a main battle tank but it would also be considerably less versatile. Tanks, whether armed with guns or with missiles, will continue to dominate the battlefield into the twenty-first century as they have in the twentieth.

Tanks are the most important weapons in a land battle but they cannot win it alone. Unless accompanied by infantry they are simply too vulnerable to close-in attack especially in built-up areas, woods, and other difficult terrain. It is hard to see anything from inside a "buttoned-up" tank (one with its hatches closed and the crew peering through periscopes) and impossible to see anything in its immediate vicinity. There is a "blind zone" of about three-quarters of an acre around a tank, within which its crew cannot see the ground, and an even larger "dead zone" within which it cannot bring its gun to bear. Enemy infantry armed with light hand-carried anti-tank weapons can readily destroy a tank if they can get close enough to it. To prevent that tanks must be accompanied by their own infantry who can clear the enemy out of buildings and underbrush, inform the tank commanders of what lies around them, and help the tanks advance through difficult terrain. "The infantry protects the tanks by acting as their eyes and ears; the tanks support the infantry with their firepower."[1] The tanks and the infantry who accompany them must work as a team, the tank–infantry team; neither can survive for very long without the other. As the Germans discovered during World War II, the best way to stop an attack is to separate the tanks from the infantry and then deal with each element separately.

If the infantry are to keep up with the tanks and to survive enemy artillery fire while on the move, they must be provided with some form of armored,

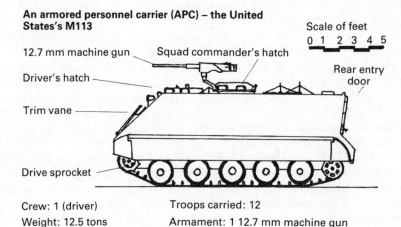

An armored personnel carrier (APC) – the United States's M113

Scale of feet

0 1 2 3 4 5

12.7 mm machine gun   Squad commander's hatch

Driver's hatch

Rear entry door

Trim vane

Drive sprocket

Crew: 1 (driver)   Troops carried: 12
Weight: 12.5 tons   Armament: 1 12.7 mm machine gun
Road speed: 42 mph   Armor: 38 mm (aluminum)

The M113 has been the United States's standard APC since the early 1960s and more than 73,500 have been built. Like other APCs, it is armored against shell fragments but is not designed to fight on the battlefield. The troops it carries must dismount from it to fight – they cannot fire from within the vehicle – and it is armed only with a single heavy machine gun.

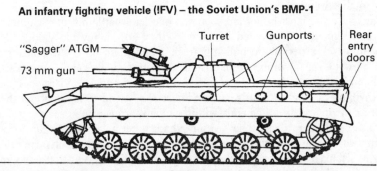

An infantry fighting vehicle (!FV) – the Soviet Union's BMP-1

Turret   Gunports·   Rear entry doors

"Sagger" ATGM

73 mm gun

Crew: 3 (commander, gunner,   Troops carried: 8
            driver)
Weight: 14.75 tons   Armament: 1 "Sagger" ATGM launcher;
Road speed: 50 mph   1 73 mm gun; 1 7.62 mm
                               machine gun

Armor: 33 mm (steel)

The BMP is the standard IFV in Soviet tank divisions. It has the armament of a light tank and the troops it carries are provided with gun ports from which they can fire from within the vehicle

Figure 11.2 *Armored personnel carriers*

mechanized transport. The infantry component of a tank–infantry team is carried in "Armored Personnel Carriers' (APCs) (figure 11.2). The United States's M113 is a typical, though now somewhat outdated, APC. It is a box-like tracked vehicle, fitted with light armor (about 1 inch thick), armed with a heavy machine gun, and capable of carrying 12 infantrymen. Deluxe model APCs, sometimes known as "infantry fighting vehicles' (IFVs), have turrets mounting ATGMs and cannon to support their infantry when dismounted. In a rapid advance, the APCs move just behind the tanks. Some APCs have gunports so that their occupants can fire from within while the vehicle is on the move, although this fire is very unlikely to hit anything. If the tanks encounter opposition they cannot quell by their own efforts, the infantry dismount from the APCs and go forward to help clear it out, either accompanied by the tanks or supported from a distance by the tanks' guns. In defense the infantry usually dismount and dig in around the tanks, protecting them from close assault while being covered by their guns.

Support Forces

Tankers, armored infantry, and other front-line troops are the cutting edge of an army, "the sharp end". They seize the ground in the attack, hold it on the defense, and suffer most of the casualties. A vast array of other forces backs them up. Artillery, close support airplanes, and gunship helicopters engage the enemy's ground forces on the battlefield, in direct support of the front-line troops. Fighter-bombers and suface-to-surface missiles attack deep behind the battlefield, striking at supply lines and reserves moving up to the front. Air superiority fighters and anti-aircraft weapons attack enemy aircraft. All of these combat forces require combat service support units to supply them with munitions, fuel, rations, and other requisites, repair damaged equipment, and treat the casualties. No one arm of the service can win a battle, any more than a single instrument can play a symphony.

Artillery caused 58 percent of the casualties in World War II and remains an extremely important weapon today. It is almost always used for indirect fire, shooting from well behind the lines at target that cannot be seen from the gun positions. Forward observers locate targets for the guns and report where the shells fall. Even if the enemy cannot see a battery, his computer-assisted radars can locate it pretty quickly once it opens fire. Guns dare not fire from the same position for very long. They must "shoot and scoot" – fire a few rounds and then quickly move to a new location.

Self-propelled guns are much more effective than towed artillery and have almost entirely replaced the latter in NATO armies. A self-propelled gun resembles a tank but it is essentially a form of artillery, a heavy gun mounted on a tank chassis (figure 11.3). It can move as fast as a tank, go anywhere a tank can go, and get into and out of firing order very quickly. This allows it to keep up with the tanks and APCs in a rapid advance and to use shoot and scoot tactics better than a towed gun could.

Gunship helicopters and close support airplanes are best seen as forms of

A self-propelled gun – France's 155 mm GCT

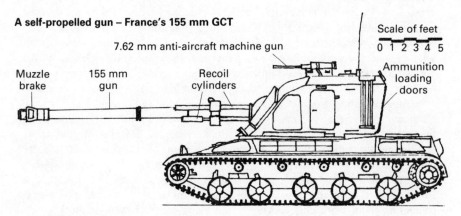

7.62 mm anti-aircraft machine gun

Scale of feet

0 1 2 3 4 5

Muzzle brake

155 mm gun

Recoil cylinders

Ammunition loading doors

Crew: 4 (commander, driver, gunner, loader)

Weight: 46 tons

Road speed: 37 mph

Armament: 1 155 mm gun; 1 7.62 mm anti-aircraft machine gun

Armor: 20 mm (steel)

The GCT (grande cadence de tir) is in service with the French Army and has also been employed by Iran in the Iran–Iraq War. Its 155 mm gun has a maximum range of 13 miles and, with its automatic loader, can fire 8 rounds a minute. The GCT carries 42 rounds of 155 mm ammunition in the rear of its turret.

A self-propelled anti-aircraft gun system – the Soviet Union's ZSU-23-4 Shilka

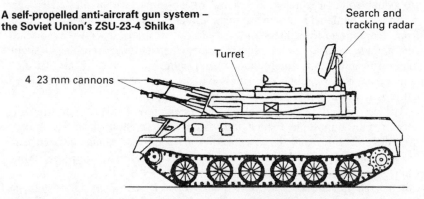

Search and tracking radar

Turret

4 23 mm cannons

Crew: 4 (commander, driver, gunner, radar observer)

Weight: 22.5 tons

Road speed: 27 mph

Armament: 4 23 mm cannons

Armor: 15 mm (steel)

The "Shilka" has been in service with the Soviet Army for more than 20 years, has also served with the armies of 27 other countries, and was used in Vietnam and in the Yom Kippur War. Its 23 mm cannons are water-cooled and all four together can fire as many as 800 rounds a minute.

Figure 11.3 *Self-propelled weapons*

flying artillery (figure 11.4). They have much greater range than guns, so they can concentrate much more quickly and effectively at the decisive point on the battlefield. They are ideal for use against mobile targets because they can seek out and destroy a moving target without the help of forward observers. Airplanes and helicopters armed with anti-tank cannons and ATGMs are especially effective against tanks and other armored vehicles, which they can attack from above and the rear where the armor is thinnest.

On the other hand, operating these weapons is extremely expensive. In a major conventional war one close-support sortie would cost nearly half a million dollars. This is not the cost of the airplane: it is the cost of flying the airplane to its target and back *once*. This includes the price of the munitions it drops or fires, the fuel it consumes, the maintenance work it requires, and one five-hundredth of the cost of the airplane to cover the odds on its being shot down.[2]. Guns can do the job far more cheaply and there are far more of them, about four times as many in Europe as there are close-support airplanes and gunship helicopters. Therefore, aircraft are best used to attack targets guns cannot reach.

Aerial attack on enemy supply lines and reserves moving along them is "interdiction". Interdiction proved to be a relatively ineffective use of United States air power in Korea and Vietnam. Communist forces in both wars showed great resourcefulness in keeping open their supply lines, sometimes repairing bombed-out segments of railroad track in as little as 8 hours. Also, their lightly-armed troops required much less resupply than modern mechanized forces would. Interdiction may be much more effective against a Soviet offensive in Europe, as Soviet strategy depends on a massive flow of reserves up to the battlefield. On the other hand, interdiction missions flown in the teeth of the Soviet air defense network may prove prohibitively costly.

Air defense is the mission of hunting the enemy's aircraft from the sky, to protect friendly troops and supply lines, and to facilitate aerial attack on enemy forces. Achieving this "air superiority" is usually the first-priority task of tactical air power. In addition to their air superiority fighters, modern armies deploy ferocious arrays of anti-aircraft guns and missiles, large numbers of many different types. The key to an effective ground-based air defense is to have a variety of weapons, each able to cover part of the air-space in which hostile aircraft might appear. Airplanes can usually evade any one type of weapon but only by flying into the teeth of another. United States pilots over North Vietnam found that it was easy enough to evade the big SA-2 missiles by going into a steep dive, but that brought them right down to where thousands of North Vietnamese light anti-aircraft guns were waiting for them.

The Soviet Union has placed great emphasis on anti-aircraft guns and missiles. The Soviet ground forces have 18 different types of anti-aircraft guns and missiles, in all sizes from the SA-10, which can reach an altitude of 19 miles, down to the SA-7, which one man can fire and is supplied to every infantry platoon. There are 40,000 of these weapons, big and small, high-, medium-, and low-altitude, mobile and stationed right up with the front-line troops, or echeloned back for hundreds of miles. Together they

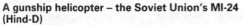
A gunship helicopter – the Soviet Union's MI-24 (Hind-D)

Scale of feet
0 _____ 10 ft

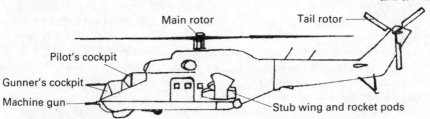

Main rotor

Tail rotor

Pilot's cockpit

Gunner's cockpit

Machine gun

Stub wing and rocket pods

Crew: 2 (pilot, gunner)
Speed: 199 mph

Armament: 1 four-barrel 12.7 mm machine gun; 4 ATGMs; 4 32-round rocket pods

The MI-24 carries eight troops in a cabin behind the cockpit, so it serves as an assault helicopter as well as a gunship. This heavily-armored helicopter has seen extensive service in Afghanistan. ("Hind" is a code name assigned to the type by NATO).

A close-support airplane – the United Kingdom's Harrier GR-3

Swivelling exhaust nozzle

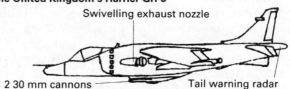

2 30 mm cannons

Tail warning radar

Crew: 1 (pilot)
Speed: 730 mph

Armament: 2 30 mm cannons; more than 5,000 lb of bombs, rockets, and other munitions

The Harrier is a VSTOL airplane; its exhaust nozzles (two on each side) rotate to point towards the ground for take offs and landings and back towards the tail for level flight. It played a prominent part in the Falkland Islands War and is in service with the United States Marine Corps as well as with the British Navy and the Royal Air Force.

A fighter-bomber – the Tornado IDS of the United Kingdom/West Germany/ Italy

Swivelling outer wing panels

Radome (housing radar sets)

27 mm cannon

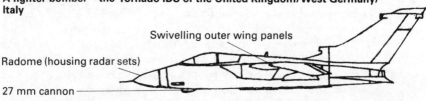

Crew: 2 (pilot, systems operator)
Speed: 920 mph (sea level)
        1,452 mph (high altitude)

Armament: 2 27 mm cannons; up to 19,840 lb of bombs and other munitions

The Tornado was designed and built by a consortium of British, West German, and Italian manufacturers and serves in the air forces of all three countries. It is a "variable geometry" type like the American B-1 bomber.

Figure 11.4 *Tactical aircraft*

can blanket with overlapping fields of fire any air-space into which enemy aircraft might venture. As long as this complex, multi-layer defensive system remains intact it will be very dangerous for anything hostile to fly in the area it defends. However, after a few days the chops and changes of mobile warfare may well open up gaps through which enemy aircraft could penetrate.

Probably the greatest vulnerability of aircraft in land warfare is their dependence on airfields. Some high performance fighter planes require hard surface runways a mile or two long. Such runways are inevitably conspicuous, fairly easy to destroy, and difficult to repair quickly. Most of the airfields in a theater of war may be out of service after a few days of high intensity conventional war. In response to this threat modern close support airplanes and fighter-bombers are more and more being designed to operate from short runways. There are even aircraft which do not need runways, such as helicopters and vertical or short take-off and landing (VSTOL) airplanes. (A VSTOL airplane is one that can take off after a very short roll or even – with a much reduced payload – straight up into the air like a helicopter.) Helicopters and VSTOL airplanes inevitably have less speed, range, and payload than conventional airplanes because a price has to be paid for the ability to jump up into the air. They do not have the range or payload that interdiction requires, but they can be effective in the close-support role.

A mechanized army cannot fight without a continual heavy flow of supplies to its combat troops. This is particularly true of the United States Army and the armies of other NATO countries. In combat an American division uses up 200 pounds of supplies per man per day or 1,500 tons for the entire division – 300 truckloads. The division carries with it enough supplies to fight for three to five days. After that it grinds to a halt and rapidly loses its ability to fight until resupplied. It will also grind to a halt if its vehicles are not continually serviced and disabled vehicles repaired. Tanks and other tracked vehicles are particularly likely to break down. In moving less than 70 miles, an armored division can expect to have a third of its tanks break down along the way even if it does not encounter the enemy at all. They can be repaired, but if the division is retreating rapidly there will not be time to do so. One of the great advantages of being on the offensive is that the attacker, if successful, gets possession of the battlefield and with it all of his and the defender's disabled but repairable vehicles.

Land War Today

The weapons and tactics used in conventional war on land have not changed greatly in the past 40 years. Contemporary tanks, APCs, towed and self-propelled guns, and close-support airplanes are recognizably the same kinds of weapons as their World War II ancestors. Nor have the ways they are used changed greatly. The Blitzkrieg tactics proposed early in the 1920s by the Englishmen B.H. Liddell Hart and J.F.C. Fuller, perfected by German tacticians during the 1930s, and then revealed to the world in the invasions of Poland and France, continue to be used today. A Soviet invasion of Western

Table 11.1 *Weapons and troops engaged in the Battle of France and deployed in Central Europe today*

	Tanks	Aircraft	Troops
Battle of France, 1940			
Germans	2,574	3,000	2,800,000
Allies	3,800	2,100	3,500,000
Total	6,374	5,100	6,300,000
Central Europe today			
Warsaw Pact	20,500	3,950	899,000
NATO	7,000	2,251	738,000
	27,500	6,201	1,637,000

Source: F.W. von Mellenthin and R.H.S. Stolfi with E. Sobik, *NATO under Attack* (Durham, NC.: Duke University Press, 1984), pp. 32–3

Europe would be another Blitzkrieg. In 1945, as it is today, tanks dominated the battlefield, they were accompanied by infantry, and the tank–infantry team was supported by self-propelled guns and close-support aircraft. ATGMs and other new weapons may someday compel a radical change in land warfare tactics but they have not yet done so.

Land war has changed in a number of ways. The most obvious is that modern conventional weapons are far more powerful than those used in World War II. Modern tank guns have considerably more accuracy and penetrating power than those in use 40 years ago. Infantry have greatly increased their firepower by acquiring ATGMs and hand-held anti-aircraft missiles. Modern close-support airplanes are faster than the World War II types and can carry several times the payload. Shells and bombs have been devised that are far more lethal than the familiar high explosive varieties: "beehive" artillery rounds to hurl several thousand small steel darts in the face of enemy infantry charging a gun position, cluster bombs that can scatter tennis-ball-sized bomblets over an entire field, bombs that sow vast numbers of tiny but effective "track-buster" minelets on the ground in front of advancing enemy tanks, and others. Even the hand grenade has been made far more deadly than before.

Not only are modern weapons more destructive but also there are far more of the heavier weapons – particularly tanks, aircraft, APCs, and self-propelled guns – today than during World War II. One can see this by comparing the forces engaged in the Battle of France in 1940 and those deployed in Central Europe today. As table 11.1 shows, the NATO and Warsaw Pact forces in Central Europe today have more aircraft and four times as many tanks as those that fought the Battle of France even though they have little more than a third as many troops. The 1940 armies had one tank for every 988 troops and one airplane for every 1,235. The present-day armies have one tank for every 60 troops and one airplane for every 264.

The vast majority of World War II combat soldiers were ordinary, non-mechanized infantry, armed with bolt-action rifles and marching on foot. The relatively few tanks, self-propelled guns and mechanized infantry were concentrated in a small number of specialized armored divisions. Of the 136 German divisions which fought in the Battle of France, only ten were Panzer divisions. The armored spearheads could break through the adversary's defenses on a narrow front and penetrate deep into his rear, but they could not hold the ground they had captured until the foot soldiers could catch up. This meant that the sustained rate of advance of a World War II army was usually limited to the speed at which foot soldiers could march.

In a modern mechanized army most divisions are in effect armored divisions with several hundred tanks apiece, their infantry carried in APCs and their artillery support provided by self-propelled guns. Of the 227 divisions in the Soviet ground forces, 202 are tank or motor-rifle divisions (a Soviet motor-rifle division has two-thirds as many tanks as a tank division). This has greatly increased the scope and tempo of offensive operations, at least on suitable terrain. A mechanized army can make a Blitzkrieg assault along its entire front, not just on a narrow sector, and if it breaks through can advance as far and as fast as its tanks can go without waiting for foot soldiers to catch up.

The greatly increased number of vehicles in a modern army also greatly increases its need for supplies and for the maintenance and repair of its thousands of vehicles. A United States armored division has nearly 4,500 vehicles in its inventory, about one for every four men, each of which has to be fueled, serviced, and kept in repair. Modern mechanized armies are immensely more powerful than their World War II predecessors but they are also far more vulnerable to any interruption of the supplies and services they need to fight. Compared with their World War II predecessors, modern armies have "fists of iron, and feet of clay."[3]

Various types of electronic equipment play a major role in war today, more so than ever before. Modern armies have extensive radio networks for communication between headquarters and the forces they control, and numerous ground-based, vehicle-mounted, and airborne radars. Without them commanders would find it difficult to command their troops and the troops would find it difficult or impossible to use many of their weapons. The importance of radios and radars has made electronic warfare (EW) a major element of war on land, at sea, and in the air. Specially-equipped EW units listen to the enemy's radio transmissions to gather information. Even if the messages cannot be decoded, the amount and direction of traffic on the radio network may be a significant clue to the enemy's plans and deployments. Direction-finding equipment can pinpoint the locations of transmitters and very often, thereby, the locations of the headquarters and combat units using those transmitters. EW also includes electronic counter measures (ECM) designed to prevent the enemy's using the airwaves. Radio transmitters and radars can be jammed, i.e. drowned out by broadcasting a very loud signal on the same frequency that they use. Jamming usually works only for a short period of time, until the enemy can take countermeasures, but disrupting his communications at the decisive moment, even just temporarily, may tip the balance between victory and defeat. Radar installations can also be put out of action by missiles designed to home in on the radio waves they emit.

All of these techniques for preventing or exploiting the enemy's use of the airwaves can themselves be countered. It is possible to deceive an enemy who is listening in on one's radio transmissions by employing some of one's own transmitters to simulate units that do not exist. Jammers can be located by direction-finding equipment and then destroyed. Jamming and direction-finding can be frustrated by transmitting messages in very short, high-energy bursts. Radar-seeking missiles can be frustrated by temporarily turning off the radars. So, to deal with ECM techniques there are so-called electronic counter-countermeasures (ECCM) and soon, no doubt, there will be counters to the counters to the counter measures (ECCCM?).

The effect of all this may be to make it almost impossible for either side to use the airwaves. Both armies may find themselves fighting with little centralized direction and with many of their weapons unusable. This would have an especially serious effect on the United States army which depends very heavily on radios and radars. The effectiveness of radar-guided weapons may fluctuate greatly in the course of even a very short war. They may be devastatingly effective at the beginning of a war and a few weeks later, when the enemy has learned how to counteract them, be almost worthless.

Armies today are much better equipped than before to fight at night. Tanks carry infrared lights or thermal imagers which enable their crews to see for several miles at night; infantry are often equipped with night vision devices; fighter-bombers can fly in the dark and even, with some difficulty, locate targets. Armies still cannot fight as well at night as during the day but an attacker who wishes to press his attack all night long can do so. If the attacker fights all night then the defender must also. Troops which are not prepared to continue fighting after the sun goes down may well suffer the same fate as the Jordanian 40th Armored Brigade in the Six Day War. It fought valiantly throughout the day on June 6, 1967 and stopped the Israeli advance on its front. The following night, as the Jordanian tankers got some well-earned sleep, an Israeli night attack swooped down and overwhelmed them.

Troops that can fight all night as well as all day can, at least in theory, fight continuously for several days on end. Military planners are examining, quite seriously, the feasibility of having soldiers fight for 72, 96, or even 120 hours without any sleep at all. Recent studies carried out for the United States army indicate that after 120 hours of continuous combat the loader in a tank crew can still function at 95 percent of his initial effectiveness while a tank platoon commander would be down to the 45 percent level. Well-trained and strongly-motivated troops can fight for a surprisingly long period of time. During the Yom Kippur War the Israeli Seventh Armored Brigade fought continuously for four days and three nights, without relief or respite of any kind. Whether the average NATO or Warsaw Pact combat unit could duplicate the Seventh Armored Brigade's feat is open to question, but the generals on both sides can be expected to push their troops to the limits of human endurance. The army whose troops can keep fighting the longest enjoys a crucial advantage.

Strategy and Tactics

Strategy is the art of employing a country's armed forces in war to attain the government's political goal. Tactics is the art of employing individual military units (platoons, perhaps, or battalions or divisions) to win a battle. The decision to invade Normandy in June 1944 was a strategic decision; the maneuvers of a battalion storming a hill are tactics.

Both in strategy and in tactics there are certain widely recognized rules for the successful conduct of military operations. The United States army recognizes nine of these "principles of war:"

1 The Objective. Just as the conduct of a war should have a clearly-defined military objective, so should the execution of any military operation, from the emplacement of a machine gun to the invasion of a continent. The man in charge of that operation must know clearly what the objective is, he must have ascertained that it is attainable, and he must make the objective clear to his subordinates.

2 The Offensive. "The offensive is the decisive form of war, the commander's only means of attaining a positive goal or of completely destroying an enemy force."[4] Even if the political goal or military objective of a war is purely defensive, the best defense frequently turns out to be a vigorous offensive. (This has been the secret of Israel's success in the Arab-Israeli wars.) By attacking or counter-attacking one seizes the initiative, controls the course of events, and forces the enemy to react to what one does instead of oneself reacting to what he does.

3 Mass. "There is in every battle-field a decisive point, the possession of which, more than of any other, helps to secure the victory."[5] The decisive point may be a hill or some other geographic feature; it may be the morale of the enemy's troops; or it may be the will to win of the enemy commander. Whatever it is, the commander must discern where that decisive point is and concentrate all the power he can on it.

4 Economy of Force. In order to concentrate strong enough forces on the decisive sector of a battlefront, it is usually necessary to practise "economy of force" in the less important sectors. That means to reduce one's strength in those areas to the absolute minimum.

5 Maneuver. Battles are seldom won by standing still. Military forces must maneuver to place themselves at the decisive point at the right time, to gain the initiative or respond to enemy maneuvers, to seize territory or to achieve other military objectives. Only a purely passive, Maginot Line defense does not require maneuver and this style of defense is seldom successful.

6 Unity of Command. A well-tried military maxim holds that to command

an army "one bad general is better than two good ones.' Military operations must be well-coordinated if they are to succeed; to insure this, all the forces dedicated to attaining the same objective should be commanded by the same person.

7   Security. Do not be taken by surprise. A commander should not become so intent on what he plans to do to the enemy that he neglects to consider what the enemy could do to him.

8   Surprise. Attempt to take the enemy by surprise. Attack when and where he least expects it and is least prepared to meet it.

9   Simplicity. The friction inevitable in war makes even the simplest maneuvers difficult and complex ones often impossible. Therefore, "other factors being equal, the simplest plan executed promptly is to be preferred over the complex plan executed later."[6]

The United States army's principles of war, and the very similar principles to which other armies adhere, have been proven by many years of experience. That does not make them laws which should or even can be followed absolutely and always. They are much more like useful rules of conduct such as "look before you leap" or "neither a borrower nor a lender be." A commander should try to observe them and, if he is breaking one, know that he is doing so and have a good reason for it. However, there are many good reasons for violating some of the principles of war on occasion, and one is that it is rarely possible to observe all of them simultaneously and in full. The commander must choose which of them he is fully to observe, frequently at the price of disregarding others.

The commander has, in fact, a choice between two possible styles of warmaking, each emphasizing certain of the principles of war while somewhat disregarding the others. These are "attrition-style" and "maneuver-style" warfare. In most situations the attrition style is best able to achieve mass, economy of force, unity of command, security, and simplicity, while maneuver-style warfare emphasizes surprise, maneuver, and the offensive. No army or commander ever relies entirely on either of these polar opposites – there are some situations which virtually dictate a battle of attrition while others require maneuver warfare – but generals and armies do tend to be primarily attrition-oriented or maneuver-oriented.

The attrition-oriented commander approaches the enemy army as if it were a large rock to be knocked to pieces with a sledgehammer. On the attack he seeks to find, fix, and destroy the enemy: locate the main strength of the enemy army, pin it down so it cannot escape his blows, and annihilate it. If he is the defender he seeks to place his forces athwart the main thrust of the enemy's advance, and concentrates their firepower against it. Attrition tactics rely primarily on firepower rather than maneuver. An attrition-oriented commander fighting on the defensive attempts to hold the line he held at the beginning of the battle and maneuvers only as necessary to plug whatever gaps the enemy makes. On the offensive he moves slowly but with all of his forces together, preferring to slow his advance rather than let any of them fall behind.

Attrition warfare has its advantages. It is a comparatively simple way to wage war and requires little initiative of lower-ranking officers. Because the maneuvers it requires are slow and ponderous and because the entire force moves together as much as possible, an entire operation can be planned in detail by the high command and lower-ranking officers need only carry out their orders as issued. All likely contingencies can be anticipated and planned for before the operation begins. Attrition warfare is also rather predictable. The larger army is pretty sure to win, just as when two rocks are scraped together the smaller will be worn away first. On the other hand, battles of attrition tend to be very costly. The great killing-grounds of World War I trench fighting – Verdun, the Somme, and Passchendaele – were attrition warfare at its brutal, unimaginative worst.

If attrition warfare treats the enemy army as an inanimate object to be broken up piece by piece, maneuver warfare treats it as a living organism with a brain and a nervous system. It wins by confusing that organism, demoralizing and ultimately paralyzing it, killing its brain (the high command) or preventing the brain from controlling its members (the combat units). Maneuver tactics attempt to disrupt the enemy army and eventually split it up into disorganized and demoralized fragments. They do this by employing what B.H. Liddell Hart termed "the indirect approach," striking the enemy where he is weakest and where he least expects to be attacked. The maneuver-oriented commander employs firepower as well as maneuver – one cannot defeat an enemy without shooting at him – but relies primarily on maneuver.

The German Blitzkrieg was a classic example of maneuver tactics on the offensive: relatively small armored spearheads breaking through on a narrow sector of the front and then fanning out in the enemy's rear, striking headquarters, supply lines, and other vulnerable targets. The armored spearheads were formidable because they could change direction rapidly, evading enemy defenses and striking where they were least expected. This enabled them to survive deep behind enemy lines (where tank columns are quite vulnerable to a flank attack if the enemy can find them) and to create the impression among the defenders of being everywhere. They triumphed as much by disrupting and demoralizing the enemy as by destroying him.

Defensive maneuver tactics consist largely of counter-attacking. The defender positions some of his forces in front of the major enemy penetrations to delay them, but also attempts to send his armored columns through the gaps between the enemy spearheads so as to attack them from the flank and rear. A maneuver defense entails waging what United States army doctrine describes as the "deep battle." This means striking at the enemy in the entire depth of his deployment, with ground forces as well as missiles and aircraft, striking at his reserves moving up from the interior and his support forces as well as his front-line troops.

Maneuver tactics create a very fluid, confused battlefield, one on which the situation changes rapidly. Two of the great strengths of maneuver tactics are flexibility and unpredictability: the enemy cannot determine where a threat is coming from in time to focus on and counter it. To exploit this advantage, lower-ranking officers must be able to act on their own initiative and seize fleeting opportunities without being told to do so by their superiors. This requires what the Germans call an *Auftragstaktik* (mission tactics) style of command. Instead of prescribing in detail what each of his units will do, a commander simply tells his subordinates what the overall objective is, what each unit's objective is, and why he has set that objective. Then he leaves each unit commander considerable freedom to decide how his unit will reach its objective. This style of command was developed by the German army during the nineteenth and twentieth centuries and is employed today most effectively by the Israeli army. *Auftragstaktik* is almost a necessity to make maneuver tactics work but it requires very able junior and middle-rank officers; leaders able to make decisions on their own in the chaos of a fast-moving battle.

It may seem that maneuver is the smart man's way of fighting a war and attrition the stupid man's. This is not the case. The smart man's way to fight a war is whichever way offers the best chance of winning at the least cost. Fighting a war of maneuver requires a high degree of intelligence and initiative of the junior officers, but attempting to fight that kind of war under the wrong circumstances shows little intelligence in the supreme commander. Other factors not being decisive, a large, clumsy, poorly-officered army should employ attrition tactics and force on the enemy a war of attrition. A small, agile, well-led army does best to employ maneuver tactics and force the enemy to fight a war of maneuver. Great victories have been won both ways. Stonewall Jackson's Valley Campaign in 1862, Rommel's battles in the Western Desert during World War II, and the Israeli conquest of the Sinai during the Six Day War (1967) are classic examples of brilliant maneuver tactics. Meade's conduct of the Battle of Gettysburg, Montgomery's conduct of the Battle of El Alamein against Rommel, and the Egyptian crossing of the Suez Canal during the Yom Kippur War are equally classic examples of attrition tactics used appropriately and successfully. Much of the skill of a commander lies in understanding what kind of battle his army is best able to win and imposing that kind of battle on his adversary.

The United States army has traditionally relied on attrition tactics but may not continue to do so. The current (1982) edition of its basic manual on strategy and tactics, *FM 100-5, Operations*, is a rather maneuver-oriented document, especially when compared with the 1976 edition. Probably the change is a wise one. The Union army in the Civil War and the United States army in the two world wars faced outnumbered but agile and well-commanded foes. It was best to fight wars of attrition against them. In a land war with the Soviet Union, the United States army almost certainly would be outnumbered, but the Soviet ground forces seem to be clumsy and rigidly centralized – the ideal enemy against whom to fight a war of maneuver. However, a strong argument can be made for employing attrition tactics in the defense of Western Europe.

Tactical Nuclear War

A major land war might not be fought with conventional weapons alone. NATO has approximately 6,000 tactical nuclear bombs and warheads, Warsaw Pact forces in Europe several thousand more and there are others deployed along the Sino-Soviet border. Tactical nuclear weapons come in all shapes and sizes. There are IRBMs, MRBMs, short range surface-to-surface missiles, and cruise missiles; nuclear-capable attack aircraft, both land-based and carrier-based; nuclear artillery in a variety of sizes; nuclear-tipped anti-aircraft missiles; and nuclear mines. The most powerful of them, the Soviet SS-20 IRBM is barely distinguishable from a strategic nuclear weapon. It has a range of more than 3,000 miles and can carry a 1.5-megaton warhead. The least powerful, the American SADM nuclear land mine, may have an explosive power as low as 0.01 kiloton – about the same as that of the most powerful conventional weapons. All this diversity can be broken down into two broad categories: theater nuclear weapons and battlefield nuclear weapons.

Theater nuclear weapons are designed primarily for use against targets well behind the battlefront but not on the territories of the superpowers. They have yields in the tens or hundreds of kilotons and ranges of 100 to 3,400 miles. The United States Pershing Ia is a typical theater nuclear weapon: a surface-to-surface missile with a range of 450 miles and a yield of 60, 200, or 400 kilotons. (The Pershing Ia's warhead, like those of most American tactical nuclear weapons, can be set to give any one of several yields.) Surface-to-surface missiles are the most important theater nuclear weapons but bombers can also play this role. Medium range bombers, such as the Soviet Union's "Backfire" or the United States FB-111, are designed primarily as theater nuclear weapons, while fighter-bombers can serve either as theater or as battlefield nuclear weapons.

Theater nuclear weapons are more vulnerable than are their strategic equivalents. IRBMs and MRBMs typically are not housed in silos, as are ICBMs. Even if they were they would be considerably more vulnerable than ICBMs. Because their range is so much less than that of an ICBM they have to be placed much closer to the enemy's territory. Ballistic missiles are more accurate the shorter the distance they have to fly so an MRBM fired at the launcher of another MRBM is much more likely to destroy it than an ICBM would be to destroy another ICBM in its silo. This means that a Soviet preemptive attack on United States or French missiles in Europe is more likely to succeed than a similar attack on ICBMs in the United States itself. The only way really to protect an IRBM or an MRBM is to move it around, which can be done. Most of them are mounted on mobile launchers which can disperse rapidly and seek concealment on warning of possible hostilities. But if the enemy can find them, he can destroy them.

Medium range bombers are very vulnerable to missile attacks on their airfields. An airfield cannot be moved or concealed the way a missile can. Medium range bombers can be kept on ground alert but this does not protect them nearly as well as it does strategic bombers. As with the missiles, the problem is the short distance between them and the enemy and thus the very short flight times of the enemy missiles that could be sent against them. Nevertheless, NATO does keep some nuclear-armed bombers on ground alert when political tensions are high.

The vulnerability of theater nuclear weapons gives the first to use them a great and perhaps decisive advantage. His first strike has an excellent chance of destroying most or all of his adversary's missiles and bombers. Theater nuclear weapons can also destroy headquarters, communications networks, air and naval bases, radar stations, anti-aircraft batteries, ports, supply dumps – in short, all the support forces an army needs to fight. The massive and effective use of theater nuclear weapons can totally isolate the battlefield and make it impossible for the combat units to keep fighting. It could also so devastate Europe that there would be little left worth fighting over.

In December 1987 the United States and the Soviet Union signed the INF (Intermediate range Nuclear Forces) Treaty which bans from Europe all United States and Soviet nuclear missiles with ranges between 310 and 3,400 miles. This treaty would leave in place medium range nuclear bombers, French and British nuclear weapons, and theater nuclear weapons outside Europe. However, if it is carried out, it will greatly reduce the number of theater nuclear weapons in Europe.

Battlefield nuclear weapons are designed for use against enemy front-line troops and the support forces immediately behind them. They have ranges of less than 100 miles and, typically, yields of one or two kilotons. The United States M110, an 8-inch self-propelled gun whose nuclear shell has a yield of one kiloton and a range of 13 miles, is a typical battlefield nuclear weapon. Nuclear artillery, fighter-bombers used to drop atomic bombs on the battlefield, short range surface-to-surface missiles, and nuclear land mines are all battlefield nuclear weapons.

The neutron bomb, more correctly known as an enhanced radiation (ER) warhead, is a recent and significant innovation in battlefield nuclear weapons. It is designed to kill primarily by direct radiation rather than by blast and heat. The explosion of an ER warhead releases ten times as much direct radiation as a standard nuclear warhead with the same yield. As crewmen inside an armored vehicle are much more vulnerable to direct radiation than the vehicle itself is to blast and heat, ER warheads make ideal anti-tank weapons. They can kill tank crewmen within a fairly large area without killing civilians or friendly soldiers outside that area. This is why the United States is now developing ER warheads, not because they are a "capitalist weapon" designed to kill people and leave property intact, as some would have it. They offer the NATO armies a means to put out of action the tremendous numbers of tanks and APCs the Soviets would use in invading Europe, while limiting the suffering that would befall the people of Europe.

ER warheads have two major disadvantages. One is that radiation kills fairly slowly so that most of the enemy soldiers who receive a fatal dose would still be able to fight for several days or even weeks. The other is that they are nuclear weapons. While they are the most discriminating and limited of their kind, their use would inevitably lead to Soviet retaliation with other, probably less discriminating nuclear weapons, and then perhaps to a nuclear total war.

In any war in which they were used, tactical nuclear weapons would provide almost all of the firepower used on the battlefield. Their firepower is so great that the power of conventional weapons dwindles into insignificance by comparison. "An average atomic weapon of 20 kilotons produces an explosive force equivalent to that of a salvo of four million 75 mm cannons."[7]

This does not mean that conventional ground forces will disappear from the battlefield once nuclear weapons have been used. Even a direct hit with a nuclear warhead will not *necessarily* destroy or prostrate well-led, well-protected troops; several times during World War II troops subjected to carpet-bombing attacks and artillery barrages equal in explosive power to a small nuclear weapon survived and held their ground. For example, on March 15, 1944 a German parachute battalion defending Monte Cassino had 1,250 tons of bombs dropped on it, followed by 200,000 shells – the equivalent of a one-kiloton nuclear explosion. As soon as the barrage lifted the survivors, now down to the strength of a company, were attacked by three New Zealand battalions. They stopped the attack. So, we must expect that some soldiers would survive and fight even on a nuclear battlefield. The nuclear strikes would, however, inflict tremendous casualties and perhaps hopelessly demoralize most of the troops they do not kill. The primary role of conventional forces on a nuclear battlefield is to seize the ground which has been cleared of enemy troops by nuclear weapons.

The only real defense against battlefield nuclear weapons is not to provide a target for them. Combat forces on a nuclear battlefield must disperse into small, independent units, perhaps no larger than a company, widely separated from each other. They must move rapidly and continually to prevent the enemy knowing where they are long enough to hit them, and also to get over areas contaminated with nuclear fallout before they sicken from radiation poisoning. They must use "hugging" tactics, staying close enough to enemy units that the enemy cannot hit them with nuclear warheads without killing his own men. They must be able to fight on their own, without directions from higher headquarters, as many headquarters will be destroyed and the EMP from nuclear explosions may cripple radio networks. And, they must carry their own supplies with them, because once the missiles fly the bases, supply dumps, and combat service support units which normally sustain an army in the field will be obliterated. Land forces on a nuclear battlefield might well be reduced to a kind of mechanized guerrilla warfare: small mobile units wandering around the battlefield and fighting random encounter battles with whatever enemy forces they come upon.

Only maneuver tactics work on a nuclear battlefield, as the dispersal and mobility required to survive preclude attrition tactics. One of the strongest arguments for maneuver tactics is that the nuclear battlefield requires them. If an army has to be dispersed to survive on a nuclear battlefield, then it probably must fight dispersed even when only conventional weapons are being used. Otherwise, the enemy could make an overwhelming surprise attack with his nuclear weapons. This creates a very difficult dilemma for commanders because even the most maneuver-oriented army has to concentrate its forces to some degree and in some circumstances to fight a conventional battle successfully. If a commander concentrates his forces he renders them vulnerable to an enemy nuclear surprise attack, but if he does not, he risks losing the conventional battle.

The most important question about tactical nuclear war is whether it could ever be justifiable to fight one. It is impossible to use large numbers of theater and battlefield nuclear weapons in a heavily populated area without killing a great many civilians. Central Europe, where such a war seems most likely to be fought, is one of the most densely populated areas in the world. The towns and villages in West Germany are said to be "only one kiloton apart," leaving few places where even a battlefield nuclear warhead could be detonated without killing civilians. ER warheads would spare the civilian population somewhat, but their use would still cause many civilian casualties and almost certainly lead to the use of more devastating nuclear weapons. An authoritative estimate offered in Jeannie Peterson's *The Aftermath* holds that the detonation of 170 theater and 1,000 battlefield nuclear weapons in the two Germanies would kill about 11 million Germans. The toll could be even greater; more than 1,100 locations in the two Germanies have been identified as suitable targets for theater nuclear weapons, and there are more than enough weapons to hit all of them. It does seem that "a people 'saved' by us through our free use of nuclear weapons over their territories would probably be the last that would ever ask us to help them."[8]

There are also chemical and biological weapons to consider. They are generally classified, together with nuclear weapons, as "weapons of mass destruction" because of their ability to kill very large numbers of people over a wide area. Biological weapons inspire perhaps even more horror than nuclear weapons, but fortunately they are very unlikely to be used. The Biological Weapons Convention of 1972 firmly forbids their development, production, stockpiling, or use; more important, they are not very serviceable weapons. They require too long to take effect and are too unpredictable. An epidemic propagated among the enemy's population might be just taking hold when the war ends – and might spread to one's own people.

Chemical weapons can be and have been used in war; they were used for the first time in World War I. Since then many new types of poison gas have been developed. Nerve gas is the most feared, but there is also blister gas, which completely destroys skin tissue; blood gas, which damages internal organs; and others. Gas can be sprayed from an airplane or a helicopter, usually the most effective way to spread it, or delivered by bombs, shells, rockets, and even mines. It can kill or incapacitate people over a wide area: not as wide an area as a nuclear warhead can devastate, but far wider than that affected by conventional munitions.

Troops caught without protection by a poison gas attack can suffer 70–90 percent casualties and 25 percent fatalities, so a massive surprise attack on an unprepared army can virtually destroy it. Fortunately, it is fairly easy to protect troops against this threat. Masks and full-length protective clothing effectively shield men on foot and the occupants of tanks and APCs are safe as long as their vehicles are properly buttoned-up and have air filters. The major effect of employing chemical weapons against well-prepared troops would not be to kill or injure them, although some would die or suffer injury before they could protect themselves, but to make it difficult for them to function. It is very fatiguing to work or fight in a gas mask and full-length protective clothing. Artillerymen who must work their guns or ground crew who have to service aircraft under the threat of chemical weapons attack will have their efficiency markedly reduced. Guns will fire many fewer rounds and aircraft make many fewer flights because of the difficulty of manning and servicing them. The great loss of efficiency suffered by an army under chemical weapons attack or threatened with it might well tip the scales in a non-nuclear land war.

Notes

1 James F. Dunnigan, *How to Make War* (New York: William Morrow, 1982), p. 47.
2 The possibly skeptical reader can find the figures in Dunnigan, ibid., pp. 360, 362–3.
3 Ibid., p.317.
4 US Army, *FM 100-5, Operations*, August 20, 1982, p. 8–1.
5 Baron de Jomini, *The Art of War*, (Philadelphia: J.B. Lippincott, 1862; reprint edn, Westport, Conn.: Greenwood Press, 1971), p. 186.
6 US Army, *FM 100-5, Operations*, p. B–5.
7 Andre Beaufre, *Strategy for Tomorrow* (New York: Crane, Russak, 1974), p. 44.
8 Bernard Brodie, *Strategy in the Missile Age* (Princeton, NJ: Princeton University Press, 1959), pp. 324–5.

12
The Soviet Ground Forces

A master plan designed by geniuses for execution by idiots.
Herman Wouk, *The Caine Mutiny*[1]

The Soviet Order of Battle

The Soviet Union has the most powerful land army in the world. It is larger
than the Chinese army and much better equipped, several times as large as
the United States army and almost as well-equipped. There are twice as
many soldiers in the Soviet ground forces as in the United States Army and
Marine Corps combined. The Soviet Union has nearly four times as many
tanks as the United States, three times as many APCs, five times as many
guns, and twice as many fighter-bombers and close-support aircraft. When
the two countries' European allies are added to the balance, so that total
numbers of NATO troops and weapons are compared with the Warsaw Pact
totals, it is a more even but still not equal balance (see table 11.1, p. 150).
Ever since the end of World War II the fear that this tremendous force might
invade Western Europe has dominated United States national security policy
almost as much as the threat of nuclear total war.

There are 227 divisions in the Soviet ground forces' order of battle; the
United States Army and Marine Corps together have 21. Soviet army divisions
have nearly as many tanks and other weapons but only about three-fifths as
many troops as their American counterparts. Most Soviet divisions are
partially-manned reserve formations. They fall into three readiness categories,
each comprising about a third of the total. Category I units are maintained
at full strength in both men and equipment and are prepared to fight at any
time or with only a few days' notice. They are as ready for action as the
regular army units of the United States and other countries. Category II
divisions are manned at half strength and have most of their equipment in
storage. They would have to be fleshed out by calling up reservists but could
go to war several weeks after being given the order to mobilize. Category III
are skeleton divisions manned at one-fifth of their war strength, and would
require up to two months to get ready. They would need to absorb large
numbers of reservists, obtain most of their weapons from reserve stocks of
obsolete equipment, and probably commandeer most of their trucks from
factories and collective farms. Category III divisions can be and have been
mobilized and sent to war; several of the units sent to fight in Afghanistan
are believed to have been Category III.

Just by fleshing out the units they already have, the Soviet ground forces could double in size in a few months. They might also be able to raise a large number of new units. There is evidence that each division can form from within itself another division the same way an amoeba can become two amoebas. The division's deputy commander and the deputy commanders of each of its subunits would take over command of the new division, and reservists bring the two divisions up to full strength. Raising all these new units would require calling up millions of reservists and supplying them weapons and equipment, but it could be done. In the Soviet Union all fit males serve in the reserves until the age of 50 and can be called up for periodic refresher training until the age of 35. The Soviets maintain great stockpiles of obsolete military equipment, rarely throwing anything away. An improvised division of elderly reservists equipped with obsolete weapons would hardly be a first-line combat force but 150 such divisions, thrown into a major war after the first-line forces on both sides had been virtually annihilated, might well win that war.

The vast majority of Soviet ground forces divisions, 202 out of the 227, are tank or motor-rifle formations, each equipped with several hundred tanks and moving its infantry in APCs. (A Soviet tank division has 333 tanks and 150 APCs, a motor rifle division 205 tanks and 375 APCs.) The Soviet ground forces have not for several decades had any non-motorized infantry. Since World War II they have been transformed from an army that moved almost entirely on foot into one that moves on tracks and wheels. Even the airborne divisions have substantial numbers of ultra-light armored vehicles.

One of the more notable differences between the Soviet and the United States armies is that the Soviets manage to get a far higher percentage of their manpower into the front-line combat units. For every three combat soldiers the United States army has ten communications, headquarters, and combat service support troops; the Soviet ground forces have two. As armies go, the Soviet Union's has a fairly high "tooth-to-tail" ratio and the United States's a very low ratio. This is why Soviet army divisions can bring into action nearly as many weapons as their American counterparts, with only three-fifths as many troops. It also helps the Soviet ground forces to keep up ten times as many combat divisions as the United States Army and the Marine Corps combined with only twice the manpower.

The Soviets achieve their high "tooth-to-tail" ratio by skimping on logistics and other aspects of combat services support. They seem to assume that a major land war would last only a few weeks and that a tank or motor-rifle division would only be good for five days of intense combat. Both assumptions, if correct, greatly reduce their supply and services requirements. A Soviet division goes into battle with enough ammunition, fuel and rations to fight for five days. It makes only the crudest attempt to repair damaged equipment and does not replace its personnel losses at all. After five days it is fought out; most of its vehicles disabled and front-line soldiers casualties. The survivors withdraw to the rear and a fresh division moves up to take their place. Safely in the rear the used-up division replenishes its supplies, fills up its ranks, and is ready to fight again later.

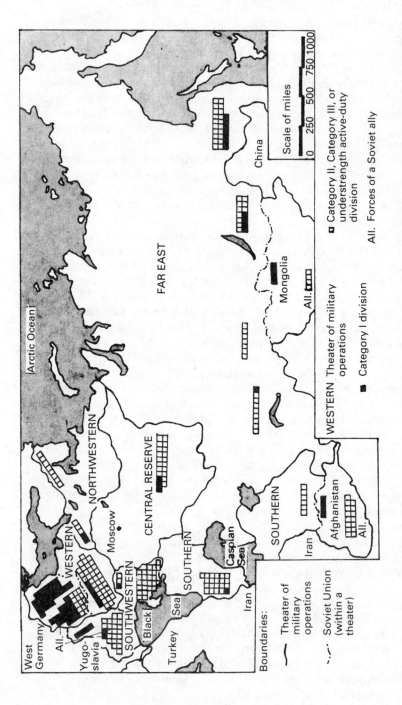

Based on map and data in *The Military Balance 1987–1988* (London: International Institute for Strategic Studies, 1987), pages 39–44.

Map 12.1 *Ground forces deployments of the Soviet Union and its allies*

The Soviets can fight this way because of their tremendous mobilization potential. During the entire course of a land war anywhere in Eurasia, they would have a continual stream of reinforcements reaching the battlefield: Category I divisions coming from the Soviet interior, then fleshed-out Category II and III units, and finally perhaps divisions organized after the war began. In contrast, the United States and its allies have to devote much greater resources to combat service support because they would have to keep their front-line units in combat as long as possible. It would be difficult to get reinforcements from the United States to the battlefield and perhaps impossible to mobilize reserves on the territories of the countries under attack. Perhaps the difference between the United States and the Soviet approaches to weapons design is that "if we can be said to treat our weapons like handkerchiefs, they use theirs like Kleenex – use them up and throw them away."[2] It is certainly the difference between the approaches of the two superpowers to the use of combat units in war.

What all this adds up to is an army designed primarily to invade Western Europe. The Soviet ground forces are deployed mostly in Eastern Europe and the European part of the Soviet Union (see map 12.1). These forces are much larger, particularly when mobilized, than they need to be to ward off a NATO attack or hold down Eastern Europe. The Soviets have 135 divisions deployed against Western Europe, and their Warsaw Pact allies another 54 – a total of 189. This includes most of the elite Category I divisions, as all Soviet units in Eastern Europe are Category I. There are 65 Soviet and Mongolian divisions, one-third as many, stationed along the Chinese border. The remainder of the Soviet ground forces are deployed against Middle Eastern countries. The 118,000 Soviet troops fighting in Afghanistan at the end of 1987 were only one-fiftieth of the total strength of the Soviet armed forces.

The tremendous preponderance of armored and mechanized forces in the Soviet order of battle is appropriate for an invasion of Western Europe, where the terrain is generally flat and the road network excellent. Soviet logistics are probably adequate for a war in Europe because the distances are short and the war would probably be resolved quickly. To move south against Turkey, Iran or Pakistan, the Soviet army would have to fight its way across high mountains and operate in territory where there are few roads. To invade China it would have to cross deserts and, if it were to occupy the heavily-populated eastern part of the country, deal with a billion hostile Chinese. Either campaign would require large numbers of non-mechanized infantry and a strong combat services support organization, the two areas in which the Soviet ground forces are weakest. The Soviet Union certainly could invade China or its neighbors to the south and might well do so. However, Soviet military planners clearly regard seizing Western Europe as the most important as well as the most formidable task which their army could be set. The Soviet ground forces are organized, equipped, and trained primarily to perform that task.

A standard exposition of Soviet military doctrine holds that "the strategic and tactical missile forces are the basis of the firepower of the Land Forces for defeating the enemy."[3] In a nuclear war the Soviet ground forces could call upon nearly 600 IRBMs and MRBMs in the Strategic Rocket Forces to attack enemy missiles, airbases, ports, headquarters, and other targets anywhere in Eurasia. The ground forces also have numerous theater and battlefield nuclear weapons of their own. Every army has 24 theater surface-to-surface missiles, typically SS-23s with a range of 220 miles and 100–500 kiloton warheads. Every tank or motor rifle division has 16 SS-21s, battlefield surface-to-surface missiles with a range of 75 miles and 20–100 kiloton warheads. (The SS-21 and SS-23 can also be fitted with conventional or chemical warheads.) Missiles are the Soviets' most important theater and battlefield nuclear weapons, but they also have several thousand bombers and fighter-bombers which can carry atomic bombs, and some guns which can fire atomic shells. Soviet capabilities for waging chemical warfare also are believed to be formidable. Reportedly, chemical rounds exist for several types of surface-to-surface missile and for all 122 mm and larger guns.

The Soviet ground forces have been carefully prepared to survive the effects of nuclear and chemical weapons. There are about 90,000 servicemen in their Chemical Troops branch – more than in the entire armed forces of Canada. The Chemical Troops' function is to protect Soviet combat and combat service support forces from nuclear and chemical weapons. Every regiment has its Chemical Troops company with decontamination equipment and means of locating and charting contaminated areas. All Soviet soldiers undergo training in defense against nuclear and chemical attack and are issued with gas masks and protective clothing. The Soviet ground forces are also well adapted to nuclear and chemical warfare in other respects. Almost all of their front-line combat troops move in closed vehicles, tanks and APCs, which afford them a high degree of protection against poison gas and radioactive fallout, enable them to maneuver rapidly to avoid enemy nuclear and chemical strikes, and allow them to cross contaminated areas rapidly. The tremendous size of the Soviet ground forces and their ability to mobilize reserves prepare them, as well as any army can be prepared, to endure the losses that would be suffered in a nuclear war. The ability of their combat units to function with a minimum flow of supplies from the rear would also serve them well if the devastation created by tactical nuclear weapons had made it impossible to get supplies to the fighting troops.

The Soviet ground forces can fight without nuclear weapons. They have done so in Afghanistan, and in the past decade have devoted more attention to conventional war than previously. Soviet military planners recognize that a war in Europe very likely would not begin nuclear and might be fought to the end with conventional weapons alone. Nevertheless, the Soviets are also well prepared to fight with nuclear or chemical weapons.

Soviet War Plans

If the Soviet army invades Western Europe, the main thrust of its attack will probably come against West Germany. West Germany is the most geographically vulnerable part of the NATO front. West Germany and the countries that lie beyond it are the richest prizes in Europe, and the Warsaw Pact's strongest forces are deployed against it. The Soviets could move southwest against Yugoslavia (not a NATO member), Greece and Turkey, or in the north across neutral Finland and Sweden to strike at Norway. There is not nearly as much to be gained from these other two aggressions and they do entail substantial, though lesser, risks. An attack on Norway, or on Yugoslavia, Greece, and Turkey might well accompany an attack on West Germany, but NATO's primary task has always been to defend West Germany.

As the attacker, the Soviet government would decide how and when the war would start. The Soviets could begin their war in any one of three ways:

1 A conventional weapons surprise attack with forces which are ready to fight and are close to the West German border in peacetime, i.e. the Soviet, East German and Czech forces in East Germany and western Czechoslovakia.
2 A conventional weapons attack with all the forces which could be mobilized and moved to the theater of operations in several months, i.e. almost everything the Soviet Union has except for the forces necessary to guard the border with China, supplemented by all the other Warsaw Pact armies.
3 A nuclear attack, using theater and battlefield nuclear weapons from the outset.

These three options seem over the years to have waxed and waned in popularity among Soviet military planners, but probably none has been definitely rejected.

The first option is sometimes called the "Hamburg grab" – a surprise attack to seize a limited objective, the city of Hamburg for example, before NATO could react to stop it. It would be essential for the attack to take NATO by surprise, and this dictates both the timing and the forces employed. Only the Warsaw Pact forces near the frontier and in constant readiness to fight could be used in the initial assault; it would be made by the 20 Soviet and six East German divisions in East Germany and the six Soviet and Czech divisions in western Czechoslovakia. Even to move up Category I divisions from Poland or the Soviet Union, let alone to mobilize Category II and III divisions, would give warning that something was about to happen. The best time for the attack would be, of course, when NATO is least prepared to meet it, perhaps a Sunday in July or August when many of the defenders would be on leave and the roads would be clogged with tourist traffic, or Christmas Eve.

Three hours after the assault force had left its barracks, its forward elements could reach and cross the West German border. NATO normally has only light covering forces on patrol at the border and these could easily be masked or brushed aside. About all they could do would be give definite warning that West Germany was being invaded. Most of the forces in West Germany are in barracks well back from the border. If they were attacked in the middle of the night, particularly on Christmas Eve, it would take at least 4 hours to wake the troops, load them into their vehicles, and start them moving towards the front. This would have to be done while the barracks were being hit by Soviet gunship helicopters and airborne detachments, so probably most units would need considerably more than 4 hours to get into action. If the Soviet armored spearheads are able to advance 19 miles an hour, as some military strategists think they can, they might be half-way across West Germany before meeting effective opposition. The forces engaged on each side would be about equal in numbers: 32 Warsaw Pact divisions against 25 larger NATO divisions in West Germany. The invaders, though, would have the advantage of surprise, and the defenders would be fighting with many of their troops absent on leave, would have to get ready to fight under very difficult circumstances, and would have little room for maneuver. The invasion force might well be able to reach the Rhine River in 48 hours; it is only 100 miles from East Germany at one point.

The Soviets would want to win this war very quickly, before NATO mobilized its full strength and above all before nuclear weapons were used. To do this, their forces must advance very rapidly and – even more important – their objectives must be limited. When they reach the Rhine they must stop and offer to make peace. If they do so, the governments of what would be left of NATO might not be able to persuade their people to continue the war; NATO would face a neatly-executed *fait accompli*. But, if the Soviet armies were to keep on going into France, France would hit back with nuclear weapons even if no other country did.

The Hamburg grab would require a level of coordination and operational skill which many think the Soviet army does not have. It would also be very risky. NATO allies might not continue the war once West Germany had been overrun but they also very well might do so *and* use nuclear weapons. Nobody can tell, least of all the people in the Kremlin, whether a successful Hamburg grab would be a *fait accompli* or the beginning of a global nuclear war.

The Soviets might not take NATO by surprise, as they must if a Hamburg grab is to succeed. They are most unlikely to execute a "bolt from the blue" invasion of West Germany without any immediate provocation, solely because they can do it. Almost certainly they would undertake such a risky operation only under the pressure of a severe crisis, and the existence of that crisis would put NATO on its guard. On the other hand, the Soviet Union has achieved successful surprise attacks before, even when the opponent had good reason to fear attack. The invasions of Manchuria in 1945, of Czechoslovakia in 1968, and of Afghanistan in 1979, all took their victims by surprise.

If the Soviets take the time to mobilize before launching an invasion, they could have available several times as many divisions as the 32 they might send into a Hamburg grab surprise attack. In ten days they could move up all the Category I divisions in their Western Theater of Military Operations and Central Reserve, including those of their Warsaw Pact allies, and have 54 divisions available. In two months they could mobilize and send into action all the Category II and Category III divisions in those two commands, giving them a total force of 120 divisions available to invade West Germany. At the same time, the 56 Warsaw Pact divisions in the Southwestern Theater of

Military Operations could be sent into Italy, Austria, Yugoslavia, and Greece, and the 13 divisions in the Northwestern Theater used to invade Scandinavia. (See map 12.1 and figure 13.1, pp. 164 and 185.)

All this mobilizing and transporting of massive land armies could not go unnoticed in Western Europe and the United States. NATO would make its own preparations. However, NATO has much smaller reserves than the Warsaw Pact. Soon after the alliances began mobilizing, NATO forces in West Germany would be outnumbered about two-to-one in combat divisions. (For further discussion of how fast NATO and the Warsaw Pact could build up their forces see chapter 13, pp. 183–4.)

The invasion would probably begin between midnight and dawn with Warsaw Pact missiles and artillery shells dropping on NATO airfields, air defense radars, and anti-aircraft batteries. (The missiles would, of course, have conventional high-explosive warheads.) Then the invader's aircraft would fly down the corridors opened through NATO's air defense belt and hit targets deep behind the lines. They would attack headquarters, nuclear weapons, airfields, supply dumps – all the targets that would be hit with theater nuclear weapons in a nuclear war. Next would come airborne and helicopter-borne detachments to seize airfields and bridges and to destroy targets the aircraft had missed. By then the battle at the frontiers would be well under way.

With NATO put on alert by Warsaw Pact preparations, the defenders could not be overwhelmed in their barracks. The invasion force would have to break through a 20-mile deep belt of prepared defenses near the border. It would hit those defenses in their entire depth and along the entire front with all of its tremendous firepower: several thousand gunship helicopters and close-support airplanes and several tens of thousands of guns. Under the cover of this fire support, the front-line tank and motor rifle divisions would advance in several widely separated waves. Typically each division would send forward a small covering force to feel out the enemy, then its first echelon to penetrate the defenses, followed by a second echelon to move through the gaps created by the first. Wherever the attackers broke through, they would keep moving ahead as far and as fast as they could, in the classic Blitzkrieg style, and all available reserves would be thrown in to widen the breach.

Once the first line of defenses had been breached, the battle would become a series of meeting engagements: unplanned collisions between the advancing armored spearheads and NATO reserves maneuvering to contain them. In this complex, chaotic struggle, Soviet operational maneuver groups (OMGs) would play a major role. These are self-contained, division-sized (usually) raiding parties able to operate behind enemy lines for several days at a time. An OMG comprises tank, armored infantry, self-propelled gun, engineer, helicopter, and air defense units – everything necessary to operate independently. It can be resupplied by air. The OMGs are to slip behind enemy lines at an opportune time, preferably on the first day of the invasion, and then roam through NATO's rear areas 50 or 100 miles behind the battlefront. They are to seek out and destroy nuclear weapons and headquarters, conduct electronic counter measures (ECM) to paralyze NATO communications, and block the retreat or redeployment of enemy forces. They would contribute greatly to keeping the offensive moving, permitting the defenders no rest to regroup.

The offensive is to continue day and night without interruption until Europe is conquered. As the front-line divisions are worn away, they are to be replaced by reinforcements coming up from the rear, and when the reinforcements can no longer fight, others will take their places. Soviet military planners expect that after the initial breakthrough their forces should be able to advance 20 miles a day. This rate of advance would take them from the West German border to the Atlantic Ocean in a month of sustained combat. However, if Soviet plans go awry, as war plans tend to do, the Soviets could face a much longer war. World War I was expected to be a short war.

The invader must seek to keep the war short, above all, to prevent it from becoming a nuclear war. A nuclear war in Europe would leave the continent in ruins even if it did not spread to the Soviet Union and become a global nuclear war. Preventing nuclear war would be the invading army's most important task. It would not be easy: NATO plans to use its 6,000 nuclear warheads if it cannot stop an invasion any other way.

Soviet military planners apparently do not expect their invasion force to be hit with nuclear weapons the moment it crosses the border. They are aware that NATO's strategy is initially to resist a conventional weapons attack with conventional weapons alone. NATO's nuclear weapons are to be held in reserve until the Soviet Union uses its own or until other means of halting the advance have failed. Even then, the various NATO governments might take several more days to reach a consensus to use nuclear weapons. The West German government, in particular, would be very slow to authorize the use of nuclear weapons on its soil. At least the first few days of the war would be fought with conventional weapons alone. The Soviets intend to put those first days to good use.

The most important missions of the Warsaw Pact forces during the eary days of the war would be to destroy enemy nuclear weapons and to prevent NATO from using those that survived attack. The missile and air strikes which would begin the war would be directed primarily against NATO nuclear weapons and the headquarters which could order their use. So would the airborne and helicopter-borne detachments landed behind enemy lines on the first morning of the war. The OMGs driving deep and rapidly into the enemy rear would also seek out and destroy headquarters and nuclear weapons. By the end of the first day, NATO might have very few nuclear weapons left to use.

Even if the NATO governments had a substantial nuclear arsenal and were losing the war, they might still hesitate to escalate to nuclear war. By the time it became clear that only nuclear weapons could halt the invasion, Warsaw Pact forces would be deep inside West Germany. The invader's airborne units and OMGs would be swarming around and be within West German cities, and his front-line combat forces would be grappling at close quarters with NATO forces. It would be difficult to hit the invaders with nuclear weapons, even using ER warheads, and not kill millions of civilians and

NATO troops. And, of course, the Soviets could retaliate with their own nuclear weapons. After several days of war much of NATO's nuclear arsenal would have been destroyed but the Soviet Union's, kept well away from the battlefield on Russian soil, would be intact. The Soviet Union and its allies would have achieved clear escalation dominance over the NATO countries. They would be winning the war with conventional weapons, but would also have far more tactical nuclear weapons available should the war escalate. All that the NATO countries could do by escalating to nuclear war would be to make the disaster which had befallen them immeasurably greater. They could not avoid defeat.

So, if the Soviets use the initial period of the war well they may be able to win it *and* prevent it escalating to a nuclear war. Or, maybe not. Nothing is certain in war and the Soviets might find themselves with a nuclear war on their hands. Even so, it would have profited them to start the fighting with conventional weapons alone. They would have gained a decisive advantage during the initial period and have destroyed so many NATO nuclear weapons that even the nuclear phase of the war would be less destructive than a war fought from the outset with nuclear weapons.

In the past few years Soviet planners seem to have decided that a war in Europe could be fought with conventional weapons alone. However, they might change their minds. If they ever come to believe that war is unavoidable and that there is little or no chance it will not escalate to a nuclear war, they might decide to begin the war themselves, with a tactical nuclear weapons attack. This option has been described in Viktor Suvorov's book, *Inside the Soviet Army*.

Suvorov, who was an officer in the Soviet ground forces during the 1960s and now lives in the West, sees gradual escalation from a low level of violence as a characteristically American, not Soviet, idea:

> As I studied American theories of war ... It became clear to me that a modern American cowboy who is working up to a decisive fight will always expect to begin by spitting at and insulting his opponent and to continue by throwing whisky in his face and chucking custard pies at him before resorting to more serious weapons. He expects to hurl chairs and bottles at his enemy and to try to stick a fork or a tableknife into his behind and then to fight with his fists and only after all this to fight it out with his gun.[4]

Suvorov thinks this is absurd. "You are going to end up by using pistols. Why not start with them?" Why take the chance that the enemy will shoot you when you can shoot him first? In contrast, he says, Soviet military leaders hold to what he calls the "axe theory:" if your most powerful weapon is an axe, hit the enemy with it the moment the fight begins before he can hit you with his. If you have nuclear weapons, hit the enemy with them the moment the war begins before he can hit you with his.

As Suvorov describes it, the offensive would begin with a half-hour long barrage from thousands of Soviet nuclear weapons: MRBMs and IRBMs of the Strategic Rocket Forces, shorter-range missiles of the ground forces, and the nuclear artillery of the front-line divisions in the attack. It would be directed against nuclear weapons, headquarters, airbases, anti-aircraft batteries, radar installations, seaports, supply dumps, and other targets of military value all over Western Europe. Then, waves of nuclear-armed aircraft would fan out over Western Europe, searching for and destroying targets which had survived the initial missile and gun salvos. This would be followed by another nuclear barrage, the missile launchers having been reloaded and the damage done by the first barrage assessed. After that the tanks and armored infantry would go forward.

"Under conditions of the employment of nuclear weapons, the attack will be characterized by extreme decisiveness, mobility, great spatial scope, high tempos, continuity of conduct day and night, unevenness of development, and rapid and abrupt changes in the situation."[5] In short, it would be like a Soviet conventional weapons invasion except even more so. The Soviets would use their battlefield nuclear weapons to destroy the NATO ground forces. Their tanks and armored infantry would occupy territory and exploit the effects of the nuclear strikes. The armored spearheads are to penetrate the gaps in NATO defenses opened up by nuclear strikes and move west as far and as fast as they can. They are to cross fallout-contaminated areas without waiting for the fallout to subside. They are to advance in march formation, seldom deploying for combat or dismounting the infantry from their APCs. NATO ground forces having been almost annihilated by the nuclear strikes, there would rarely be enough resistance to justify stopping to overcome it. However, NATO might still have some nuclear weapons left, so the attackers would move in very dispersed formations. Each division would be spread out over 200 square miles. Soviet planners expect their troops to be able to advance 50 miles a day in a nuclear war.

Nobody knows what a Soviet preemptive nuclear attack on Western Europe would lead to. Probably, though, it would lead to nuclear warheads exploding on Soviet territory. They might be the warheads of NATO theater nuclear weapons which had survived the preemptive attack or they might be from British and French SLBMs. They might even be from United States strategic nuclear weapons. If anything could incite an American first strike on the Soviet Union, it would be a Soviet nuclear attack on Western Europe. Even if, against all odds, the Soviet Union escaped direct attack, there would not be much left of Western Europe for its armies to occupy. While the other two options offer some hope of avoiding a nuclear war, this one sacrifices that hope from the outset. On the other hand, if the Soviets come to believe that a nuclear war in Europe is inevitable, they would certainly wish to be the ones to start it.

How Formidable is the Soviet Army?

Seen from the outside, the Soviet armed forces appear extremely menacing: very large, adequately-equipped and trained, and guided by a well-crafted military doctrine. Seen from the inside, by those who have served in them, they look rather less impressive. In the past few years a number of people who had served in the Soviet armed forces have come to the United States and recounted their experiences. Some of the incidents they relate make amusing reading:

1 Lieutenant Viktor Belenko was a pilot in one of the elite fighter wings of the Soviet National Air Defense. He flew a Mig-25, the Soviet Union's most potent fighter plane. Everybody in his elite fighter wing drank heavily. Officers and enlisted men alike drank great quantities of the hydraulic fluid supplied for use in the aircraft, falsifying records and stealing in order to obtain it. (The Mig-25's hydraulic systems use half-a-ton of high-grade grain alcohol.) Word came down that the Minister of Defense was going to visit the airbase. Terrified at the prospect, the wing's commanders decided that the best way to make a good impression on him would be to build a paved road from where his helicopter was going to land to the base and line it with trees. Everybody was ordered out to build the road and dig up fully-grown trees to transplant along it. The Minister's visit was delayed and the trees all died, so everybody went out to dig up more trees. After several sets of trees had been transplanted and died the visit was cancelled and the airmen went back to operating fighter planes. Having spent the summer transplanting trees, the pilots seemed to have lost some of their flying skills and during the first several days of renewed operations two MiG-25s crashed. Shortly after that Lieutenant Belenko defected to Japan.

2 The centerpiece of the "Dnieper' maneuvers in 1967 was an underwater crossing of the Dnieper River by over 5,000 tanks. It is possible to drive a tank underwater if it is made air-tight and equipped with a snorkel. The problem is that a tank underwater is very hard to steer; a shade too much pressure on the steering levers will make it spin around in circles. To make sure that all the tanks got across safely, several thousand soldiers spent 4 months building a 100-lane concrete-and-steel-mesh highway on the bottom of the Dnieper River. The tank crossing must have been an impressive sight, but it had no connection at all with realistic preparation for war. "In wartime ... the enemy would hardly allow you to spend four months splashing around in a river paving its bed."[6]

3 During Warsaw Pact maneuvers in 1984 four Russian soldiers were found sleeping in a forest, totally unable to explain what had become of their tank. The tank was later found in the possession of a tavern-keeper who was slowly dismantling it and selling the pieces as scrap metal. The Russian soldiers had traded it to him for 24 bottles of vodka and some herring and pickles.

Such anecdotes lead one to wonder how such a "Keystone Kops" army could ever seem a threat to anybody. Then one remembers that the Keystone Kops army in question won World War II.

The Soviet armed forces have their defects. As the anecdotes related above show, much of their training is designed more to impress senior officers and government officials than to prepare for combat. Senior officers demand unattainable levels of performance and their subordinates react by pretending to comply. The deception is often tacitly accepted by the senior officers themselves, who must answer to their equally unreasonable bosses. However, this is not just a Soviet trait. Other peacetime armies tempt their officers to look good at the expense of realistic preparation for combat, although seldom to the extent attributed to the Soviet armed forces. One of the United States armed forces' chronic problems is that officers frequently fail to acknowledge and seek help with a problem in their units; they fear that revealing the problem will earn them a poor efficiency rating. It is hard for a peacetime army to know how well it is preparing itself for combat. Troops cannot be tested by combat if their country is not at war. Officers in a peacetime army necessarily are evaluated on how well prepared their troops appear to be rather than on how well they actually would perform. That is why the tendency to put appearance above reality is hard to resist.

Another serious defect of the Soviet armed forces is the lack of initiative and sense of responsibility displayed by their junior and middle-level officers. Lower-ranking officers wait for orders from above, carry out those orders exactly as issued regardless of the consequences, and fail to act on their own initiative. This would be a defect in any army, but it is especially serious for one which relies so much on rapid and flexible maneuver. The Soviet ground forces are heavily committed to maneuver tactics but they do not employ the *Auftragstaktik* style of command which is necessary to make maneuver tactics work. Instead, they employ a rigidly-centralized style of command which is much more suited to attrition tactics. Every detail of an operation is planned in advance at a high level and subordinate commanders are required to execute the plan rigidly.

The simpler battlefield decision-making becomes, the more feasible it is for the commander himself to control every aspect of a fast-moving battle of maneuver and not delegate any of the decision-making to his subordinates. Soviet military planners devote considerable effort to rationalizing and simplifying the battlefield decision-making process. They have established "norms" for almost everything an army does: norms dictating how much room a motor-rifle battalion must have to execute a flank attack, how many ATGMs are required to destroy a given number of enemy tanks in a given time, how fast a tank division should be able to advance against light opposition, how much road space a motor rifle regiment should occupy on the march, etc., etc. Using these norms should make the commander's job easier – at the price, most likely, of making his tactics stereotyped and predictable. Also, while these norms have been tested extensively on maneuvers, they cannot have been used in combat. They may prove to be quite unrealistic once the shooting starts.

The Soviet army of today has these defects and many others but it is still a formidable fighting force. The Russian armies of 1812 and 1941 had equally grave defects but they managed to destroy the most powerful and proficient military machines of their day. Their numerical superiority, the ruthless determination with which they were used, and the self-sacrificing endurance of the Russian soldier more than made up for Russian weaknesses in equipment, organization, and leadership. Similar strengths cancel out similar weaknesses today.

The Russians are not "ten feet tall," they can be defeated, and Russian armies have been defeated often enough in the past. If the NATO armies were man-for-man as proficient as the Israeli army then NATO would have little reason to fear a Soviet invasion. However, most of them are not anywhere near that proficient, and a big mediocre army can usually defeat a small mediocre army. To defend Western Europe successfully, NATO's armies must be man-for-man not just superior to the Soviet army but sufficiently superior to cancel out Soviet numerical superiority and the other advantages the Soviets enjoy.

Notes

1 This is how one of the characters in Wouk's novel described the United States navy.
2 Steven Zaloga, quoted in Andrew Cockburn, *The Threat: Inside the Soviet Military Machine* (New York: Random House, 1983), p. 158.
3 Colonel General N.A. Lomov, *Scientific-Technical Progress and the Revolution in Military Affairs* (Moscow: Voyenizdat, 1973; Washington, DC: Government Printing Office, 1975), pp.105–6.
4 Viktor Suvorov, *Inside the Soviet Army* (New York: Macmillan, 1982), p. 160.
5 Colonel A.A. Sidorenko, *The Offensive* (Moscow: Voyenizdat, 1970; Washington, DC: Government Printing Office, 1975), p.221.
6 Viktor Suvorov, *The "Liberators"* (New York: W.W. Norton, 1981), p.70. The other two anecdotes are drawn, respectively, from: John Barron, *MIG Pilot: The Final Escape of Lt. Belenko* (New York: Avon Books, 1980), pp. 94–108; "Russians Swap Vodka For Tank, Paper Says", *New York Times*, 5 August, 1985.

13
The Defense of Europe

NATO military planners have found it as hard to decide how to defend Western Europe as Soviet planners seem to have found deciding how to attack it. Since World War II, three very different strategies for the protection of Western Europe have been proposed, debated, and to varying extents implemented. They are:

1 Massive Retaliation – deterrence of any Soviet attack on Western Europe by the threat of a massive United States nuclear strike on the Soviet Union.
2 Flexible Response – the defense of Western Europe by conventional weapons alone unless the invader uses nuclear weapons.
3 European Nuclear Deterrent – deterrence by the threat of retaliation with European nuclear weapons. These might be weapons owned by individual countries, or they might belong to a single European strategic nuclear force owned by several countries.

Massive Retaliation and Flexible Response have been the strongest competitors for the favor of NATO strategists. Both have been, at one time or other, the accepted NATO doctrine; both have been reflected in NATO war plans, weapons procurement decisions, and force deployments. Neither is clearly the accepted NATO doctrine today; some documents and force deployments support one and others support the other. The European Nuclear Deterrent option has only from time to time been proposed.

The lack of consensus within NATO is not surprising. Like other alliances, it is a limited partnership which each of the partners entered to secure its own national interests. The alliance does have common objectives, otherwise it would not exist, but each member prefers to attain those objectives as much as possible at the risk and expense of the others. Besides this normal weakness of alliances, NATO has its own particular problems. It unites one very large country, the United States, with 15 much smaller countries. This largest partner is a long distance away from the enemy the alliance was formed to counter while most of the smaller countries are quite near that enemy. These circumstances make it difficult for the NATO countries to unite behind any of their three options. Any option inevitably places heavier risks and burdens on some members than on others, while another strategy would distribute the risks and burdens differently. Naturally, each country advocates the strategy that will protect it the most and burden it the least.

It might be possible to resolve these differences if one of the options were so clearly superior to the others that nobody could oppose it – but none of them is. Each has sufficient merit to appeal to some NATO strategists and enough defects to be rejected by others. Making a clear choice among them is also difficult because of the protean nature of the threat the strategies are to ward off. Soviet invasion planners have three options of their own and nobody can know which they would choose. How one proposes to defend Europe depends very much on how one thinks the Soviets are most likely to attack it.

Massive Retaliation

When United States military planners first began to think seriously about how to stop a Soviet invasion of Western Europe, they saw that each of the superpowers had one great military asset. The Soviet Union had by far the most powerful land army in the world; the United States had nuclear weapons. They decided that matching Soviet strength on the ground was futile and unnecessary. Instead, the United States should rely on its nuclear weapons; it should respond to a Soviet invasion by a nuclear strike on Russian cities. The first American plan for a war with the Soviet Union, drawn up in 1948, projected a month-long atomic bombing campaign against 70 Russian cities. This attack was expected to kill or wound a great many Russians, some 7 million, but it would not have destroyed the Soviet Union and it probably would not by itself have halted an invasion. NATO military planners soon realized the American atomic bombing attack would have to be supplemented by strong resistance on the ground. In 1952 the NATO governments agreed to deploy 96 active-duty divisions in Europe. At the same time, the plans for American nuclear retaliation against the Soviet Union were never abandoned.

In 1953, when Dwight D. Eisenhower became president of the United States, he found the comparative advantages of the United States and the Soviet Union essentially unchanged. The Soviet Union still had the most powerful land army in the world. The NATO countries had strengthened their land armies, but they never deployed more than a third of the force they had decided in 1952 would be necessary to stop the Soviet army. The United States remained overwhelmingly superior in nuclear weapons. The Soviet Union had tested its first atomic bomb in 1949, but it still had only a handful of nuclear warheads. The few dozen Soviet strategic bombers available could reach the United States only on a one-way suicide mission. Meanwhile the United States had increased its nuclear warheads stockpile from 50 in 1948 to 1,350 in 1953. It also had nearly 1,000 strategic bombers, most of them far superior to the Soviet Union's. An American attack in 1953, unlike the 1948 version, probably would have stopped a Soviet invasion – and destroyed the Soviet Union as well. A 1954 Strategic Air Command briefing on what a full-scale American attack would do conveyed this impression: "virtually all of Russia would be nothing but a smoking, radiating ruin at the end of two hours."[1]

It seemed that so formidable a capability should be used to defend Western Europe. That was the strategy chosen by the Eisenhower administration and christened by it "Massive Retaliation.' The United States would respond to a Soviet invasion of Western Europe the same way it would to an attack on the United States itself – with an all-out nuclear strike on the Soviet Union. (Lesser aggressions were to be countered by less extreme military measures, employing either tactical nuclear weapons or conventional weapons, depending on the provocation.)

Massive Retaliation is a deterrence strategy just as Mutual Assured Destruction is (see Chapter 9). It places entire reliance on preventing a Soviet attack and holds out little hope of a successful outcome for anybody should deterrence fail. Like MAD, it rests on a threat that it would be utter folly to carry out, but which may provide excellent protection if the Soviets believe it would be carried out. It is, in fact, Mutual Assured Destruction employed to protect Western Europe as well as the United States. It treats Western Europe, for purposes of deterrence, as if Western Europe were an extension of the United States and endeavors to make the Soviets see things the same way. Massive Retaliation is thus sometimes described as a strategy of "extended deterrence," in contrast to MAD's "basic deterrence" of any attack on the United States itself.

In the Massive Retaliation strategy, the same strategic nuclear force that deters attack on the United States also protects Western Europe. The United States must have some conventional forces in Western Europe, but far fewer than the Flexible Response strategy would require. This makes Massive Retaliation a comparatively inexpensive way to protect Western Europe – a major reason why the Eisenhower administration adopted this strategy and to this day one of its strengths. Its great defect is the weak credibility of the threat on which it rests.

In the 1940s, when the United States owned all the nuclear weapons in the world, its threat to use them on the Soviet Union was thoroughly credible. In the 1950s, when the United States had overwhelming nuclear superiority, the threat was still very credible. It is not at all credible now. The Soviet Union long ago accumulated a nuclear stockpile large enough to inflict swift and total destruction on the United States. That is why ever since the early 1960s the United States government has been looking for some other way to protect Western Europe. However, the Massive Retaliation pledge has never been rescinded, and no generally accepted alternative to it has been found.

The Massive Retaliation pledge is not (to use Thomas Schelling's terminology) inherently persuasive. If it is to protect Europe at all it must be made persuasive. There are several ways to do this, most of them requiring that the United States give up its freedom of action or its ability to control the situation that would result from a Soviet invasion. It is almost impossible for the United States credibly to threaten that it would deliberately start a nuclear total war. What it can do is create a situation in which a nuclear total war would be the most likely, if unintended, result of a Soviet invasion.

The easiest, most obvious, and least dangerous way to make the Massive Retaliation threat persuasive is to make a public pledge to carry it out. The United States did this when it joined NATO. The NATO Treaty obliges it to regard any attack on one of its NATO allies as if that were an attack on the United States, and thus presumably as requiring the same response – nuclear retaliation. Just signing a treaty, however, does little to make the Massive Retaliation threat persuasive. The existence of a treaty *in itself* will not have much impact on what the United States government does at the moment of truth. Any American president would rather see the United States get a reputation for breaking treaties than have 70 million Americans die in a nuclear war. The Soviets know this as well as anybody.

If the Massive Retaliation pledge is to mean very much, it must be reinforced. Stationing United States troops in West Germany does this. In a pure Massive Retaliation strategy, their function is to be killed in the event of a Soviet invasion. The American Seventh Army, 205,000 strong and stationed directly across the best avenue of attack for an invasion force, would inevitably suffer heavy casualties. (The servicemen's 300,000 civilian dependents would also suffer heavily unless they were evacuated before the fighting started.) Once large numbers of Americans had been killed in action, the United States government might be forced to declare war, just as it was after Pearl Harbor. And, if the combined United States and Western European ground forces were too weak to stop the Soviet onslaught, the United States would then have to use its nuclear weapons. Knowing that an invasion would almost certainly kindle a nuclear war, the Soviets will not invade Western Europe – such is the theory.

The Seventh Army's true function in a Massive Retaliation strategy is not to defend but to deter. It is a "trip-wire" connected to the trigger of the United States strategic nuclear force. By crossing the border in force and killing United States servicemen, the Warsaw Pact forces would in effect stumble over the trip-wire – and American nuclear weapons would fire back at them. A proper trip-wire force must be large enough but not too large. It must be large enough so that its destruction would compel the United States to fight: at the same time, it must not be so large that the United States might be able to defend Western Europe with conventional weapons alone. If it were, the Soviets might think the United States planned to respond to an invasion with conventional weapons alone and thus they would not face the risk of a nuclear war. Perhaps the Seventh Army's slogan should be:

If we are mark'd to die, we are enow
To do our country loss.[2]

The less damage a nuclear total war would do to the United States the more likely the United States government is to escalate a local war into one. Adopting a Damage Limitation strategy (see Chapter 9) for the protection of the United States and taking the measures necessary to implement it would therefore make the Massive Retaliation pledge more credible. If the United States had a civil defense system, strong active defenses, and a powerful first-strike, counterforce capability, it could plausibly threaten to start a nuclear total war in defense of Europe. While Mutual Assured Destruction cannot possibly achieve anything more than basic deterrence, Damage Limitation might secure substantial extended deterrence. This is one of the best arguments for a Damage Limitation strategy. On the other hand, if Damage Limitation is not the best way to protect the United States – which it is not – Massive Retaliation is much less attractive.

Another way to reinforce the Massive Retaliation pledge is to place American theater nuclear weapons in Western Europe. These weapons are much more likely to be launched in response to an invasion than are strategic nuclear weapons in the United States.

Admittedly, the United States is very unlikely deliberately to launch its nuclear weapons in Western Europe in response to a Soviet invasion. If it did, the Soviets would probably hit back directly at the United States. They might not even be able to tell where some of the warheads exploding on Soviet soil had come from. The United States has a number of Poseidon missile submarines assigned to its nuclear weapons force in Western Europe. Therefore, to know that it had been hit "only" with American nuclear weapons assigned to Western Europe, "Moscow would somehow have to determine that incoming Poseidon missiles were launched from officially authorized SACEUR-controlled stocks, rather than from the U.S.'s central strategic submarine force."[3] Even if the Soviets could make that distinction it would not matter very much to them. American theater nuclear weapons in Europe have sufficient range to hit Soviet territory and if they were used many of them would do so. Once nuclear weapons explode on Soviet territory the last step on the escalation ladder would have been taken and a nuclear total war would be in progress.

This will all be very clear once the fighting starts, if it is not clear now. Once they really have to decide, American leaders would almost certainly choose to withhold their nuclear weapons and see Europe conquered rather than use them and see the United States destroyed. However, by then the decision would be out of their hands. During a war, or while the armies are being mobilized for a war, American nuclear weapons in Europe would be removed from their heavily-guarded storage depots and put into the hands of the combat units which are to use them. Once they are so dispersed, it would be almost impossible to prevent the commanders of those military units from using them. That means those weapons would be used in all but a very limited war in Europe. With the invaders driving relentlessly ahead and the American nuclear weapons under unremitting attack, some field commander somewhere would decide to use his weapons before they were all destroyed. That is exactly how American theater nuclear weapons in Western Europe may deter a Soviet invasion; simply by existing they constitute a threat that a major war in Europe will escalate to a nuclear total war, regardless of anybody's intentions. However, to apprehend this threat requires grasping some rather subtle concepts which Soviet leaders may not accept. Subtle threats are difficult to convey and easy to disregard in a serious crisis.

Henry Kissinger has very cogently stated the case against Massive Retaliation:

> It is absurd to base the strategy of the West on the credibility of the threat of mutual suicide ... European allies should not keep asking us to multiply strategic assurances that we cannot possibly mean, or if we do mean, we should not want to execute because if we execute, we risk the destruction of civilization.[4]

Massive Retaliation has the same fault as all other purely deterrent strategies, that of resting on a bluff that might be called. Deterrence can fail. If it does, Massive Retaliation offers even less useful guidance as to what actually to do than does MAD. To destroy the world because Western Europe had been invaded would, from an American point of view, be even more insane than to do so because the United States had been destroyed.

Massive Retaliation's bluff is more likely to be called than MAD's. The treat to destroy the world in defense of Western Europe is not credible, and no amount of ingenuity and resolution can make it so. De Gaulle was right; no country will commit suicide in defense of an ally. Trying to convince the Soviets of something that is not in fact true is hardly a sound basis for the defense of a vital interest. To be sure, even a small chance that a war in Europe would escalate to a nuclear total war may be enough to deter an invasion. On the other hand, the Soviets are very much aware of the conflicts of interest between the United States and its allies. They are much more likely to overestimate than to underestimate the seriousness of those conflicts because of the Marxist-Leninist lenses through which they see the world. This makes them harder to convince and deter than another ruling elite might be.

The Massive Retaliation pledge may suffice in quiet times. If the Soviet Union continues to be led by cautious men, if NATO stays together, and if the Soviet empire remains reasonably stable, the Soviets may never be prepared to run great risks to conquer Western Europe. However, a deterrent must also deter in stormy times. Future catastrophes or opportunities might well drive the Soviets to take risks they have not taken in the past. They might even risk a nuclear total war if they judge the risk to be slight and all of Europe were the prize. Massive Retaliation cannot with certainty keep Western Europe secure. It may be the best protection possible, but nobody should assume it is without looking at the alternatives.

Flexible Response

Almost as soon as the Massive Retaliation policy was announced by the Eisenhower administration it was rejected by many American and European defense professionals. They pointed out that the United States had already lost its nuclear monopoly, that the Soviet Union was rapidly building up its strategic nuclear force, and that soon the Soviets would be able to destroy the United States even after suffering a massive nuclear strike. Once they could do that, the Massive Retaliation pledge would not be credible enough to afford much protection. Furthermore, the critics asserted, the most probable threat was not a full-scale invasion of Western Europe. The Soviets were more likely to make a limited, carefully-controlled incursion, perhaps a Hamburg grab. The United States would not start a nuclear war over Hamburg, but the Massive Retaliation strategy left it only a choice between that and doing nothing. Therefore, critics of Massive Retaliation proposed much more emphasis tactical nuclear and conventional forces, forces that the United States could actually use in the kind of war it was likely to fight.

The Kennedy administration, coming to power in 1961, took these arguments to heart and soon espoused a different strategy for the defense of Europe. The basic concept of the new Flexible Response strategy was that NATO's response to a Soviet aggression should depend on the nature and level of that aggression. The punishment should fit the crime. If the Soviets were to invade Western Europe with conventional forces alone, NATO would defend itself with conventional forces alone. Only if the enemy were to resort to nuclear weapons would NATO use its own nuclear weapons. The advocates of Flexible Response assumed that the Soviet Union would not be the first to use nuclear weapons. In the early 1960s the United States had far more and better tactical nuclear weapons, so Soviet escalation would simply have led to a much more destructive war, not to a Soviet victory. Therefore, if NATO could defend itself without using nuclear weapons, a war in Europe need not escalate to the nuclear level. Flexible Response would be, it was hoped, a more effective deterrent than Massive Retaliation because the action it threatened was more credible. It would also be an effective defense if deterrence should fail, keeping the Soviets out of Western Europe without destroying the entire world.

The European governments were not particularly enthusiastic about the new United States strategy. They had been fairly comfortable with the old one – the British actually took up Massive Retaliation before the Americans did. None of them saw the United States government's changing its mind as a sufficient reason for them to change theirs. Not until 1967, under considerable American pressure, did the European governments accept Flexible Response as the common NATO strategy. Even then, "Europeans endorsed the formal US position while making it clear that they would not devote great resources to implementing it."[5]

They had good reason for their skepticism. There are three important questions which those who advocate Flexible Response must answer:

1 Can Western Europe be defended with conventional forces alone against the Warsaw Pact's powerful land armies?
2 If an invasion were defeated with conventional forces alone, would the Soviets resort to nuclear weapons?
3 Does Flexible Response make a Soviet invasion more likely?

The advocates of Flexible Response can answer the first question with some confidence. It is the second and third questions that reveal the real flaws in their strategy.

There are 25 divisions stationed in West Germany to meet a Warsaw Pact onslaught. Half of them belong to the West German army. The others come from the United States (forces equivalent to six divisions), the United Kingdom (three divisions), France (three divisions), Belgium (one division), the Netherlands (one brigade), and Canada (one brigade). (The French forces in West Germany are not under the NATO military command, but almost certainly would help defend West Germany against an invasion.) These units are organized and equipped much the way their Warsaw Pact opponents are: armored and mechanized infantry formations with tanks, APCs, and self-propelled guns.

Most authorities consider these forces to be man-for-man superior to their opponents. None of the NATO armies has problems as severe as the poor training practices and lack of initiative prevalent in the Soviet armed forces. Probably all of them would fight well. The West Germans, in particular, have inherited a formidable reputation and would be fighting in defense of their own territory. Russian troops would also fight resolutely, but some of their allies might have little stomach for the war. However, nobody can really know how well the various armies would perform as most of them have not been to war for more than 40 years.

In addition to the quality of its troops, NATO would enjoy two more important advantages. One is that it would fight on the defensive. As von Clausewitz pointed out, "defense is easier than attack."[6] The defender can dig in and use the terrain to conceal and protect his forces, while the attacker must expose his as he moves them forward. Tacticians generally reckon that the attacker must have at least a three-to-one numerical superiority at the point of attack. This does not require a three-to-one superiority across the entire front. The attacker has one great advantage: he decides where the fight is to be. He can mass his forces where he has chosen to break through, thus achieving the necessary numerical superiority at that point, even though he is outnumbered along the rest of the front. Even so, a widely-accepted rule of thumb is that the attacker must have a three-to-two numerical superiority across the entire front. When both alliances are fully mobilized, the Warsaw Pact *does* have more than this degree of superiority – but not so much more that other factors could not compensate for it.

Another important advantage NATO has is the kind of terrain it defends. West Germany is not the ideal place in which to execute a Blitzkrieg. The southern part of the country is mostly hilly and heavily forested, and while armored forces can move across such countryside – the Germans themselves did so very successfully when they invaded France in 1940 – it is not easy. A resolute, well-conducted defense can hold such terrain against even very powerful armored forces. There are several good invasion corridors in the south, but NATO military planners have massed NATO's forces to block them. In northern Germany the North German Plain is flat, but it is also cut up by numerous canals and small rivers which would slow down an invader quite considerably. The Soviets have devoted much thought and effort to the problem of crossing rivers quickly, but it seems there are no easy solutions. When it rains, as it does rather frequently in that area, much of this plain becomes waterlogged and difficult for vehicles to travel across. Also, much of northern Germany is now built-up, with suburban sprawl extending over thousands of square miles. This, too, would be a considerable obstacle to an invader. A determined defense could entangle the invader in house-to-house fighting most of the way to the Rhine River.

NATO's greatest weakness is the numerical inferiority of its forces in West Germany. It does not have as many troops as the Warsaw Pact available for combat immediately and NATO has far fewer reinforcements than the Warsaw Pact to send to the battlefield.

NATO does have reinforcements available. In the first few days of a NATO mobilization the Dutch and Belgian forces, most of which are held back in their homelands during peacetime, would move up to their assigned positions in the NATO border defenses. The six United States army divisions based in the continental United States but earmarked to fight in Europe would be flown across the Atlantic. These United States army divisions have all the vehicles and other heavy equipment they would need stockpiled in West Germany; only the troops themselves would have to be transported. The French would send forward at least some of the 12 active duty divisions they have stationed in France. The West Germans would mobiize their Territorial Army troops. These are lightly-armed reserves intended mainly to secure NATO rear areas but they are also capable of harassing enemy combat units to some extent. West Germany can mobilize several hundred thousand Territorial Army troops in 72 hours, a force equal in numbers to its standing army. Ten days after it starts mobilizing NATO could have as many as 43 divisions in the field. In two months it might have 18 more divisions in action. All these calculations assume that only those countries which have troops on West Germany's territory in peacetime would come to her defense in a war. The other NATO countries would have enough to do to defend themselves if Warsaw Pact forces were to invade them too. Even if they were not invaded they would probably choose to hold their forces back to protect their own territories.

NATO, then, can build up its strength very considerably. But the Warsaw Pact can build up its strength even more and do so more rapidly. By the time NATO has 43 divisions in place, the Pact could have 54; by the time NATO has 61, the Pact could have 120 (see figure 13.1). That is if the armies mobilize in peacetime. If the Soviets begin the war with what they have ready for action immediately and then both sides start mobilizing, NATO will have real trouble marshalling its forces. American reinforcements flying across the Atlantic very likely would find the airfields they were to land at and the equipment they were to use destroyed or in Soviet hands. Later American reinforcements coming by sea would have to survive Soviet submarine and aerial attacks on the convoys carrying them. With every square mile of West Germany under attack, calling up the Territorial Army and preparing it to fight would pose many difficulties. French mobilization might also be impeded by aerial attacks on French territory. Meanwhile, the Warsaw Pact's mobilization

NATO

1 NATO divisions ready immediately (▰▶): 25–12 West German, 6 United States, 3 British, 3 French, 1 Belgian, all in West Germany

2 NATO divisions ready 1–10 days after mobilization: (▱▶): 18–2 West German Territorial Army; 4 United States, 1 British, 5 French, 1 Belgian, 3 Dutch, 2 Danish in their respective homelands

3 NATO divisions ready 11–60 days after mobilization (▱▷): 18–10 United States (some National Guard), 1 British, 7 French

NATO forces build-up

Day 1: 25 divisions
Day 10: 43 divisions
Day 60: 61 divisions

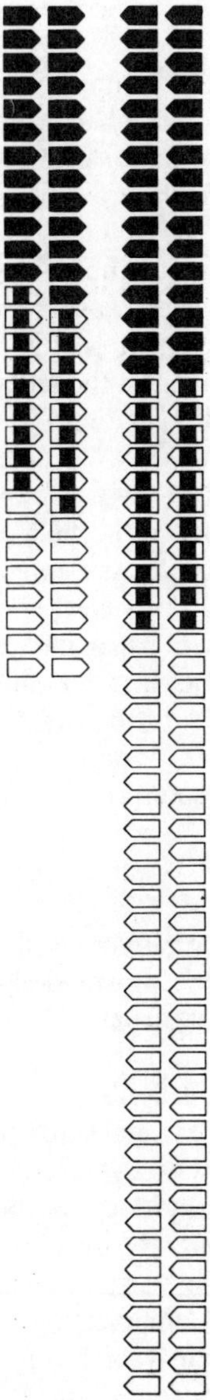

WARSAW PACT

1 Warsaw Pact divisions ready immediately (◀▰): 32–20 Soviet and 6 East German in East Germany; 4 Soviet and 2 Czechoslovakian in western Czechoslovakia (all Category I)

2 Warsaw Pact divisions ready 1–10 days after mobilization (◁▰): 22–2 Soviet and 8 Polish in Poland, 1 Soviet and 2 Czechoslovakian in Czechoslovakia, 6 Soviet in the Soviet Union's part of the Western theater of military operations, 3 Soviet from the Central Reserve (all Category I)

3 Warsaw Pact divisions ready 11–60 days after mobilization (◁▱): 66–54 Soviet, 5 Polish, and 7 Czechoslovakian (all Category II and Category III in the Western theater and the Central Reserve)

Warsaw Pact forces build-up

Day 1: 32 divisions
Day 10: 54 divisions
Day 60: 120 divisions

Figure 13.1 *Ground forces deployments and reinforcements for war in West Germany.*
The United States has in West Germany four divisions and separate brigades and regiments equal to two more divisions. Author's estimates, based on *The Military Balance 1987–1988* (London: International Institute for Strategic Studies, 1987); Andrew Hamilton, "Redressing the Conventional Balance: NATO's Reserve Manpower", *International Security*, 6 (Summer 1985), pp. 112–20

would be proceeding relatively smoothly unless Russian territory were attacked – which would likely lead to a global nuclear war. Under these conditions, the Warsaw Pact's advantage in mobilized strength might be much greater than two to one.

Another NATO weakness is the comparatively short distance the enemy would have to advance to inflict mortal damage on the alliance. The total distance enemy troops would have to advance to the Atlantic Ocean is some 600 miles, but NATO would be in severe difficulty by the time Warsaw Pact forces had covered a quarter of that distance. The lines of communication of its forces in West Germany run mainly north–south along or near the Rhine River, within easy reach of an invader. Losing Frankfurt would virtually split the NATO forces in two, and Frankfurt is little more than 60 miles from the border. This is a strong argument against the maneuver tactics which otherwise might be the best way for NATO to fight. A maneuver-oriented defense cannot help but give ground and the NATO defenses in West Germany have very little ground they can afford to give.

It is hard to predict who would win a major conventional war in Europe. Any judgement rests on very debatable assumptions about how well the armies would fight, how fast each side can mobilize, and how much of an advantage the defender has over the attacker. Most likely, though, the forces NATO has today could stop a Warsaw Pact invasion short of the Rhine River *if* two crucial conditions are met:

1 NATO begins to mobilize as soon as the Warsaw Pact does or at most a few days later.
2 France joins in the defense of Western Europe.

French participation in the defense would add formidable ground and tactical air forces to NATO's order of battle. It would also allow reorienting NATO's lines of communication east–west, from Germany to the Atlantic Ocean, making them less vulnerable.

Many of the strategists who think NATO must have stronger conventional forces suggest relatively inexpensive measures to accomplish this. Some of them propose that NATO augment its reserves. They point out that the Warsaw Pact's numerical superiority does not become really menacing until the alliances have been mobilizing for several weeks. NATO already has enough active duty forces; what it needs are more reserve divisions, to be available a month or two after the mobilization starts. Providing these should not be prohibitively expensive, as a reserve unit costs much less to keep up than an active duty unit. Other strategists want the NATO armies consistently and thoroughly to adopt maneuver tactics and the *Auftragstaktik* command style that goes with them. They see this as the best way to exploit the Warsaw Pact's defects and to overcome its numerical superiority. Still others pin their hopes to high technology conventional weapons, particularly weapons designed to destroy enemy reserves moving up to the battlefront, or to building fortifications along the West German border.

Even those who think that NATO must greatly increase the number of troops it keeps on active duty believe that the members of the alliance have the resources to do so. The 1952 goal of 96 active duty divisions is not out of reach if the peoples and governments of NATO are persuaded that it must be attained. NATO has half again the population of the Warsaw Pact and three times its economic resources. Western Europe can be defended in a conventional war, perhaps with the forces it has now, certainly with the forces it could have.

The problem is, what does NATO gain by being able to stop an invasion without using nuclear weapons? The Soviet Union could still use them. "We cannot and ought not go on assuming a Soviet Union bent on major aggression but afraid of using nuclear weapons."[7] The Soviets could begin their invasion with a massive tactical nuclear weapons attack. It would be a desperate venture, but the better able NATO is to fight a conventional war the more attractive this option looks compared with the Soviets' other two.

Even if the Soviets begin the invasion with conventional forces alone, they could still resort to nuclear weapons if their conventional forces were defeated. Soviet military doctrine holds out the hope that Western Europe can be conquered without using nuclear weapons, but it makes no commitment whatever not to use them if necessary. On the contrary, it asserts that a conventional war in Europe could escalate to a nuclear war at any time. If it does, the Soviets would want to be the ones who escalate it.

The Soviet leaders would launch a full-scale conventional invasion of Western Europe only if they were resolved to run a desperate risk for a very high stake. Once an attack were under way they would probably not expect their regime – or themselves – to survive defeat. One of the strongest lessons of Russian history is that regimes that lose major wars are overthrown. That was how the Tsarist regime was finally toppled in 1917 and it seems that Stalin's greatest fear in the summer of 1941 was that the initial German victories would drive the Russian people to overthrow him. The Soviet leaders probably could and would accept defeat in a limited Hamburg grab operation rather than escalate to nuclear war. They probably could not accept defeat in a full-scale war in Europe, even a conventional one. Instead, they would resort to nuclear weapons. Flexible Response holds out some hope that a full-scale conventional war in Europe would not escalate to nuclear total war, but it is not much more than a hope.

At the same time, Flexible Response does not guaranteee that there will be no war in Europe. It may even encourage the Soviets to aggress by letting them think the risk they run is limited. If they are convinced that NATO will not be the first to use nuclear weapons, they can hope to conquer Western Europe without a nuclear war: whatever happens in Western Europe, nuclear warheads will not explode on Russian cities. They might launch their invasion feeling sure it would not lead to a nuclear war – and then find out they were wrong.

European Nuclear Deterrent

Massive Retaliation and Flexible Response have opposite but equally grave flaws as deterrence strategies. The Massive Retaliation pledge is extremely powerful but probably not sufficiently credible. The Flexible Response threat, that if attacked with conventional weapons NATO will resist with conventional weapons, is quite credible but probably not powerful enough. It tempts the Soviets to think they could conquer Western Europe and not pay too heavy a price. A powerful strategic nuclear force controlled by the European countries themselves would be a much better deterrent. It would be far more powerful than any conventional deterrent NATO could have and the threat to use it would be far more credible than the Massive Retaliation pledge.

There already are two European-controlled nuclear deterrent forces – those of France and the United Kingdom. They protect France and the United Kingdom fairly well: it is doubtful whether they give much protection to the rest of Western Europe.

The United Kingdom, as a full member of NATO, is just as much bound as is the United States to consider an attack on any NATO member as an attack on itself. British leaders have always maintained that they would launch their nuclear weapons against an invader; they have made the same Massive Retaliation pledge that United States leaders have. (Considering, though, the much smaller size of the British nuclear deterrent, it might better be termed "mini-retaliation.") No doubt the United Kingdom has made this pledge sincerely and its promises are as much to be relied on as the promises made by the United States. However, if the American Massive Retaliation pledge is not fully credible to Soviet leaders – and that is the rationale of the British nuclear force – it is hard to see why a similar British pledge is any more credible.

The French nuclear deterrent was built to protect France. French political and military leaders have never promised that French nuclear weapons might be used to defend any other country. Even if they had, a French Massive Retaliation pledge would not be very credible. If no United States president would sacrifice Washington in defense of Paris, would any French president sacrifice Paris in defense of Bonn?

Perhaps the countries of Western Europe might club together to build a single strategic nuclear force to protect all of them. Something like this, the Multilateral Force (MLF), was proposed during the early 1960s. The MLF was to have been a strategic nuclear force within the NATO framework, manned and paid for by all the NATO countries (including the United States) which chose to participate in it. It would have had a fleet of 25 surface ships, each armed with eight ballistic missiles. Each ship would have been manned by contingents drawn from at least three different countries with no more than 40 percent of a crew coming from any one country. Presumably the commanders of all three national contingents on a ship would have had to turn their keys or press their buttons to launch the ship's missiles. The order to launch the missiles could only have been given with the unanimous consent of all the countries participating in the MLF.

This plan never got very far. Some NATO governments refused from the beginning to have anything to do with it, others pursued it reluctantly, none was really enthusiastic. Many rejected it because it would have given Germans a certain access to nuclear weapons, even though the West German government could have done nothing without the unanimous consent of its partners. A more cogent objection was that it would not have been a credible deterrent. If any one of a dozen or more countries can veto the use of a military force, it will never be used; at least one country almost certainly will cast a veto, no matter what happens. Despite what its critics said, the MLF would not have put 15 fingers on the nuclear trigger; it would have put 15 fingers on the safety catch. This is, however, the only basis on which any group of sovereign states can club together to form a common strategic nuclear force. No government will voluntarily yield the decision to start a nuclear war to a committee of its allies. It will insist on the right to veto that decision. There may some day be a United States of Europe and if so it could have its own strategic nuclear force. Until that day comes, as NATO's experience with the MLF shows, there almost certainly will not be a common Western European strategic nuclear force.

The political conditions that might make a common European strategic nuclear force possible do not exist and may be unattainable. Nuclear forces belonging to individual countries protect only those individual countries. However, there is one country which, if it had nuclear weapons, could protect most of Western Europe as well as itself. That country is West Germany. The Soviets simply cannot get at most of the population and resources of Western Europe without crossing West German territory. West Germany needs nuclear weapons more than any other country in Western Europe. It is geographically the most vulnerable and faces the heaviest concentration of Warsaw Pact military power. If West Germany had a powerful strategic nuclear force, the Soviets could not launch an all-out invasion of Western Europe without risking massive damage to their own country.

Considered purely as a military strategy, equipping West Germany with nuclear weapons may be the best way to protect Western Europe. Unfortunately it faces great – probably insuperable – political obstacles. World War II has not been forgotten. Even today, more than 40 years later, many people in Europe and elsewhere are not ready to trust any German government with nuclear weapons. The West Germans themselves show no strong desire to have their own strategic nuclear force. These attitudes are strong enought to prevent West Germany acquiring nuclear weapons for some time to come. However, if the flaws in both the Massive Retaliation and the Flexible Response strategies become more evident and less tolerable, NATO military planners will be looking for alternatives. They may well give the option of equipping West Germany with nuclear weapons more attention than it has received so far.

How to Defend Europe

As a general rule to which there are exceptions on both sides, American strategists favor Flexible Response while the Europeans prefer Massive Retaliation. To Americans the primary objective of NATO, besides keeping the Soviet Union from conquering Western Europe, is to prevent a strategic nuclear war. They would very much regret having to fight a conventional war in Europe and not only because of the American soldiers who would be killed in it. They would regret even more having to fight a tactical nuclear war in Europe. But the only war that could destroy the United States would be one fought with strategic nuclear weapons. Flexible Response seems to most Americans the best way to ensure that, whatever happens in Europe, nuclear warheads will not explode on American cities.

American strategists see very clearly that Massive Retaliation calls on the United States government to do what it would not do at the moment of truth. Furthermore, it relies almost entirely on what the United States does, threatens to do, and is prepared to do. It is the United States which is to maintain the strategic nuclear force which protects Europe and to launch the attack which will lead to its own destruction if Europe is invaded. It is the United States which is to maintain the trip-wire conventional forces which make the Massive Retaliation pledge credible, offering the lives of its young men as a guarantee that it will carry out its treaty commitments. European armed forces play no *essential* role in a pure Massive Retaliation strategy. It is all too easy for Massive Retaliation to serve as, or be seen as, an excuse for the Europeans to throw the burden of defending their continent on to the United States. This may have been acceptable in the early 1950s, when Western Europe was still weak from the after-effects of World War II, but not today.

The Europeans see things differently. For them the primary objective must be to prevent any war, not just a strategic nuclear war. They are very well aware of the damage even a conventional war would do to their crowded continent; World War II taught them that. A "successful" conventional weapons defense of West Germany would be a catastrophe for the West Germans. A tactical nuclear war might kill as large a percentage of the population of Western Europe as a strategic nuclear war would of the population of the United States. For the Europeans, Flexible Response offers the Soviets too much temptation to launch an invasion. Aware as they are of the defects of Massive Retaliation, they still see it as the best way to prevent war entirely.

Another defect of Flexible Response, in European eyes, is that it requires powerful conventional forces. The Europeans must provide most of them. The European countries have most of the manpower in the alliance, nearly twice as much as the United States, and their manpower is more usable in a European war. European armed forces are right where they would be needed, or nearby, while United States conventional forces would have to be carried across the Atlantic at great effort and risk. The division of labor suggested by the Flexible Response strategy is that the United States provides the tactical nuclear weapons which prevent the Soviet Union escalating a conflict and Western Europe provides the ground forces. The United States provides a small, elite, relatively inexpensive force which would be kept out of the fighting as much as possible. Western Europe provides the most expensive part of the NATO military machine, which is also the part that would do most of the dying.

The European Nuclear Deterrent option appeals mostly to Europeans who believe in nuclear deterrence but doubt the efficacy of the American Massive Retaliation pledge. They may, like the British, wish to reinforce that pledge or, like the French, not rely on it at all. This option also appeals to some Americans who wish to reduce the United States's involvement in the defense of Europe.

It is remarkable that the alliance has survived these differences of interest and military doctrine. It has done so by refusing to make a definite choice among the three possible strategies for the protection of Western Europe. Each has been implemented to a degree. The Massive Retaliation pledge is still in effect and the United States still has its Seventh Army in Germany as security for that pledge. On the other hand, Flexible Response has been the official NATO strategy since 1967 and the alliance has built up powerful conventional forces to support it – much more powerful than a purely trip-wire force would need to be. France and the United Kingdom have their nuclear deterrents but the European nuclear deterrent that would best protect Western Europe, a West German force, has not been built.

Usually failure to choose clearly and decisively among alternatives is a bad mistake when making strategic decisions. This is one of the few cases in which it is not. If NATO strategists do not know how they would respond to a Soviet invasion then the Soviets cannot possibly know either – which makes planning that invasion or resolving to undertake it quite difficult. Furthermore, Soviet strategists have their own choice of three different invasion plans. Having a variety of strategies and the forces to implement them gives NATO a reasonably good defense against or deterrent to each of the Soviet Union's options.

Flexible Response offers the best way to deal with a Hamburg grab in Central Europe or a limited Soviet aggression on the northern or southern flank of NATO. These threats can be defeated with conventional weapons alone if NATO has powerful enough conventional forces, in the right places and on alert. The Soviets could not employ in any of these operations more than a small part of their armed forces: in a Hamburg grab for fear of sacrificing the element of surprise, and in the other cases because of the Soviet Union's logistic limitations and the unsuitability of most of its army for fighting in mountainous areas. These are threats that can be defended against. On the other hand, they cannot reliably be deterred. The Massive Retaliation pledge simply is not credible against them. A United States president might, just conceivably, sacrifice Washington in defense of London or Paris, but it is utterly absurd to think he would do so in defense of Trondheim or Salonika.

If the Soviet Union hits Western Europe with a massive nuclear weapons barrage, the only possible response would be to hit back with nuclear weapons (or surrender). It would not much matter whether they were strategic or tactical weapons because strategic nuclear weapons would be used eventually. There is no way Europe could be defended against this attack, nor would there be much left to defend. A nuclear attack can only be deterred. Fortunately it is the easiest threat to detèr because it is the most likely to provoke NATO nuclear retaliation. France and the United Kingdom would certainly hit back with their nuclear weapons. The United States, also, would be much more likely to launch a nuclear strike on the Soviet Union if the Soviets had used nuclear weapons first. The United States's Massive Retaliation pledge and the nuclear deterrents of France and the United Kingdom are the most effective deterrents of a Soviet nuclear attack now available.

The most difficult option to counter is the full-scale attack made with conventional weapons alone. That attack cannot be defended against because a successful defense would probably lead to a nuclear total war. Once the Soviets mobilize their entire strength for a decisive effort, they cannot accept defeat. NATO *must* deter a full-scale conventional weapons attack, but no one strategy can by itself deter such an attack as reliably as it needs to be. Massive Retaliation is not credible enough, Flexible Response not powerful enough, and the nuclear deterrents of France and the United Kingdom protect only part of Western Europe. The ideal deterrent would be a powerful West German strategic nuclear force. Perhaps West Germany eventually may have one, but not for many years, if ever. In the meantime the peoples of North America and Western Europe must hope that the combination of fairly powerful NATO conventional forces, a not very credible United States Massive Retaliation pledge, and the nuclear deterrents of France and the United Kingdom will protect Europe adequately.

Notes

1 David Alan Rosenberg, "'A Smoking Radiating Ruin at the End of Two Hours': Documents on American Plans for Nuclear War with the Soviet Union, 1954–55", *International Security*, 6 (Winter 1981/82), p.25.
2 William Shakespeare, *Henry V*, IV,iii, 20–1. This slogan would not be a morale-raiser but it would be honest.
3 Paul Bracken, *The Command and Control of Nuclear Forces* (New Haven, Conn.: Yale University Press, 1983), p. 156. SACEUR (Supreme Allied Commander Europe) is the military officer in command of all the forces under NATO control in Europe.
4 Henry A. Kissinger, "NATO: The Next Thirty Years", in *Strategic Deterrence in a Changing Environment*, ed. Christoph Bertram (Westmead, England: Gower Publishing, 1981), p. 109.
5 Lawrence Freedman, *The Evolution of Nuclear Strategy* (New York: St Martin's Press, 1981), p.285.
6 Carl von Clausewitz, *On War* (Princeton, NJ: Princeton University Press, 1987), p.357.
7 Bernard Brodie, *Escalation and the Nuclear Option* (Princeton, NJ: Princeton University Press, 1966), p.131.

PART V
Naval War and Power Projection

We have fed our sea for a thousand years
  And she calls us, still unfed,
Though there's never a wave of all her waves
  But marks our English dead:
We have strawed our best to the weed's unrest,
  To the shark and the sheering gull.
If blood be the price of admiralty,
  Lord God, we ha' paid in full!

Rudyard Kipling, "The Song of the Dead"*

* *Rudyard Kipling's Verse: Definitive Edition* (Garden City, NY: Doubleday and Company, Inc., 1940), p. 172. Reprinted with permission.

14
Command of the Sea

He that commands the sea is at great liberty, and may take as much and as little of the war as he will.
    Francis Bacon, "Of the True Greatness of Kingdoms and Estates"

The Uses of Sea Power

When countries go to war they use their navies to help their land armies fight and deny enemy armies the ability to fight. "Men live on land and it is there that the final decision must be sought."[1] On the other hand, 71 percent of the earth's surface is covered with water and every continent is essentially an island surrounded by some part of the great ocean that girdles our planet. Most countries have sea coasts and can be invaded from the sea; a few, such as Japan and the United Kingdom, are islands and cannot be invaded overland. Many depend on seaborne trade to sustain their peoples and supply their armies; without it they cannot fight for very long. The sea is often the best means of access to some place on land and sometimes it is the only way. That is why navies have been built and sent out to seek "command of the sea."

Having command of the sea means being able to use the sea for one's purposes and prevent the enemy using it for his. One's warships, troopships, and merchantmen can sail in safety or at least without prohibitive risk. Enemy ships cannot. As long as navies have existed the instrument by which they have contested command of the sea has been the battlefleet, the largest force that can be assembled of the most powerful warships available. In the Ancient World battlefleets consisted of oared galleys, in the Age of Sail of galleons or ships of the line, in the Age of Steam of battleships or aircraft carriers. A battlefleet won command of a sea by driving the enemy's battlefleet from that sea; destroying it, blockading it in its harbors, or blocking the straits it had to sail through to reach that sea. The Spartans won command of the Aegean Sea in 405 BC by destroying the entire Athenian fleet at Aegospotami. The British won command of the sea during the Napoleonic Wars by blockading the French battlefleet in its harbors. In World War I the British navy kept command of the Atlantic Ocean by controlling the English Channel and the exits from the North Sea through which the German High Seas Fleet would have had to sail to reach the Atlantic.

Today surface ship battlefleets of aircraft carriers and missile-armed cruisers continue to be important instruments of sea power, but they have powerful rivals. Submarines and land-based aircraft may be able to drive enemy warships from the surface of the ocean even without the help of a battlefleet. It may also be possible to destroy an enemy battlefleet with a combination of reconnaissance satellites and ballistic missiles: satellites to detect and target the ships, and missiles to barrage them with nuclear warheads. These new weapons probably make getting and keeping undisputed command of the sea more difficult. The enemy may continue to dispute it even after his battlefleet has been driven from the sea. However, it is still possible to command most of an ocean most of the time and important benefits flow from doing so.

A navy that commands the seas around its homeland keeps the homeland safe from attack by sea. Until the twentieth century that was the only way an island country such as the United Kingdom could be attacked. As long as the Royal Navy dominated the English Channel, the North Sea and the North Atlantic, no invader could set foot on British soil. Their powerful navy gave the British a degree of national security their continental neighbors could not hope to enjoy. The development of bombing aircraft weakened the protection which command of the sea gave the British. In both world wars German bombers flew unimpeded across British-dominated waters to attack British cities directly. They could not, however, conquer the United Kingdom; only a seaborne invasion could have done that. Nuclear warheads and ballistic missiles have made island sea powers even more vulnerable. ICBMs can fly across any ocean in the world and destroy any country they hit. Nevertheless, in a conventional war the ability to prevent a seaborne invasion is still important.

A powerful navy also protects its possessor's maritime trade from attack. That country can ship its exports all around the world, draw imports from all over the world, and transport its armies across the oceans. For some countries the ability to do this is almost as vital as defending their territories against invasion. For centuries the United Kingdom's prosperity has depended on its maritime trade. If an enemy had ever managed to drive that trade from the seas he might have brought the country to its knees without landing a single soldier on its territory. The United States also must protect the sea-lanes its ships use, although for different reasons. If the United States had not been able to transport its armies and its goods across the Atlantic and Pacific Oceans during the two world wars, it would have had little impact on those struggles. If it cannot do so today there is little it can do to help most of its allies. Transport aircraft can substitute for shipping but only to a very limited degree; they lack the carrying capacity to move more than a small part of what is needed to sustain an economy or fight a war.

Command of the sea is the first line of defense of maritime trade. Destroying or penning up the enemy's battlefleet eliminates the most formidable threat to the sea-lanes. It does not, though, eliminate all possible threats. Even a tightly blockaded enemy can get a few surface raiders to sea and attack merchantmen and troopships from the air. Above all, he can use submarines. In the twentieth century the submarine has been perhaps the most effective of all naval weapons. A navy that cannot contest the command of the sea may still be able to use submarines with deadly effect against its enemy's merchant marine. In the two world wars the German surface fleet never seriously threatened to take command of the sea away from the British – but German U-boats nearly won both wars.

The protection of maritime trade requires, then, a second line of defense: numerous specialized anti-submarine warfare (ASW) ships and aircraft. There are three ways to use these weapons:

1 To seek out and destroy enemy submarines, either in an entire ocean or just along the sea-lanes used by one's own shipping.
2 To form the transport ships into convoys and use ASW forces to protect each convoy.
3 To keep enemy submarines away from the sea-lanes by blocking the straits they must pass through to reach them.

In the past, seeking out and destroying enemy submarines was not feasible enough to afford maritime trade much protection. The immensity of the sea defeated efforts to search it adequately. One British admiral stated the problem graphically:

> Imagine England and Scotland and Wales rolled flat, all hedges, towns, lakes, rivers removed until it was merely a sandy waste. One single rabbit in that vast area would be relatively the same in size as a submarine in the North Atlantic. Moreover, it would be a rabbit which could disappear under the sand in half a minute without leaving a track behind it.[2]

Today hydrophone arrays and sonar-equipped aircraft have made searching an ocean for submarines much easier, although on the other hand, modern submarines are harder to find than World War II submarines were.

The beauty of convoy tactics is that they turn the immensity of the sea against the submarine. It has to search out the convoy in all those tens of millions of square miles of water. The ASW forces protecting a convoy do not have to search out the submarine; it has to come to them. The closer it comes, the more vulnerable the submarine itself is to the ASW forces. On the other hand, convoying reduces the number of voyages a transport can make in a given period of time. It also gives enemy submarines a very concentrated, lucrative target if they can defeat a convoy's escorts.

In the two world wars convoying, more than anything else, defeated the U-boats. Today convoy tactics are still probably the best way to protect maritime trade, but they must deal with the greatly increased range of modern anti-ship weapons. Torpedo-armed World War II submarines generally had to get within a mile of a ship to sink it; modern submarines armed with anti-ship cruise missiles can launch them from several hundred miles away. The ASW forces protecting a convoy must therefore dominate a very large area around the convoy.

If there is a strait or even a relatively narrow arm of the sea which enemy submarines must pass through to reach the sea-lanes, blocking it can be a very effective ASW tactic. The most effective barrier is a minefield reinforced by patrolling submarines, surface ships and ASW aircraft. This tactic works well only if geography allows. The narrow body of water which the enemy's submarines have to pass through must be too far away from his bases for his minesweepers to keep cleared. This tactic would not have kept German U-boats out of the North Atlantic during World War II once the Germans had control of France.

Because a country that commands the sea is safe from any seaborne invasion and can protect its maritime trade, it "may take . . . as little of the war as [it] will." It is much less vulnerable than its opponent. Command of the sea also gives a seapower the means to attack its opponent in ways that would not otherwise be possible and, in that sense, to "take as much . . . of the war as [it] will." A country that commands the sea can destroy the enemy's maritime trade and invade his territory from the sea. These are the offensive uses of command of the sea, just as security against invasion and protection of maritime trade are its defensive uses.

Destruction of the enemy's maritime trade is an important fruit of command of the sea, and winning command harvests that fruit almost without further effort. If a navy is strong enough to drive the enemy's warships from the sea it can do the same *a fortiori* to his merchant ships. This can almost entirely destroy the foreign trade even of a country with land frontiers across which it can draw supplies. By 1918 the British blockade had reduced Germany's foreign trade to one-seventh of its pre-war level. A commercial blockade may or may not contribute decisively to winning a war. Its effectiveness depends very much on the length of the war, the nature of the enemy's economy, and his access to sources of supply on land. Blockades accomplish little in short wars, because it takes time for the effects of a stoppage of trade to weaken a country's ability to fight. The French blockade of Prussia during the Franco-Prussian War (1870–1) did little to avert France's defeat because the war was over in seven months. Nor can blockades do much damage to a great continental power. The British blockade in World War II did little to weaken Germany's ability to make war because the Germans could draw on the economic resources of most of Europe. On the other hand, in a long war, against an enemy which must import across the seas the weapons his army must have or the food to feed his people, blockade is a powerful weapon. It is slow to act, but in the end perhaps the surest of all means of winning a way. The Union navy's blockade of Confederate trade and the Royal Navy's blockade of Germany during World War I both contributed decisively to ultimate victory.

Attacking targets on land from the sea is known as "power projection." It can take many forms. Ships can bombard targets on land with guns and missiles or can launch aircraft to bomb them. They can also land troops on the enemy's shore, in any number from a dozen commandos sent to destroy a radar station to an army sent to invade a continent. The enemy can protect his coastline against seaborne power projection efforts, but not easily. A fleet can appear over the horizon, often with little or no warning, anywhere along his shores. The defender must place fortifications, airfields, garrisons, minefields – whatever means of defense he chooses to employ – all along a coastline that may stretch for thousands of miles. A relatively small army, working with a navy that commands the seas, can thus tie down enemy forces many times larger than itself. Napoleon is said to have complained that "With 30,000 men in transports at the Downs, [an anchorage off the southeastern tip of the United Kingdom] the English can paralyze 300,000 of my army."[3] So even if a navy does not strike a single blow against an enemy's coastline, its ability to do so may notably affect the outcome of a war.

In land warfare, seapower's ability to turn the enemy's flank by landing ground forces behind his lines is an important asset. The Union navy did it many times during the American Civil War, along the seashore and even more frequently along the rivers in the interior of the country. Almost a century later the United States gained its greatest victory of the Korean War by an amphibious landing at Inchon.

If two countries at war are separated by a body of water only the one able to send its army across the water and invade the opponent's territory can hope to win a total victory. Because they did not command the English Channel, neither Hitler nor Napoleon could invade the United Kingdom. Because the British could send their armies across the Channel, they helped to defeat both Hitler and Napoleon.

Command of the sea is not all that is required for a successful landing. There must also be ground forces available to be landed and ships to put them ashore. The troops employed are preferably, although not necessarily, specially trained, specially equipped marine infantry. The landing craft needed to put them ashore are specialized craft, designed solely for that purpose, and it takes a large fleet of them to land very many troops.

On land, an outnumbered army has the option, which it usually takes, of standing on the defensive. The defender can hope to defeat the attacker, even if heavily outnumbered, because of the advantage the terrain confers on him. At sea, the defender has no such advantage because there really is no terrain. There are no hills to shelter behind, nor are there rivers an attacker must cross. Therefore, in the past naval strategists held that a badly outnumbered battlefleet had little hope of winning command of the sea. Once brought to batttle it would be sunk; the best it could do was avoid battle, wait for the enemy to split his fleet up into detachments, and then attack one of the detachments. The enemy was seldom obliging enough to permit this.

The development of aircraft and anti-ship missiles has challenged this doctrine. Surprise has become as important as numbers. Aircraft and missiles can do so much more damage so much more rapidly than guns that the fleet which gets its blow in first has an excellent chance of winning even if it is badly outnumbered. That was how the United State navy won the Battle of Midway in 1942. Superior strength is, though, still a very important advantage. A large fleet has a much better chance of surviving a small fleet's surprise attack than a small fleet has of surviving attack by a large one.

A country that does not hope to win command of the sea can choose not to build a battlefleet and instead concentrate its naval efforts in two other areas: coast defense and commerce-raiding. By doing this it may be able to deprive its opponent of two of the benefits that he would otherwise obtain from command of the sea: the ability to attack targets on land from the sea, and the ability to move trade and troops across the oceans.

A coast defense navy devotes its efforts to keeping the enemy fleet away from its coastline and to fighting along the shore on the flanks of the land armies. It relies primarily on a "mosquito fleet" of numerous small, short range ships – missile boats, small submarines, and landing craft – on minefields, and on land-based aircraft. These forces work in very close cooperation with the ground forces, often being used to attack enemy ground forces from the sea or to move friendly ground forces along the coast and even up major rivers. These were the tactics of the Soviet Navy during World War II.

Commerce-raiding is a "sea-denial" strategy. It depends on building large numbers of ships designed to attack enemy merchantmen and slipping them past the enemy's blockade to do so. In the Age of Sail commerce-raiding was the business of frigates, corvettes, and other light, fast-sailing ships. In the twentieth century submarines have been by far the deadliest commerce-raiders but long range patrol bombers have also proven effective. The United States during the War of 1812, Germany during the two world wars, and France in several eighteenth century wars used their navies primarily for sea-denial. It has never yet won a war, but twice it came close to doing so for the Germans.

The classic doctrines of sea power were formulated long before the advent of nuclear weapons. They seem, however, to remain valid even today *in conventional wars*. What navies can do in a nuclear war depends on what kind of nuclear war it is. Sea power, in the traditional sense, would probably be meaningless in a nuclear total war. On the other hand, in a limited nuclear war at sea there would be battles for command of the sea, battlefleet against battlefleet, as well as power projection and commerce-raiding operations. They would simply take different forms and perhaps have different outcomes than if nuclear weapons were not used.

In a nuclear total war one form of sea power, the missile submarine, would be supremely important. Particularly if the war lasted more than a few hours, it would be vital to ensure the safety of one's missile submarines and very useful to destroy as many of the enemy's as possible. Other forms of sea power might not play much of a role. Even if there were warships still afloat to contest command of the sea, the outcome of that contest would not be important. With almost all of the world's ports and other cities destroyed, there would be little maritime trade to protect or attack. Nor would there be many targets worth the attentions of naval power projection forces. Amphibious landings would be very dubious enterprises, even if there were still forces available to attempt them or objectives worth seizing. They are probably not feasible even in the most limited of nuclear wars. A beachhead is a very concentrated target and just a few small nuclear warheads can destroy it.

The United States and the Soviet Union have made considerable preparations for a limited nuclear war at sea. Their navies have several thousand tactical nuclear weapons between them: missiles, bombs, torpedoes, depth charges, etc. Approximately one-third of the United States navy's ships are equipped to carry nuclear weapons, as are most of the Soviet navy's larger ships. A naval war between the superpowers is likely to become a nuclear war because these nuclear-armed ships would be in the thick of the fighting from the outset. The captain of a warship that is about to be destroyed would be very strongly tempted to use his most powerful weapons – regardless of his orders. Another reason why a nuclear war at sea is more likely than one on land is that it would have less severe consequences. "A nuclear explosion does not leave a hole in the water," as naval strategists say. A nuclear war confined to the high seas would not kill large numbers of civilians or cause the devastation it would on land.

In a nuclear war at sea the means of attack would be overwhelmingly superior to any defense against them. A very small ship can launch a nuclear missile and the very largest can be destroyed by a single warhead. A 200-ton missile boat can destroy a 90,000-ton aircraft carrier. Large ships can carry longer-range missiles, and more of them, but they have far less advantage over small ships than in a conventional war at sea. A fleet of many small ships would be far more survivable, and thus more effective, than one of a few large ships. Battlefleets and convoys would sail widely dispersed, each ship several miles away from its nearest neighbors so that hitting one would not sink several others as well. (A 200-kiloton explosion can sink a ship more than a mile away.) Even a numerous, widely dispersed, strongly defended battlefleet might not survive more than a few hours of nuclear war at sea. Every hour that it did survive, it could deal out fearfully heavy blows to enemy ships or targets on land. In this type of war, success would depend on taking the enemy by surprise. The first blow would probably be decisive.

Warships

The most powerful weapons of naval warfare are aircraft carriers. The 15 super-carriers in the United States navy are the largest warships ever built. The most modern of them, the nuclear-powered "Nimitz" class ships, are more than 1,000 feet long and have a full load displacement of more than 90,000 tons (see figure 14.1). (The displacement of a warship is a measure of its weight. It represents the weight of water the ship displaces when floating in it.) Each "Nimitz" carries 89 aircraft: 24 fighter planes to defend against enemy aircraft and missiles, four airborne early-warning airplanes to guide the fighters, 34 attack planes to hit enemy ships and targets on land, eight tanker and ECM airplanes to help the attack planes reach their targets, 18 ASW airplanes and helicopters, and a transport airplane. Whatever a battlefleet has to do, the carrier has aircraft that can do it.

An aircraft carrier is, then, essentially a floating, movable air base. Its mobility confers on it many qualities which an airbase on land does not have. It can move to any part of the world where there is salt water at a steady 700 miles a day. This allows the United States navy to get a carrier within a week or two to almost any place a war is likely to be fought. An air base takes months to build. The carrier does not need anybody's permission to go where it is sent (as long as it stays at least 12 miles offshore) while building an air base requires the permission of the host government. The United States has

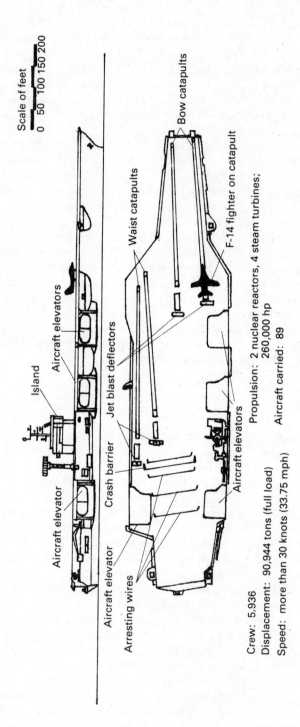

Scale of feet
0 50 100 150 200

Aircraft elevator

Island

Aircraft elevators

Aircraft elevator

Crash barrier

Jet blast deflectors

Arresting wires

Waist catapults

Bow catapults

F-14 fighter on catapult

Aircraft elevators

Crew: 5,936
Displacement: 90,944 tons (full load)
Speed: more than 30 knots (33.75 mph)

Propulsion: 2 nuclear reactors, 4 steam turbines;
260,000 hp
Aircraft carried: 89

The "Nimitz" class super-carriers are the largest warships ever built. Four are now in service and two more under construction. The "angled" flight deck enables the carrier to launch airplanes from the forward portion of the deck and at the same time land other airplanes on the aft portion. When airplanes are being landed on a carrier, the arresting wires are stretched across the deck; an airplane coming in to land trails behind it a hook which catches on an arresting wire and slows it to a stop. Airplanes are launched by steam-powered catapults.

Figure 14.1 *A super-carrier – the United States's Nimitz*

not been given permission to build any bases on the Persian Gulf or in several other areas vital to American interests. Finally, if it becomes necessary to retreat, the carrier simply sails away while an air base must be abandoned to the enemy.

The carrier's greatest defect is sometimes said to be its vulnerability. This is arguable. Carriers certainly can be sunk but air bases can be bombarded, raided, or overrun. During the Vietnam War the United States lost more than 400 aircraft destroyed on the ground while its aircraft carriers were never attacked. The Six Day War provided another illustration of the vulnerability of land-based airpower. On June 5, 1967 the Israeli air force destroyed the entire Egyptian air force on the ground in one morning – and then destroyed the Jordanian and Syrian air forces that afternoon. Carriers are probably more vulnerable than air bases in a nuclear war, but they seem to be less so in many conventional wars.

Carrier-based airpower is very expensive. A "Nimitz" class carrier costs more than two billion dollars. Each carrier must be accompanied by half a dozen other surface warships, several supply ships, and usually by an attack submarine. The other ships are there primarily to protect the carrier. Half of the carrier's aircraft also serve primarily to protect the carrier and its escorts. The entire battle group, costing about 8 billion dollars brings 34 attack aircraft to the scene of the action and enables them to fight. Furthermore, the United States navy is able to keep only one-third of its carrier task forces deployed in the areas where they are most likely to be needed. The other two-thirds are based in the United States, being overhauled and conducting training exercises. (Half of these could be sent into action on short notice if necessary.) So, to have 34 attack aircraft ready to fight in the Indian Ocean or the Mediterranean, the United States navy must maintain 267 aircraft of all types on three aircraft carriers, and about 30 other ships, all at a total cost of 24 billion dollars. This works out to nearly a billion dollars for each forward deployed attack plane. Land-based airpower is less expensive.

Only the United States navy has super-carriers (the Soviet Union will have its first in service by the end of the 1980s), but several navies have small VSTOL carriers. These are ships designed to operate helicopters or VSTOL airplanes, but too small for modern conventional airplanes. The United Kingdom's "Invincible" class ships are typical VSTOL carriers: 677 feet long and with 19,500 tons displacement (see figure 14.2). Each "Invincible" carries five VSTOL fighters and nine ASW helicopters. VSTOL cariers are designed primarily to defend convoys and battlefleets against enemy submarine or air attack. Helicopters are excellent for hunting down and destroying submarines and VSTOL fighters can deal with patrol bombers. But compared with super-carriers, VSTOL carriers are not much use in power projection or winning command of the sea, although they can play those roles to a limited extent.

There are only 30 aircraft carriers in all the world's navies but more than 1,000 other ocean-going surface warships: frigates (1,000–3,000 tons), destroyers (3,000–5,000 tons), cruisers (5,000–25,000 tons), and even a few battleships. Surface warships escort aircraft carriers, merchant ship convoys, and amphibious squadrons, attack other surface ships, and bombard targets

A battleship – the United States's *New Jersey*

Stacks   Bridge   Scale of feet

Aft 16-inch gun turret   Forward 16-inch gun turrets

Crew: 1,518 (65 officers, 1,453 enlisted men)
Displacement: 58,000 tons (full load)
Speed: 35 knots (39.375 mph)

Propulsion: 8 boilers, 4 steam turbines; 212,000 hp
Main armament: 9 16-inch guns
Armor: 17.5-inches (steel)

The "New Jersey" was the first of the World War II "Iowa" class battleships to be returned to service and was in action off the coast of Lebanon in 1983. All four are to be back in service by 1989.

A rocket cruiser – the Soviet Union's *Kirov*

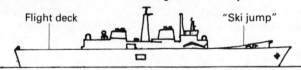

Accommodation for 3 helicopters   Forecastle

Crew: 900
Displacement: 28,000 tons (full load)
Speed: 33 knots (37.125 mph)

Propulsion: 2 nuclear reactors, boilers, and steam turbines; 150,000 hp
Main armament: 20 SS-N-19 cruise missiles

The "Kirov" went into service in 1980, the first of what is to be eventually a class of four ships. The "Kirovs" are the Soviet Union's first nuclear powered surface warships. The SS-N-19 missiles are housed in the forecastle, in tubes slanting forward at a 45-degree angle.

A VSTOL carrier – the United Kingdom's *Ark Royal*

Flight deck   "Ski jump"

Crew: 954
Displacement: 19,812 tons (full load)
Speed: 28 knots (31.5 mph)

Propulsion: 4 gas turbines; 94,000 hp
Aircraft carried: 20 (8 VSTOL airplanes, 12 helicopters)

The "Ark Royal" is a sister-ship of the "Invincible" which fought in the Falkland Islands War in 1982. The "ski-jump" at the end of the flight deck enables VSTOL airplanes to take off with heavier loads than they could otherwise manage.

Figure 14.2 *Surface warships*

on shore. To carry out these missions and to protect themselves they must have weapons and sensors capable of performing four quite different tasks:

1 Anti-air warfare (AAW) – shooting down enemy aircraft and cruise missiles.
2 Anti-submarine warfare (ASW) – locating and destroying enemy submarines.
3 Anti-surface warfare (ASUW) – attacking enemy surface ships.
4 Shore bombardment – attacking targets on land with guns and missiles.

As a general rule each task requires weapons and sensors designed to perform that task alone. ASW weapons pose no threat to aircraft or surface ships, and AAW weapons cannot destroy submarines. Therefore, every surface warship must carry at least a few AAW weapons, and usually some ASW weapons, for self-defense.

To be effective a fleet's AAW defense must be multi-layered; it must force incoming aircraft and missiles to run a gauntlet of layer after layer of defensive weapons. If there is an aircraft carrier in the fleet, the outer layer is the combat air patrol maintained by its fighters. A United States F-14 Tomcat fighter can patrol more than 300 miles from its carrier, detect enemy aircraft 190 miles away, and track 24 targets simultaneously. The second layer of defense is area defense surface-to-air missiles (SAMs). These are large, long-range, shipboard missiles. Some area defense SAMs have ranges as great as 80 miles. These weapons are bulky and very expensive, and their fire control systems even more so. Usually only one or two ships in a fleet carry area defense SAMs, these being specialized AAW ships which use their missiles to protect the entire fleet. The third layer of defense is point defense SAMs – much smaller weapons with ranges of 5 miles or so. Point defense SAMs are small enough and cheap enough so that every ship in a fleet can be given a battery of them to defend itself. Finally, whatever gets past the fighters, the area defense missiles, and the point defense missiles encounters the fourth layer of defense. This consists of multi-barrelled machine cannon, such as the United States 20 mm "Phalanx" weapon, able to fire several hundred rounds in a few seconds. No one layer of defense can stop all of the attackers, but very few of them should get through all four.

The most difficult aspect of ASW is finding the enemy submarine. Sonar is the means to do this. Passive sonar uses hydrophones, such as those in the SOSUS network, to listen for the sounds a submarine makes. Active sonar locates a submarine by sending out sound waves and bouncing them off the submarine back to the sonar operator. Sonar works very well sometimes, particularly at short range, but not always. Sound waves do odd things underwater. For example, there is the "layer depth" problem: where there are two layers of seawater, one colder than the other, sound waves bounce off the boundary between them. A submarine can hide from a sonar-equipped surface ship by keeping below the boundary. There are various ways to compensate for this and the other peculiarities of sound transmission underwater, but the use of sonar is still an art rather than a science. The effective range of even a powerful active sonar is about 30 miles although occasionally it can locate a submarine at a greater distance. (Passive sonar can sometimes detect the presence of a submarine several thousand miles away, but cannot tell its exact location.)

Attack planes, cruise missiles, and heavy guns can all be used as ASUW or shore bombardment weapons. Cruise missiles are probably the best ASUW weapons. Anti-ship cruise missiles can hit targets 300 miles away, 13 times the range of the heaviest guns. They cannot maneuver the way airplanes can but in all other respects they are harder to stop: much smaller, less vulnerable to battle damage at close range, and some of them considerably faster. Unlike airplanes, they can be launched from almost any surface warship, even one as small as 200 tons. They can also be carried by a submarine (and launched underwater), by a patrol bomber, or by an attack plane. The Soviet "Kirov" class rocket cruisers, at 23,000 tons displacement, are the most powerful ASUW warships in the world (see figure 14.2). A "Kirov" carries 20 very large SS-N-19 missiles; one hit from such a missile would probably sink most warships, and the missile has a range of about 300 miles.

Heavy guns, 5-inch and larger, are primarily useful for shore bombardment. The World War II battleships the United States is bringing back into service each carry nine 16-inch guns. These "New Jersey" class ships, twice the size of the "Kirovs," are by far the most powerful shore bombardment ships in the world. Guns do not have nearly as much range as airplanes or cruise missiles – the guns of the "New Jersey" class have a range of 23 miles – but there are many important targets within 23 miles of some coastlines. Battleships are better for attacking those targets than aircraft carriers or rocket cruisers. Their great strength as shore bombardment weapons is the tremendous weight of high explosives they can hurl at an enemy. In a given length of time the guns of a "New Jersey" can deliver ten times as many tons of munitions as a super-carrier's aircraft. A "Kirov" carries only 20 SS-N-19 missiles; a "New Jersey' carries more than 1,000 16-inch shells.

The past 40 years have revolutionized the design and capabilities of the submarine. World War II submarines were essentially surface ships that could submerge for a few hours. A German Type VII U-boat, the most widely-used ocean-going type of the period, could cruise underwater at 4.5 miles an hour for 18 hours. Such boats spent most of their time on the surface. A modern nuclear submarine normally stays submerged for several months at a time and can move 30 or 40 miles an hour underwater. The only real limit on its range is the endurance of its crew. Unlike conventional submarines, it is as fast as most surface warships and thus can attack them effectively. A nuclear attack submarine is also the ideal weapon for hunting down and destroying enemy submarines. It can go wherever they go and its sonar can detect them when sonar installations on surface ships cannot. Nuclear submarines are expensive, though, and at high speed they are vulnerable to detection by enemy passive sonar.

There are still several hundred conventional submarines in service. A conventional submarine has both electric and diesel motors: electric motors to drive the submarine and diesel motors to recharge the batteries from which the electric motors draw their power. A modern conventional submarine can stay submerged for several weeks on end by using its snorkel – a tube that it sticks up above the surface to suck in air. A submarine is very easy to detect, and hence vulnerable, while snorkeling – which it must do to run its diesel motors – but it only has to snorkel for an hour or two a day. The rest of the time it can be far quieter and harder to find than a nuclear submarine. Conventional submarines are not fast enough to attack surface warships or enemy submarines very effectively. They can, however, attack enemy merchantmen. As they are much cheaper than nuclear submarines, it makes sense for a navy that has a commerce–raiding mission to build some conventional submarines for this purpose. The other important mission of conventional boats is coast defense. As they can be built in much smaller sizes than nuclear submarines – as small as 500 tons, while 2,500 tons is the smallest a practical nuclear submarine can be – they are suitable for use in shallow seas and coastal waters where a larger boat cannot operate safely.

Amphibious warfare vessels are essential instruments of naval power projection. They have two tasks to perform; one is to transport marine infantry and their vehicles to the coastal waters where they are to land, and the other is to put them on the beach. Sometimes both tasks are performed by the same ship. The World War II LST (landing ship tank) was an ocean-going ship with water tight doors in its bow. It could be run right up on to a beach and then disgorge a cargo of tanks through the bow doors. Although they could perform both tasks, the LSTs were not ideal for either one. Since World War II the world's navies have concentrated more on building much larger ships, able to cross oceans more comfortably than an LST and to cruise for months on end with marines on board. These large ships do not run up on to the beach themselves. Instead, they carry a number of small landing craft which they can load up, launch, and send in to make the landing. The entire aft portion of an LSD (landing ship dock) is a large docking-well sheltering three to nine landing craft (see figure 14.3). The well is flooded and the gate on the LSD's stern lowered to float them out. Other amphibious warfare vessels carry helicopters instead of landing craft. These ships closely resemble small aircraft carriers and operate 25 troop-carrying helicopters apiece. The latest and largest American amphibious assault vessels carry both helicopters and landing craft; they have a flight deck on top and a docking well underneath it. These are truly massive ships, each as large as a World War II battleship.

The Purposes of United States Naval Power

Defense of many of the United States's vital interests depends on command of the sea. "One U.S. states, several U.S. territories, and 41 of the 43 nations with which we have defense treaties and agreements, all lie overseas."[4] So do all of the neutral countries important enough for the United States to defend or intervene in. So, also, do all the countries which the United States has fought since 1945, threatened to fight, or might some day have to fight:

the Soviet Union, China, Cuba, Nicaragua, North Korea, Vietnam, and Iran, among others. Wherever United States ground forces fight their next war, it will be separated from the United States by a body of salt water. They will have to rely on American seapower to transport them safely across the ocean and perhaps to help them fight their way on to an enemy-held shore. The navy may even be able to win a very small war entirely by its own efforts, without the help of the ground forces at all.

Maritime trade is another vital interest which the United States navy must protect. The United States imports every year, mostly by sea, 26 percent of the oil it uses, 36 percent of its iron ore, 97 percent of its bauxite (used to make aluminum), and 90–100 percent of certain rare and utterly indispensable metals – cobalt, manganese, columbium, and tantalum. There are stockpiles or alternative sources of supply of most of these raw materials but a naval blockade would still do great damage to the United States's economy. The loss of oil imports would be particularly damaging because oil is absolutely essential to an industrial economy and the United States has stockpiled only enough for several months' consumption. It would also be fairly easy to cut off the United States's oil imports. The oil must be shipped in bulk, and much of it must come halfway around the world from a very turbulent area close to the Soviet Union. The United States must also help its major allies protect their maritime trade. Many of them, notably Japan and the United Kingdom, are even more dependent on use of the sea-lanes than is the United States.

If the United States were to lose command of the sea, it could not conceivably move by air the troops and supplies needed to fight. Transport aircraft simply do not have much carrying capacity. The entire United States air force long range transport fleet can move about 50,000 tons of cargo at a time. This is the cargo capacity of *one* large bulk cargo ship. It is enough to move half of a mechanized division in one lift. Sealift is absolutely essential to fight any war that lasts more than a week or two or involves more than a few thousand troops. Ninety percent of the supplies the United States would need to fight a major war in Eurasia must come by sea.

Airlift is also important, particularly in the first few days of a war, to get the most vitally-needed troops and supplies on the ground as fast as possible. The successful United States effort to resupply Israel during the Yom Kippur War (1973) relied totally on airlift. "The first ship that dropped anchor ... contained more outsize cargo than all aircraft combined, but by then the war was over."[5] Nevertheless, airlift is not a substitute for sealift; the United States needs both.

There are many kinds of war the United States navy might be called upon to fight. One of the most important virtues of sea power is its flexibility, the ability of warships to move to and fight in any part of the globe washed by salt water. The various kinds of war the Navy might be sent to fight can, though, be broken down into two broad categories: wars with the Soviet Union, and wars with any lesser power.

In a war with any country other than the Soviet Union, the primary mission of the United States navy is power projection. It bombards targets on land

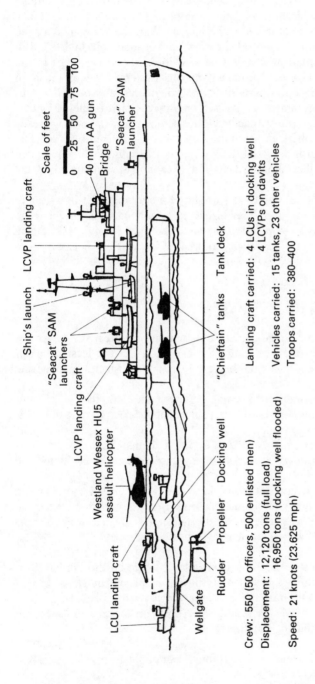

Scale of feet

0   25   50   75   100

40 mm AA gun

Bridge

"Seacat" SAM launcher

LCVP landing craft

Ship's launch

"Seacat" SAM launchers

LCVP landing craft

Westland Wessex HU5 assault helicopter

LCU landing craft

"Chieftain" tanks

Tank deck

Docking well

Propeller

Rudder

Wellgate

Landing craft carried: 4 LCUs in docking well
                        4 LCVPs on davits

Vehicles carried: 15 tanks, 23 other vehicles

Troops carried: 380–400

Crew: 550 (50 officers, 500 enlisted men)

Displacement: 12,120 tons (full load)
              16,950 tons (docking well flooded)

Speed: 21 knots (23.625 mph)

The "Fearless" played a major role in the Falkland Islands War. The ship is shown with the docking well fully flooded, which brings her deep in the water by the stern, and with the wellgate down. An LCU is shown entering the docking well and another is inside. For ocean cruising the wellgate is raised and the docking well pumped dry so that the LCUs rest on the bottom of the well. The ship can carry five helicopters on the deck over the docking well.

Figure 14.3   *An amphibious warfare ship – the United Kingdom's Fearless*

with the guns of its surface ships, sends carrier aircraft to attack targets beyond the range of the guns, and puts ashore and supports United States ground troops. That is what it did in the Korean War, the Vietnam War, the invasion of Grenada, in 1983–4 off the coast of Lebanon, and in 1987–8 in the Persian Gulf. Tomorrow it might be sent to do the same things in the Caribbean against Cuba or Nicaragua, in the Eastern Mediterranean in defense of Israel, or in other parts of the world. In this kind of war it does not have to fight for command of the sea or protect the sea-lanes. None of the lesser powers can seriously challenge the command of the sea it already enjoys nor threaten American shipping. Not even China could do so. The Chinese navy is primarily a coastal defense force. It is a formidable opponent within 200 miles of the Chinese coast but it cannot win command of the Pacific Ocean.

A naval war with the Soviet Union would be a very different proposition. The Soviet Union has a very powerful navy, the only one able to challenge the United States navy's command of the sea. The most difficult tasks facing the United States navy are to prevent, to prepare for, and – if necessary – to wage a naval war with the Soviet Union.

The Soviet Navy

Geography does not encourage the Soviet Union to be a great sea power. Of the major countries in the world, the Soviet Union has the least need for command of the sea and faces the biggest geographic obstacles to achieving it.

The Soviet Union, unlike the United States, does not have salt water between itself and wherever it will probably fight its next war. It does not have to cross an ocean to bring help to its allies. (Cuba and Vietnam are the only significant exceptions.) Nor is the Soviet Union anywhere near as dependent on sea-borne raw materials imports as are the United States and her major allies.

Most of the Soviet Union's coastline faces the Arctic Ocean and is almost totally unusable for military or commercial purposes. This coast is iced-in for all but 3 months of the year. The rest of the coastline is divided by intervening parts of Eurasia into four widely-separated base complexes:

1   In the far northwest – the Kola Peninsula, fronting on the Barents Sea.
2   To the west – Kaliningrad and ports further south on the Baltic Sea.
3   To the southwest – Sevastopol and other ports on the Black Sea.
4   In the Far East – Vladivostok and other ports on the Sea of Japan, the Sea of Okhotsk, and the Pacific Ocean.

Thus, the Soviet Union must maintain four separate fleets, one for each base complex (see map 14.1).

The Northern Fleet, based on the Kola Peninsula, is the most powerful of the four fleets, particularly in ocean-going submarines. It commands all Soviet

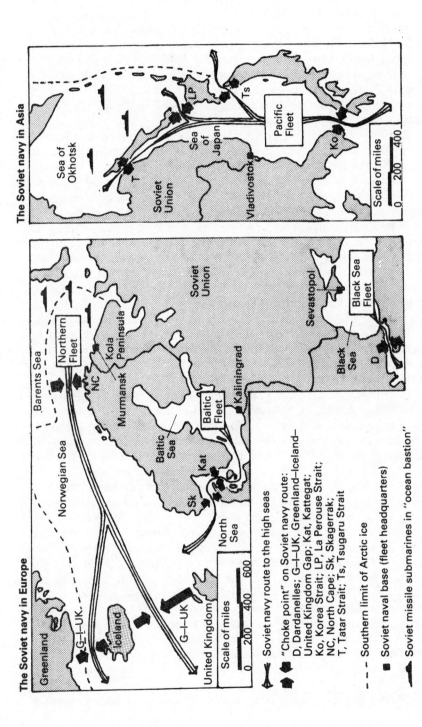

The Soviet navy in Europe

Greenland

G–I–UK.

Barents Sea

Norwegian Sea

Iceland

G-I-UK

North Sea

United Kingdom

Northern Fleet

NC

Murmansk

Kola Peninsula

Soviet Union

Baltic Sea

Sk

Kat

Baltic Fleet

Kaliningrad

Sevastopol

Black Sea

D

Black Sea Fleet

Scale of miles

0   200   400   600

The Soviet navy in Asia

Sea of Okhotsk

T

LP

Ts

Sea of Japan

Vladivostok

Pacific Fleet

Ko

Soviet Union

Scale of miles

0   200   400

⚓ Soviet navy route to the high seas

🢁 "Choke point" on Soviet navy route:
D, Dardanelles; G–I–UK, Greenland–Iceland–United Kingdom Gap; Kat, Kattegat; Ko, Korea Strait; LP, La Perouse Strait; NC, North Cape; Sk, Skagerrak; T, Tatar Strait; Ts, Tsugaru Strait

– – – Southern limit of Arctic ice

■ Soviet naval base (fleet headquarters)

⚓ Soviet missile submarines in "ocean bastion"

Map 14.1 *Geography and Soviet sea power*

warships in the Barents Sea, the Norwegian Sea, and the Atlantic Ocean, as well as Soviet submarines in the Mediterranean. The Baltic Fleet is primarily a coastal defense force, configured to support Soviet ground forces operations along the shores of the Baltic. It may even be under the operational command of the ground forces rather than the navy. The Black Sea Fleet also controls Soviet surface warships in the Mediterranean. The Pacific Fleet, in the Far East, is nearly as powerful as the Northern Fleet and is especially strong in surface warships. It commands all Soviet warships in the Pacific Ocean and the small squadron in the Indian Ocean.

Bringing any two of these scattered fleets together has always been a daunting task, even in circumstances in which the waters between them are not controlled by an enemy. In October 1904 the Baltic Fleet was sent to rescue the Pacific Fleet which had been driven into harbor by the Japanese navy. Seven months and 18,000 miles later, it arrived off the coast of Japan, having recoaled at sea more than 30 times, put down several mutinies, and nearly started wars with France and Great Britain. Utterly exhausted by its epic voyage, the Baltic Sea Fleet was very quickly sunk at the Battle of Tsushima. The fleet it had been sent to rescue had already been destroyed. Similar efforts to concentrate the Russian navy have been more successful than the "Voyage of the Damned," but even today the Soviets would find it difficult to reinforce one of their fleets in wartime.

Each of these base complexes is virtually an inland sea. The naval bases on the Kola Peninsula are ice-free year-round but the Arctic ice pack extends to within several hundred miles of the coastline. Ships using these bases must sortie through a narrow corridor between the ice and the coast of Norway, and then through the G–I–UK (Greenland–Iceland–United Kingdom) gap to reach the Atlantic Ocean. Ships in the Baltic Sea must go through the Great Belt, the Skaggerak and the Kattegat to get out of the Baltic – a passage that is only 6 miles across at one point. Black Sea Fleet ships must get through the Dardanelles, less than 2 miles wide, to reach the Mediterranean. Most of the ships in the Pacific Fleet must use the Korea Strait or the La Perouse Strait to reach the Pacific Ocean. All of these passages are adjacent to or even controlled by allies of the United States. So, before the Soviet navy can even begin to fight for command of the high seas, it must break through some serious geographic obstacles to reach them.

Fortunately for the Soviets, their base complexes are nearly as inaccessible to rival sea powers as the high seas are to the Soviet navy. An enemy fleet attacking the Kola Peninsula must sail up the same narrow corridor that the Soviet ships based there must use. The Baltic Sea and the Black Sea are virtually Soviet lakes. Most of the land around them belongs to the Soviet Union and its allies and on neither sea is there a sea power strong enough to challenge Soviet command of that sea. In the Far East, the Sea of Okhotsk is almost completely surrounded by Soviet-held territory. The same geographic features that make it difficult for the Soviet navy to seize command of the high seas also considerably reduce the effectiveness of sea power directed against the Soviet Union.

Defying the dictates of geography, the Soviets have created for themselves a powerful navy – rather more powerful than seems reasonable to their opponents. The Soviet navy is the second most powerful in the world. It actually has more ships than the United States navy. (In manpower, total displacement of all ships, and overall capability – better measurements of naval strength than numbers of ships – the United States navy is still the strongest in the world.) The Soviets have built up this very powerful force partly because they feel that as a superpower the Soviet Union should have a powerful navy; without it they could not claim equality with the United States. They also have some very important missions for their navy in war.

In a nuclear total war with the United States, the first priority task of the Soviet navy would be to protect its ballistic missile submarines. Soviet military planners expect a nuclear total war to last for several weeks or months or even longer. They plan to hold their missile submarines in reserve, the SLBMs unlaunched, to deal the final, decisive blows of that war. Missile submarines can be held in reserve better than ICBMs or bombers because of their great survivability, but in a protracted nuclear war even they would need protection. The Soviets intend to protect their missile submarines by concentrating them in two easily guarded "ocean bastions": the Barents Sea and the Sea of Okhotsk. These waters are difficult for enemy ASW forces to enter, are protected by the Soviet navy's two strongest fleets, and are close to powerful land-based Soviet air forces. The Soviet Union's "Delta" and "Typhoon" class missile submarines carry SLBMs with enough range to hit the United States from the Barents Sea or from the Sea of Okhotsk. These boats already carry most of the striking power of the Soviet missile submarine force and are rapidly replacing the "Yankee" class boats armed with shorter range missiles.

Second in importance to protecting the Soviet Union's missile submarines would be destroying those of the United States. The Soviet navy has devoted considerable resources to anti-submarine warfare. Its six VSTOL carriers and helicopter carriers are specialized ASW ships and most of its cruisers and destroyers are armed primarily with ASW weapons. It also has a large fleet of nuclear-powered attack submarines. Many of these would be held back to protect the ocean bastions against United States attack submarines but others, particularly the carriers, would be sent out on the high seas to destroy United States missile submarines. The Soviet navy is not yet able to neutralize the American missile submarine force. However, as the discussion of Soviet nuclear strategy in chapter 8 showed, that is one of its missions. If and when the technology required to hunt down missile submarines becomes available, the Soviets will exploit it to the maximum.

In a conventional or a limited nuclear war, the primary mission of the Soviet navy would be to win command of the sea. This requires destroying United States surface ships, particularly aircraft carriers and their escorts. The Soviets have nearly 70 cruise missile submarines, 30 ASUW cruisers and destroyers, and several hundred patrol bombers. Soviet surface ships and cruise missile submarines make a practice of trailing United States carrier task forces. Wherever there is a forward-deployed United States task force, there is likely also to be a Soviet squadron lurking nearby, within missile range. When the order comes to start the war, every Soviet surface ship, cruise missile submarine, and patrol bomber within range would fire off its cruise missiles as nearly simultaneously as possible. A few minutes later the United States task force would face a hundred or so cruise missiles coming at it over the horizon. It the task force survived the first 15 minutes of this "D-Day shootout" it would probably be master of the situation. Soviet warships are designed to put all their firepower into that first, overwhelming salvo and carry few or no reloads for their missile launchers. It would be an exciting 15 minutes.

Denying the United States use of the sea-lanes to Europe, Japan, or the Persian Gulf is a distinctly lower-priority mission. It would have little impact on a nuclear war or a short conventional conflict, but might decide the outcome of a protracted conventional war. If the Soviet navy wins the "D-Day shoutout," and thus command of the seas, it can close the sea-lanes without much further ado. If not, the Soviets are prepared to mount a serious sea-denial campaign. They have 124 medium range conventional submarines suitable for attacking merchantmen and troopships. Most of these were built several decades ago when the commerce-raiding mission seemed relatively more important to Soviet planners than it does today, but the Soviets are still building conventional submarines. Any cruise missile submarines or surface ships that survive an unsuccessful "D-Day shootout" could also be used for commerce-raiding, as could land-based patrol bombers. Another way to halt seaborne trade is to destroy the ports it must use. The Soviets probably would send bombers and missiles against every port in any part of Eurasia they invade. These attacks could close the sea-lanes very effectively without any help from the Soviet navy.

Another of the Soviet navy's missions is power projection. Each of the four fleets has a small but highly-trained marine infantry force and some landing craft. Each fleet also has a number of older, gun-armed warships suitable for shore bombardment and the support of amphibious landings. These forces could be used to support the advance of Soviet land armies along the coast. They could also help the Soviet navy overcome the geographic barriers that keep it back from the high seas. Amphibious forces with the Northern Fleet might seize Iceland or northern Norway to make it easier for the main fleet units to operate in the North Atlantic. Baltic Fleet and Black Sea Fleet forces could be sent to force open the Skaggerak and the Dardanelles. Amphibious forces with the Pacific Fleet might take control of the Korea and La Perouse Straits. They could also raid the Chinese coast in a war with China.

United States Naval Strategy and Naval Policy

In a war with the Soviet Union, nuclear or conventional, the United States could not take command of the sea or free use of the sea-lanes for granted. At best the United States navy would face a hard fight at the beginning of the war to seize command of the sea. Even after (and if) the Soviet navy had been driven from the surface, it could still wage a formidable sea-denial campaign with its submarines and land-based aircraft. The United States navy might have to fight convoys through to Western Europe, the Persian Gulf, or Japan without first having won command of the sea. These would be desperate, costly battles against Soviet bombers, surface ships, and submarines – almost every naval weapon the Soviet Union possesses.

The navy could take a hand in the fighting on land. In a war in Europe, amphibious squadrons and carrier task forces would probaby be used on the northern and southern flanks of the NATO front. Elements of the Atlantic Fleet might be sent into the Norwegian Sea to counter Soviet amphibious assaults on Iceland or northern Norway. The Mediterranean fleet might be sent into the eastern Mediterranean to help close the Dardanelles to Soviet ships and to defend Greece, Turkey, and perhaps Israel. United States naval forces in the Indian Ocean might be sent into the Persian Gulf if the Soviet Union were to invade Iran or Pakistan. If Japan were attacked the Pacific Fleet might be sent to operate in Japanese waters. All of these would be high-risk operations. They would bring United States surface ships within range of powerful Soviet land-based bomber forces. United States warships venturing into the Norwegian Sea or Japanese waters would also be exposed to attack by the most powerful surface ship and submarine forces the Soviets have. (The United States Sixth Fleet in the Mediterranean would face only the Soviet Union's Mediterranean squadron, a much smaller force.) Despite all risks, these power projection operations would have to be undertaken. They would be the most feasible way, perhaps the only way, for the United States quickly to deploy substantial forces in those theaters of war.

Perhaps the United States navy could do more, particularly if it had more ships. Under the leadership of John F. Lehman, Jr., Secretary of the Navy in 1981–7, navy strategists formulated a "maritime strategy" that enjoins it to do so. This maritime strategy proposes that the United States navy respond to a Soviet invasion of Western Europe by striking directly at the Soviet mainland. This is to be done at two widely-separated points: the Kola Peninsula in the extreme northwest, and the Soviet coastline in the Far East. (Hardly anyone proposes sending United States fleets into the Black Sea or the Baltic.) In Europe, at the outbreak of the war, American attack submarines are to sweep across the Norwegian Sea and on into the Barents Sea, intercepting and destroying Soviet submarines attempting to get out and attack NATO shipping. After these seas have been cleared of Soviet submarines, powerful American carrier task forces are to sail in and attack Soviet bases on the Kola Peninsula. Then the United States Marines land and seize beachheads on Soviet territory. United States naval and amphibious operations in the Far East are to follow a roughly similar pattern. Navy strategists hope that these attacks will achieve three objectives:

1 Force the Soviet navy to come out and fight, leading to its destruction and enabling the United States navy quickly to seize command of the sea.
2 Weaken the Soviet threat to Western Europe by forcing the Soviets to divert powerful tactical air and ground forces to defending the Kola Peninsula and the Far East.

3   Impel the Soviets to end the war by destroying their missile submarines.

Adoption of the maritime strategy has required, and served to justify, greatly strengthening the United States navy. Lehman set as his goal "a 600-ship navy" – up from the 479 ships in service in January 1981 – and had nearly achieved his goal by the time he left office in 1987. This has probably been an unwise use of American resources. The maritime strategy imposes unacceptably heavy risks on the United States surface fleet, does not help greatly to deter or defeat a Soviet invasion of Western Europe, and may lead to a nuclear total war.

United States carrier battle groups sent to attack the Kola Peninsula or the Soviet Far East would very likely not come back:

> By the time the carriers were within 1,600 miles of Soviet air bases, they would be within range of over 90 percent of the U.S.S.R.'s land-based bombers. Yet, the Soviet bases would still be over 1,000 miles beyond the range of carrier aircraft ... the carrier force would be subject to Soviet air bombardment for nearly two days before it was close enough to strike Soviet bases. The force would also be subject to attack by submarines and surface ships with long-range missiles that would have been deployed along the route. In short, we would be fighting the Soviets on their turf at times and places of their choosing, well before we could assume the offensive.[6]

These battles in the Barents Sea and the Sea of Japan would decide command of the sea, and fighting the decisive battles there gives the Soviet navy every advantage. The United States task forces would have to fight the entire Northern and Pacific Fleets: surface ships and attack submarines held back to protect Soviet missile submarines, all the Soviet warships that would otherwise have been sent out on the high seas, swarms of coastal light craft, and land-based bombers. If, on the other hand, the decisive battles were fought in the North Atlantic and southeast of Japan, the Soviets could use only those ocean-going ships they do not assign to protecting their missile submarines. They could not use their "mosquito fleet" or most of their land-based aircraft, and American or allied land-based aircraft could take a hand in the fight. It seems much more sensible to fight the Soviet navy on the high seas than to thrust one's head into its lair.

Even if the carriers *can* fight their way through Soviet defenses, they cannot do enough damage seriously to affect a full-scale land war. Not even the United States navy is powerful enough to threaten Soviet land forces. The six aircraft carriers the United States has in the Pacific carry slightly more than 500 aircraft; the Soviet Union has about 1,500 aircraft in the Far East. The United States has two Marine divisions in the Pacific (and enough amphibious warfare vessels to carry one of them); the Soviet Union has 27 divisions (seven of them Category I or II) in its Far Eastern Military District. The ratio of forces in an attack on the Kola Peninsula would be about the same. United States power projection efforts against these two great fortresses can be contained quite adequately by the forces already there; the Soviets would not have to divert any troops from conquering Europe.

If American attack submarines are sent to destroy Soviet submarines off the coast of the Soviet Union, they will undoubtedly sink some Soviet missile submarines. The American boats would be operating where Soviet missile submarines are concentrated, in the Barents Sea and the Sea of Okhotsk, and missile submarines cannot reliably be distinguished from other types. Advocates of the maritime strategy suggest that this will persuade the Soviets to end the war before such an important part of their strategic nuclear deterrent is destroyed. It may do so, but it also might persuade them to launch their SLBMs – and their other strategic nuclear weapons – rather than see those weapons destroyed.

Command of the sea cannot in itself defeat a Soviet overland invasion. The Soviets can invade Europe, overrun the entire Middle East, or conquer China without sending a single ship to sea. If the United States ever goes to war with the Soviet Union, it will not win that war at sea.

On the other hand, the United States could lose the war at sea. Without free use of the sea-lanes it cannot do much to help any of its Eurasian allies. The United States must have command of the sea so that its land armies and air forces can fight where they are needed. Its seapower must supplement, though not substitute for, its land power.

United States naval policy consists primarily of deciding what weapons the United States navy is to have. Military planners cannot know where or how the navy will fight, so they must see that it is equipped to meet a variety of contingencies. They can be sure, though, that aircraft carriers will play a major role in any American war at sea. "The super carrier is the heart and soul of the U.S. surface fleet."[7] If a war must be fought in the Third World, carrier-based aircraft are the best power projection weapons the United States has. Most of the countries in which a small war could happen are within flying range of waters in which aircraft carriers can sail. Carriers are the only way to bring United States tactical air power quickly to bear in almost all of the Third World – wherever the United States does not have air bases already built. Aircraft carriers must also play a major role in any conventional war with the Soviet Union. They would be vital to any effort to defend the Persian Gulf, northern Norway, or the Dardanelles. Perhaps a third of them might be sunk in the "D-Day shootout," but the others should survive to carry out their missions – if they are not sent to attack the Kola Peninsula. Even in a nuclear total war, a carrier which survives long enough to launch its aircraft can do as much damage as a missile submarine.

The United States navy has 99 cruisers and destroyers, primarily to escort and protect its super-carriers. The carriers need their escorts even in a war with a Third World country. American warships probably will not in the future be able to cruise off enemy-held coasts with the impunity they enjoyed during the Korean and Vietnam Wars. A number of Third World countries can now mount powerful ASUW attacks, as the Argentinian air force showed during the Falkland Islands War. Those attacks can be defeated, but only by a strong, well-coordinated fleet defense. Carrier-borne fighter planes can help defend a fleet, but it needs also SAM-armed surface ships. In a war with the Soviet Union, the carriers need even more to have cruisers and destroyers protecting them.

The cruisers and destroyers can also supplement the carriers' offensive capabilities with the "Harpoon" cruise missiles most of them now carry. It is no longer possible, as it once was, to wipe out almost the entire offensive capability of a United States task force by sinking its aircraft carrier.

Other types of naval weapon are important only in some naval wars, not in others. The United States navy's attack submarines would play several vital roles in a war with the Soviet Union. They would be absolutely essential both to protect American missile submarines and other ships from attack by Soviet submarines by sinking those Soviet submarines, and to attack Soviet surface warships. They might be even more important in a nuclear war than in a conventional war. On the other hand, it is hard to envisage them participating effectively in a war against any of the lesser naval powers (except perhaps China). Only one American submarine was used in either the Korean or the Vietnam War, and that was in a very minor role, landing commandos behind enemy lines.

The United States navy has 97 attack submarines, all but four of them nuclear-powered. It has not built a conventional submarine since 1959. That is the right policy, even though conventional submarines play useful roles in other navies. The United States does not need submarines for coast defense or commerce-raiding, the two missions conventional submarines can carry out well. All of the missions given to American submarines require high performance nuclear-powered boats.

Amphibious warfare vessels are most useful for wars with countries other than the Soviet Union. They would have a very hard time putting a landing force ashore on Soviet territory, but have shown many times their ability to do so in the Third World.

American battleships are also primarily power projection weapons for use in the Third World. Each carries cruise missiles as well as guns, so they can participate to good purpose in ship versus ship duels, but most of their firepower is in their heavy guns. They are more limited power projection and ASUW instruments than aircraft carriers but also much less expensive. The cost of restoring a "New Jersey" class ship to service is about one-fifth the cost of building a super-carrier. Their usefulness in certain scenarios justifies the comparatively modest cost of putting them into service.

Notes

1 Geoffrey Till et al., *Maritime Strategy and the Nuclear Age* (New York: St Martin's Press, 1982), p. 40.
2 Admiral Sir Reginald H. Bacon, *The Life of John Rushworth Earl Jellicoe* (London: Cassell, 1936), p. 349.
3 Quoted in Bernard Brodie, *A Guide to Naval Strategy* (Princeton, NJ: Princeton University Press, 1958; reprint edn, Westport, Conn.: Greenwood Press, 1977), p. 155.
4 James L. George (ed.), *Problems of Sea Power as We Approach the Twenty-First Century*

(Washington, DC: American Enterprise Institute for Public Policy Research, 1978), p. 326.

5 John M. Collins, *U.S.–Soviet Military Balance: Concepts and Capabilities, 1960–1980* (New York: McGraw-Hill, 1980), p. 271.

6 Admiral Stansfield Turner and Captain George Thibault, "Preparing for the Unexpected: The Need for a New Military Strategy", *Foreign Affairs*, 61 (Fall 1982), pp. 126–7.

7 Sherwood S. Cordier, *U.S. Military Power and Rapid Deployment Requirements in the 1980s* (Boulder, Col.: Westview Press, 1983), p. 110.

15
Light Infantry for Power Projection

The British Army should be a projectile to be fired by the British Navy.
Edward Grey, Viscount Falloden (attributed)

There is a limit to how much power projection can be achieved by shore bombardment or carrier-based aircraft. Winning more than a very small war requires ground forces. These ground forces may have to travel and fight several thousand miles away from their permanent bases; they may also need to land on enemy-held territory. Such power projection missions require either airlift or sealift to get the ground forces where they are needed. The ground forces themselves are usually specialized power projection units: lightly-armed infantry which are easy to transport and have undergone very rigorous training for their distinctive and very demanding roles. The United States and the Soviet Union have by far the largest power projection ground forces in the world, but France, the United Kingdom, and several other countries also have significant specialist forces of this type.

It is difficult to transport even lightly-armed troops to a distant battlefield and support them in combat. Sealift is expensive and cannot transport anything very rapidly. A cargo ship takes at least 3 weeks to reach the Persian Gulf from the east coast of the United States. Airlift is fast but is far more expensive and has much less carrying capacity than sealift. The Lockheed C-5 transport plane required to carry one main battle tank costs 200 million dollars – one hundred times as much as the tank itself. The cost and difficulty of long-range power projection compel military planners to make some hard choices. Not even a superpower can send a large, heavily-armed force quickly to a distant theater of war. It can send a few lightly-armed troops in several days, and a larger, more heavily-armed force in several weeks, but sending an entire mechanized army overseas requires months.

There are several ways to land ground forces on enemy-held territory. They can parachute out of airplanes or be put down by helicopters behind enemy lines; they can also make an amphibious landing on an enemy-held coastline. These are complex, demanding operations even when unopposed; against serious opposition they are also hazardous and sometimes very costly.

Airborne Troops

Airborne forces are power projection light infantry which can be carried to the battlefield by transport aircraft and dropped with their equipment behind enemy lines by parachute. Airborne units can respond very quickly to emergencies and have great strategic mobility. The United States army's 82nd Airborne Division always has a company ready to move out at 2 hours' notice and a brigade ready to move in 24 hours. It regularly practices flying across the Atlantic Ocean and dropping paratroopers in West Germany. Formations such as the 82nd Airborne are ideal for getting a small force into action very quickly. Whenever a battalion on the spot within a few hours can accomplish more than a division there a week later, an airborne unit is the most useful.

Airborne forces pay a high price for their strategic mobility. They are the most lightly-armed type of ground forces. No airborne division has ever possessed a main battle tank; air-dropping a 50-ton weapon is simply not practical. Airborne divisions usually do have a few light tanks, but nothing that could stand up to a main battle tank. They also tend not to be very mobile once they hit the ground. The 82nd Airborne and other Free World airborne units have nothing but foot infantry; they move only as fast as a man can walk. Soviet airborne divisions each have a regiment equipped with ultra-light APCs, but the other two infantry regiments in a division walk. Also, airborne units cannot fight behind enemy lines for very long. The paratroopers carry with them enough supplies to fight for two or three days. After that they must be resupplied by air, always a difficult and uncertain undertaking. Airborne operations are usually planned on the assumption that other forces will be able to link up with the airborne forces after a few days.

Airborne units compensate to some degree for the limitations of their equipment by the high quality of their men. In any army, paratroopers are an elite – among the toughest, most daring, best trained soldiers available. Max Hastings describes the British Parachute Regiment as "a unit which, when it is needed, is needed very badly indeed: to attempt the impossible."[1] During the Falkland Islands War the British 2nd Parachute Battalion attacked and routed a well-entrenched Argentinian force of four times its strength. There is a limit, though, to how much courage and training can substitute for heavy weapons. In September 1944 at Arnhem another British airborne unit, the 1st Airborne Division, was annihilated in a battle with an SS Panzer division. Even very good soldiers, armed with submachine guns, could not prevail against equally good soldiers armed with Tiger tanks.

Dropping several thousand soldiers with all their equipment behind enemy lines has never been easy. World War II parachute drops were almost always costly and some of them were disasters. Transport planes – inherently large, slow and unhandy – are very vulnerable to anti-aircraft fire and enemy fighters. Also, even a well-executed mass parachute drop inevitably leaves the paratroopers scattered over a rather large drop zone, badly disorganized, and very vulnerable. An enemy mobile force alert enough to counterattack before the paratroopers sort themselves out and prepare to fight can cause havoc.

No other military operation, not even an amphibious landing, can fail as quickly and totally as a mass parachute drop. Airborne units must be employed skillfully if they are not to fail disastrously. Successful airborne operations require an astute combination of decisiveness and caution: decisiveness to use airborne units when the opportunity arises, and caution to avoid sending them on a mission beyond their strength.

Helicopter-borne units, such as the United States 101st Air Assault Division or the ten Soviet air assault brigades, are similar to airborne units in many ways. They too are lightly-armed elite units able to land behind enemy lines. They can react to an emergency as quickly as airborne units and project power very effectively to a distance of several hundred miles. However, they are not useful for long range power projection because helicopters do not have much range. A typical assault helicopter has a range of 200–300 miles – one-tenth that of a typical transport airplane.

Air-portable units are yet another form of power projection light infantry. Their mission is rapidly to reinforce friendly forces anywhere in the world. The United States army now has four air-portable light divisions. These units are relatively easy to airlift and can travel as far and as fast as an airborne division but are not expected to land on enemy-held territory. Transport airplanes carry them to landing fields on friendly territory. These light divisions do not have the main battle tanks and other heavy equipment of the armored and mechanized divisions, but they are fully motorized and very mobile once landed.

Airborne, helicopter-borne, and air-portable units can all come to the aid of a threatened ally much more quickly than any other type of ground force. Airborne and helicopter-borne units are also ideal for a *coup de main*. The Soviet invasions of Czechoslovakia (August 1968) and Afghanistan (December 1979) each began with an airborne division landing on the airport of the victim's capital city. In each case, the airborne division was able to seize the capital city and prevent any attempt to resist long before other Soviet forces could have arrived on the scene.

Airborne, helicopter-borne, and air-portable units are no more than a small part of any country's armed forces. Even the Soviet Union's seven airborne divisions (more than all the rest of the world put together) are only one-thirtieth of the divisions in the Soviet ground forces. Transport aircraft and helicopters are too expensive for even a superpower to airlift very large ground forces.

Marines

Marines travel by ship and land on beaches. In all other respects they are very much like airborne units and other power projection light infantry. Like paratroopers, they are lightly-armed but highly-trained elite troops used to conduct extremely demanding and hazardous military operations. Again like the paratroopers, they are prepared to travel long distances to the battlefield and then land on enemy-held territory. Marine forces resemble airborne forces in one more respect; they tend to be rather small. The United States has by far the most powerful marine force in the world: three large divisions and a powerful tactical air force. Half the marines in the world belong to the United States Marine Corps. Even so, the Marine Corps is a quarter the size of the United States Army.

The differences between marine infantry and airborne infantry derive mainly from the differences between ships and airplanes. Because ships have much more cargo capacity than airplanes, marine infantry can be more heavily armed. Both the United States Marine Corps and the Soviet naval infantry have fairly large numbers of main battle tanks. They also have amphibious APCs, vehicles that can be launched from amphibious warfare vessels offshore, carry marine infantry to the beach, and then move inland. Ships are also much slower than airplanes, so it takes the marines much longer to cover the same distance. The United States can fly an airborne division to the Persian Gulf in 14 days but it would need twice as long to ship a marine division there. Amphibious forces, though, can stay at sea for months at a time near where they might be needed, and thus be available for rapid deployment. The United States has two Marine Amphibious Units at sea all the time, each embarked on a squadron of five or six amphibious warfare vessels. (A Marine Amphibious Unit consists of an infantry battalion, reinforced by tank, artillery, helicopter, and supply detachments. It has a strength of 1,800 Marines.) One of these units is currently deployed in the Mediterranean and another alternates between the Pacific Ocean and the Indian Ocean. When United States troops went into Lebanon in 1958, into the Dominican Republic in 1965, and into Grenada in 1983, these small marine units were on the scene before any airborne units were. But to get more than 1,800 troops quickly to a trouble spot, or to get them anywhere there is not a Marine Amphibious Unit handy, one must call on the airborne.

The first challenge to any amphibious landing is getting the invasion fleet to the area, through the shoals and minefields it is likely to encounter. (The beaches at Inchon, during the Korean War, could be approached only by a narrow, twisting channel 30 miles long; the sinking of a single ship in that channel could have bottled up the entire fleet.) Once the marines are in their landing craft, they face a very hazardous run in to the beach. "There are no foxholes in the surf." The landing craft are slow-moving while in the water and are lightly armored. They must creep in under enemy fire all the way. Once ashore, the landing force must move inland as far and as fast as possible, despite the relatively small number of vehicles it has and regardless of casualties. Crowded on to the invasion beach it makes an ideal target for enemy air strikes and artillery and is very vulnerable to a counterattack because it has nowhere it can retreat to.

Something must be done to even the odds. The accepted rule of thumb is that the landing force must be at least three times as strong as the defenders. If at all possible, the invader must use sea power's mobility and flexibility to strike the enemy where he least expects it and is least prepared to meet it. "The preferred business of an amphibious force is not to land where the enemy is, but to land where he is not."[2] Another expedient, on which the

United States Marines rely heavily, is "vertical envelopment." Many of the Marines used in an American amphibious landing are flown in by helicopter. They land some distance behind the shore defenses, attack them from the rear, and block enemy reinforcements moving towards the landing area. This may be a very costly operation because helicopters are quite vulnerable to anti-aircraft fire, but without it the landing might fail entirely.

The invasion fleet must achieve clear air superiority and command of the sea where the landing is to take place. It must carry out a heavy pre-invasion bombardment of the shore defenses; this is the primary role of its heavy guns and an important role of its carrier-based aircraft. It must isolate the battlefield, prevent the enemy bringing up reinforcements. It must supply the landing force "over the beach" without the use of permanent docking and unloading facilities on shore. (In most invasions those facilities either never existed or were destroyed in the pre-invasion bombardment.) Finally, in most major amphibious operations, the invasion force must be able to bring up powerful reinforcements to exploit the bridgehead that has been won. Otherwise, an alert, mobile defender can pen the invader into his bridgehead and frustrate even a successful landing, as the Germans did at Anzio in 1944. The later stages of an amphibious landing are essentially a race between the defender and the invader, each trying to concentrate stronger forces in the area of the landing.

There probably will be amphibious assaults in future wars, as there were in Grenada in 1983, the Falkland Islands in 1982, and Cyprus in 1974. The amphibious capability of marine forces and their ability to supply themselves over the beach permit them also to carry out other important power projection missions. They can be sent to the aid of any country which has a sea coast, even if that country does not have available the ports and airfields other types of ground forces would need. Marines are ideal for combat in remote, inhospitable areas with few roads or ports, such as Iceland or northern Norway. The presence of a Marine Amphibious Unit offshore has strengthened the United States government's hand in more than one crisis. And, if worst comes to worst, the Marines can evacuate United States citizens from a country where their lives are in danger. Three-quarters of all the Americans living abroad are close enough to a sea coast for the Marines to be able to get them out.

Notes

1 Max Hastings and Simon Jenkins, *The Battle for the Falklands* (New York: W.W. Norton, 1983), p. 233.
2 Frank Uhlig, Jr, "Amphibious, Mine, and Auxiliary Forces", in *The U.S. Navy: The View from the Mid-1980s*, ed. James L. George (Boulder, C: Westview Press, 1985), p. 127.

PART VI
Guerrilla War

The Arab army ... had won a province when the civilians in it had been taught to die for the ideal of freedom; the presence or absence of the enemy was a secondary matter.
Thomas Edward Lawrence [Lawrence of Arabia], "The Lessons of Arabia"

Political power grows out of the barrel of a gun.
Mao Tse-tung, "Problems of War and Strategy"

16
Revolutionary Guerrilla War

All modern revolutions have ended in a reinforcement of the power of the State.

Albert Camus, *The Rebel*

The Causes of Guerrilla Wars

Guerrilla war is "war in the shadows." It is waged by small, lightly-armed detachments living off the land. Guerrillas conceal themselves in difficult terrain and among the civilian population and avoid combat unless they have an overwhelming advantage.

> The guerrillas must move with the fluidity of water and the ease of the blowing wind. Their tactics must deceive, tempt, and confuse the enemy. they must lead the enemy to believe that they will attack him from the east and north, and they must then strike him from the west and south. They must strike, then rapidly disperse. They must move at night.[1]

. . . otherwise, they do not live very long.

Guerrillas must fight this way because they are much weaker than their opponents. Guerrilla war is the way the weak fight the strong. It is the weapon used by political forces that do not control a sovereign state against those that do: the people of a conquered territory against an invader, the people of a colony against a colonial regime, members of a minority nationality against the government of a country they wish to secede from, and revolutionaries against a government they wish to overthrow. No political movement, however much popular support it has, possesses the same means of raising money, obtaining weapons, and recruiting soldiers that a sovereign state does. When it takes up arms against the state it must wage guerrilla war until it is strong enough to fight a conventional war.

In the late twentieth century the most important of the various types of guerrilla war has been the "revolutionary guerrilla war." This is a war waged by a political movement within a country to overthrow that country's government. It is as much or more an instrument of politics as any other kind of war, but of a different kind of politics: domestic politics rather than foreign policy. A revolutionary guerrilla war is a civil war.

There are other ways to overthrow a government, notably the *coup d'etat* and the mass uprising. A *coup d'etat* is the way that an army seizes power. A mass uprising is the way that the people of the capital city (usually) and other cities overthrow an unpopular regime. Neither of these forms of violence commonly rises to the dignity of a war. Neither in a *coup* nor in an uprising do two military forces fight each other for more than a few days. Instead, the government either keeps control of its army and very quickly suppresses the threat, or it loses control and is overthrown.

Coups and mass uprisings are quick, urban struggles. A guerrilla war of any type is a prolonged, rural war. The guerrillas begin as a small band of fugitives and, if successful, they become through years of struggle a powerful army exercising authority over a large population. This can only happen in a rural area. The countryside shelters the guerrillas while they are weak and provides them with the resources they need to become strong. No primarily urban guerrilla insurgency has ever yet overthrown a government, although urban guerrilla warfare has been attempted from time to time. Almost all guerrilla insurgencies, and all the successful ones, have originated in rural areas.

Revolutionary guerrilla wars usually happen in Third World countries. Most of the Third World is still far less urbanized than Europe or North America and thus provides a more favorable environment for the conduct of guerrilla war. The grievances that inspire violent revolutions are stronger in the Third World than elsewhere. Finally, Third World governments tend to be weaker and easier to topple than either the liberal democracies of North America and Western Europe or the Communist-ruled dictatorships.

Revolutionary guerrilla wars are, then, peasant wars. Based as they are in the rural areas of Third World countries, they must have the support of the peasant farmers, who constitute the vast majority of the population in those areas. Even if peasants do not provide the guerrilla army's leadership – and they seldom do – they must provide most of its manpower and sustenance. They must also keep the guerrillas informed about what the government forces are doing and deny information about the guerrillas to the government. The peasantry are, in Mao Tse-tung's famous analogy, the element within which the guerrillas must live as fish swim in the water. A guerrilla army can no more survive among a hostile peasantry than fish can live in poisoned water.

Peasants can be persuaded to support a guerrilla insurgency. In most Third World countries there is considerable latent, and even overt, hostility towards the government. Almost everybody in the Third World lives in poverty. In some countries the average per capita income is less than 200 dollars a year, and in most it is less than 1,000 dollars. What wealth there is in the Third World is controlled by a small part of the population; income distribution is far more uneven than in the industrialized countries. It is also very much concentrated in the cities. Rural incomes are on average less than half as high as urban incomes. Land, the basic source of wealth in the countryside, is also divided up very unevenly. Hundreds of millions of Third World peasants have less land than they need to live on or none at all. They must rent land, commonly paying half of what they grow to the landowner, or hire themselves out as laborers. At the same time much of the land, particularly in Latin America, is held in huge estates, often by absentee landlords who make little use of it.

Third World governments get blamed for the poverty and inequality of the societies over which they preside. But they are unpopular for other reasons as well. Most of them are dictatorships and primarily serve the interests of small, often corrupt, elites. Some resemble nothing else so much as gangs of criminals selling "protection" to the people they rule. They tend also to favor the cities, where the dominant elite live, and make little effort to serve the rural areas. The Third World peasant usually has little reason to trust or respect the government of his country.

Poverty and injustice do not by themselves lead to revolutions. The absolutely destitute seldom have the strength or courage to rebel. Revolutionary guerrilla wars do not happen in the most backward societies. Instead they usually occur where there is still much poverty and injustice but where there has also been considerable economic progress. They happen during the early and middle stages of the modernization process: the long, complex transition from a traditional agrarian society to a modern industrial state. In the very long run modernization ensures political stability – there is not likely to be a violent revolution in Sweden or Canada – but in the short run it destabilizes most social systems. Countries that are just setting out on the road toward a modern industrial state are very vulnerable to all forms of revolutionary violence.

Modernization creates new possibilities, new expectations, and new insecurities. It demonstrates that poverty and injustice are not inevitable and that people can shape their own destinies. It weakens and discredits traditional elites and ways of life. It makes the peasant feel, justifiably, that his livelihood and his spiritual universe are threatened by profound and mysterious changes which he can neither understand nor control. It usually brings an increase in the cultural, economic, and political impact of the United States and Western Europe on the country undergoing modernization: an impact which provokes an angry nationalist backlash, not least among the peasantry. Finally, modernization breaks down the isolation of rural villages and brings their inhabitants into contact with the growing and often increasingly disaffected urban intelligentsia. All of this serves the purposes of revolutionaries. W. W. Rostow describes Communists as "the scavengers of the modernization process."[2] So are other revolutionary guerrillas.

Revolutionary guerrilla wars are begun and led by revolutionary political parties – small, well-articulated organizations whose members are mostly not of peasant origin. The revolutionary party mobilizes, organizes, and guides peasant energies. A guerrilla insurgency cannot succeed without both an effective revolutionary party and widespread rural discontent. The party is the seed from which the insurrection grows, while the peasantry are the soil from which it draws its nourishment. If the seed falls on barren ground it will not grow and bring forth fruit, but without a seed there will be no fruit either.

Most twentieth-century revolutionary parties have been formed and led by members of the urban, educated middle class – lawyers, teachers, students, civil servants, etc. These members of the intelligentsia may not suffer poverty and hunger, but they feel all the other grievances that make revolutionaries: resentment of foreign domination, contempt for corrupt governments, and a hunger for meaning and order in their lives. They tend, in fact, to feel these grievances even more strongly than the peasantry. Education frees them from the fatalism and the network of customary beliefs that benumb and protect the peasant. It opens their eyes to new possibilities. They bring to a guerrilla insurrection the ability to articulate an ideology and the ability to form and run an organization.

Organizational ability is vital to the conduct of any revolutionary guerrilla war. The most important task the revolutionary party faces is to organize and build up a guerrilla army and the political apparatus to support it. Articulating an ideology is scarcely less vital. A revolutionary ideology wins the peasants' support for the revolutionary party by offering them solutions for their problems. It explains to them the causes of their grievances and offers them scapegoats to blame. It holds up before them a vision of what life could be like and shows how that future can be attained. By doing so it responds, not only to the tangible grievances, but also to the great intangible injury peasants suffer from the modernization process: the feeling that a once-familiar world has become incomprehensible and meaningless.

Help from the outside is not necessary to a guerrilla insurgency, but it is very useful. The government of a foreign country can aid guerrillas in a number of ways, particularly if that country has a common border with the one where the guerrillas are operating. It can provide a sanctuary where they can train their recruits, shelter their headquarters, and treat their casualties. It can also supply them with weapons and other necessities. It can even send its armed forces across the border to fight alongside the guerrillas. (This last is, of course, a very overt act of war so it must be disguised or reserved for the final, decisive stage of the conflict.) However, nobody can export a revolution. No amount of outside aid can create grievances where they do not exist or build a strong, well-led revolutionary party.

Communist parties have led many guerrilla insurgencies. Sometimes they win – as they did in China and Vietnam – sometimes they lose – as they did in Malaya, Greece, and the Philippines – but overall their record has been formidable. The Marxist–Leninist ideology, particularly in its Maoist form, has considerable appeal in the Third World. The way Communist parties are organized serves the needs of guerrilla insurgencies very well indeed. A party run on Leninist principles combines strong leadership at the top with the necessary degree of flexibility at the grass roots. It conceals its activities effectively and can survive repression better than almost any other kind of political organization. Also, it can mobilize non-Communist allies, through front organizations, while denying them any influence on party policy. Non-Communist revolutionaries all over the world have learned from the Communists how to conduct a revolutionary guerrilla war.

Communist-led guerrilla insurgencies also get considerable help from Communist-ruled governments. Mao Tse-tung's guerrillas in China got a certain amount of aid from the Soviet Union; Ho Chi Minh's forces in the Indochina War (1946–54) and later in the Vietnam War (1960–75) were aided by China and the Soviet Union; the Communist guerrillas in the Greek Civil War (1946–9) depended on aid provided by Albania, Yugoslavia, and Bulgaria.

The Politics, Strategy, and Tactics of Revolutionary Guerrilla Warfare

At the beginning of a guerrilla insurgency the government it opposes has three very great assets:

1 An army.
2 An administrative apparatus – all the organs of national and local government, particularly including the police force, which enable it to raise money, enforce its laws, collect information, and provide services to its citizens.
3 The citizens' habit of obedience to its decrees – they have paid their taxes and obeyed the laws in the past and expect to do so in the future.

The government has these assets simply because it *is* the government, however weak and incompetently managed it may be. Its army may be poorly trained and demoralized, its administrative apparatus corrupt, and the people's habit of obedience fragile, but they do exist.

When they begin their struggle, the guerrillas do not have an army, an administrative apparatus, or the habitual obedience of the people. (There may be considerable latent sympathy for them, but that is not the same thing as active support.) Guerrillas seize power by acquiring these assets for themselves and depriving the government of them. They must defeat the government:

1 Militarily – by wearing down its army, building up one of their own, and eventually destroying the government's army on the battlefield.
2 Organizationally – by destroying the government's administrative apparatus, building up their own, and bringing more and more of the country under their control.
3 Psychologically – by destroying the people's faith in and fear of the government, inducing them to disregard its authority, and winning popular support for themselves.

No one of these aspects of the struggle can be neglected or separated from the other two. Success in each requires, and also facilitates, success in the others. All three are important throughout the entire course of the war. As a general rule, though, the psychological aspect of the struggle is most important in the early stages of a guerrilla war, the organizational aspect in its middle stages, and the military aspect towards the end of the war. When the guerrillas have triumphed in all three aspects of the struggle, they will no longer be guerrillas; they will be the government of the country.

Mao Tse-tung was the great guerrilla leader of the twentieth century, the man who led the Chinese Communist Party to victory in the most populous country on earth. For that and other reasons, guerrilla insurgencies all over the world have followed his doctrine of "protracted war."

Mao held that, as the name of his doctrine implies, winning a revolutionary guerrilla war takes a long time. Revolutionaries must be prepared for decades of political agitation, organizational work, and armed struggle. (It took him nearly a quarter of a century to overthrow the Kuomintang regime in China.) It is futile to begin the military struggle without first having created the necessary political preconditions. The revolutionaries must begin by preparing and organizing themselves. They must form a revolutionary party, attract carefully-chosen recruits into it, and train them to be disciplined, dedicated professional revolutionaries. As the party grows it will become an underground "parallel hierarchy," modelled on the administrative apparatus of the government it seeks to overthrow. Each village will have its party cell, composed of the party members in that village. Each cell will have a leader, chosen by and responsible to a district leader, who in turn reports to a province leader, all the way up the chain of command to the national leadership. Each level of the hierarchy will have specialized branches for finance, propaganda, military training, etc., just as the government has a police force or a department of public works. The "parallel hierarchy" not only leads the fight to overthrow the government, but also prepares itself to run the country after the government has been overthrown.

The revolutionary party's first task is to conduct propaganda against the regime. There are many themes it can sound, but nationalism is usually the most effective because it is the only one that appeals to everybody. Even a genuinely sovereign and independent regime can plausibly be described as a servant of foreign interests. Many Third World governments have alliances with the United States and other powerful countries, and almost all allow foreign-owned corporations to do business on their territories. There are also arguments that appeal to particular sectors of the population. The revolutionaries can promise land to poor peasants and landless laborers, compensation and revenge to those who have been brutalized or shaken down by agents of the government, and careers in the revolutionary party to the ambitious. They can even win over people who have little to gain by a revolution: landlords and merchants, for example, who will be told they can keep their property if they support the revolution. (The revolutionaries are free to ignore these promises once they are in power.)

Violence plays a relatively minor role in the early stages of a revolutionary guerrilla insurgency. It most often takes the form of assassinations. The guerrillas kill police officers and informers so that their own agitators can work in the villages without being arrested. They kill other agents of the government to demonstrate their power and to weaken the administrative system. However, at this early stage revolutionary terrorism is still very selective. The guerrillas do not have the strength to attack the entire administrative system, even in remote rural areas. They concentrate on assassinating the officials who do them the most harm.

Once the revolutionaries have won a certain amount of popular support, they can begin the military struggle. They build up a guerrilla army. This enables them to start attacking the entire administrative system in the rural areas, not just a few of its most dangerous or most vulnerable members. It is impossible for the government's army to be everywhere at once. With armed guerrilla bands roaming the countryside at night, village officials, tax collectors, and landlords soon flee to the larger towns where the army can protect them. The government thus loses the ability to collect taxes, recruit soldiers into its army, keep itself informed about guerrilla activities, and generally to assert its authority in the countryside. It also loses a great deal of respect. People see that it has become more dangerous to be loyal to the government and less dangerous to help the guerrillas. Even those who were loyal before begin to hedge their bets.

The revolutionary party now begins to fill the organizational vacuum it has created. Its agents begin to collect taxes from the peasantry and to recruit young men for the guerrilla army. They can also use their newly-won authority to carry out land reform measures the government should have carried out while it had the chance. They confiscate government-owned land and the holdings of landlords who have fled to the towns and distribute this land to poor peasants and landless laborers. This gives the party a group of people in each village who will do what it says lest the landlords return and take back the land.

> It can be explained very simply to the peasants: if you want to keep your land, you must fight the imperialists, and if you want to fight the imperialists, your son must go into the army and you must pay taxes.[3]

The guerrilla army is, at this stage, much weaker than its opponent. The revolutionary party does not have the resources to build a strong army. It has some resources – otherwise it could not raise an army at all – but the manpower, money, and supplies available to the party are much less than what even a weak government can command. The party must equip its army primarily with weapons captured from the government and man it with volunteers. Because of its lack of strength, the guerrilla army must devote most of its effort simply to surviving the governments's attempts to destroy it.

A guerrilla army survives by being hard to find. It operates in small detachments, typically squads and platoons of a dozen or a few dozen men, occasionally in companies or battalions of a few hundred. These detachments conceal themselves in the most inaccessible terrain available: swamps, jungles, mountains, or deserts. They move on foot, mostly at night, and employ few or no heavy weapons. They avoid combat, if at all possible, particularly with large enemy formations. Guerrillas almost always prefer to abandon a position rather than expose the force holding it to destruction.

It is not enough, though, just to survive. A guerrilla army must also wear down and demoralize the government's army bit by bit. The guerrillas do this by pouncing on and destroying small, isolated enemy detachments. Guerrillas cannot afford to fight a battle that lasts more than a few hours, nor can they fight a battle of attrition. They must concentrate rapidly and in overwhelming strength against an enemy detachment, overwhelm it before it can be reinforced, and disperse just as rapidly. Guerrilla tactics are maneuver tactics carried to the extreme; they rely on mobility, surprise, and the initiative of lower-ranking commanders. Aggressively and astutely employed, these tactics pose a very difficult dilemma to the enemy high command. If it disperses its forces across the countryside to protect the government's administrative apparatus in the rural areas, they will be devoured piece by piece. On the other hand, if it concentrates its forces in the cities and towns where they will not be vulnerable, it gives up control over most of the countryside.

The rural areas in a country afflicted with a guerrilla war tend to be a "no man's land" which neither side completely controls. The government may rule them during the day. Under the protection of its army, civil servants enforce government demands on the peasantry and try to root out the revolutionary party's administrative apparatus. Then at night the government army holes up in its fortified outposts and the guerrillas take over. The hapless peasantry find themselves paying taxes and furnishing recruits to support both armies, and anybody who fully satisfies one side's demands is likely to be punished by the other.

The guerrillas will want to establish "base areas" as soon as possible. These are areas into which government forces do not penetrate and where the guerrillas exercise complete control over the local population (see map 16.1). The first base areas appear in the most remote, rugged, thinly-populated parts of the country, the terrain which it is easiest for a still-weak guerilla army to defend. Even these bases can at least shelter the guerrilla army's headquarters, training grounds, and hospitals. This is especially important to guerrillas who do not have a friendly neighboring government which will allow those essential elements of the guerrilla army on its territory.

As the guerrillas become more powerful, they can expand their base areas to take in densely-settled agricultural areas as well. These are harder to defend than the remote wastelands but also more valuable as base areas. The people of a well-populated area can be organized to serve guerrilla purposes as completely and efficiently as if they were the citizens of a sovereign state which the guerrillas control. They can provide the guerrillas with more money and recruits than could the peasants in a contested area who must also supply the government. They can be put to work caring for wounded guerrillas or making weapons and ammunition in improvised workshops. They can also be organized into a village militia. It was Mao's policy during the Chinese Civil War to bring into the militia everybody between the ages of 16 and 45 in a base area, both men and women. The militia defends the base area, provides stretcher-bearers and carriers, and serves as a manpower reserve for the first-line guerrilla units.

Best of all, although hardest to win and defend, is a base area which includes a sizable town. Here the guerrillas can establish a provisional government, designate the town as its capital, and seek diplomatic recognition from other countries. Most countries will not grant it, but a close ally might. If the guerrillas can set up a plausible-looking provisional government they deprive their opponent of his greatest legal and psychological asset: his status as *the* government of his country.

Towards the end of a successful revolutionary guerrilla insurgency, most of the country is a guerrilla base area. The government holds the cities, the

Long An Province, South Vietnam, late 1968

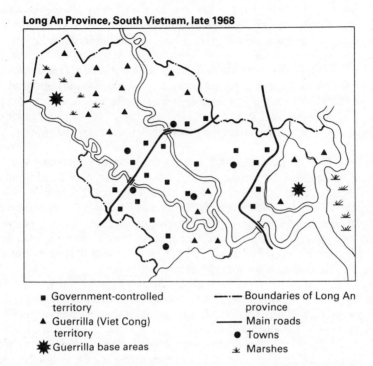

- ■ Government-controlled territory
- ▲ Guerrilla (Viet Cong) territory
- ✳ Guerrilla base areas
- —··— Boundaries of Long An province
- —— Main roads
- ● Towns
- ⚶ Marshes

[a]Based on map in Jeffrey Race, *War Comes to Long An*, rear endsheet.

**A search-and-destroy operation – "Operation Camargue,"
Indochina (Vietnam), July 28–August 4, 1953**

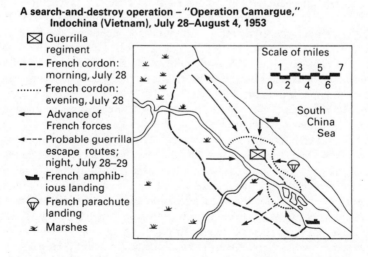

- ⊠ Guerrilla regiment
- – – – French cordon: morning, July 28
- ······· French cordon: evening, July 28
- ◄— Advance of French forces
- ◄– – – Probable guerrilla escape routes; night, July 28–29
- 🚢 French amphibious landing
- ⊕ French parachute landing
- ⚶ Marshes

Scale of miles
1 3 5 7
0 2 4 6

South China Sea

[b]Based on maps in Bernard Fall, *Street Without Joy*, pp. 146, 166.

Map 16.1 *The geography of guerrilla war*

larger towns, and (most of the time) the roads and railroads between them. All else is guerrilla territory. The government, though, still has an army. That army probably has been greatly weakened and demoralized, but it still has substantial numbers of tanks, guns, and aircraft. At this point the guerrilla army has accomplished as much as it can by fighting as a guerrilla army. To capture the cities and conquer the country it now must destroy the government's army on the battlefield, and to do that it must fight as a conventional army fights. It now must have tanks, guns, and aircraft of its own, be organized into divisions and larger formations, and employ conventional land warfare tactics.

Learning how to fight this way can be hard for the guerrillas and it is a serious mistake for a guerrilla army to make the transition to conventional warfare before it has gained a decisive advantage over the enemy army. A formidable Communist-led guerrilla insurgency in Greece was vanquished largely because it tried to make the transition from guerrilla to conventional warfare too soon. The Vietnamese Communists made the same mistake in 1951–2 and again in 1972 and suffered their only serious defeats in a quarter century of successful revolutionary guerrilla warfare.

The revolutionary party must employ a different strategy for the final stage of the war if it is fighting to liberate a colony or to overthrow a regime that is supported by a powerful ally. The guerrillas cannot possibly conquer the colonial power or the ally itself; as long as that country wants to keep fighting it can do so. A colonial power or an ally can, though, be induced to withdraw from the war. Holding on to a colony or supporting an ally is not important enough to justify fighting to the death. Therefore, the guerrillas' objective in such a war is not to destroy the enemy army on the battlefield but to convince the holders of political power in the enemy country to withdraw. The guerrillas must conduct their military operations so as to have the maximum political impact on the enemy country. For example, when the United States was fighting in Vietnam the Vietnamese Communists launched an all-out offensive in every election year – in 1964, 1968, and 1972. The most successful of them, the Tet offensive of 1968, was a military disaster. The Vietcong suffered very heavy casualties and did not capture any territory they were able to hold for more than a few days. Nevertheless, Tet was a political success – it convinced President Lyndon Johnson that the United States could not win the war – and that was what counted.

Protracted war is the hard way to take over a country. It requires careful preparation, a long struggle, and the mobilization of great psychological, political, and military resources. Some guerrilla leaders think there is an easier way to do it. Fidel Castro and his disciples hold that a revolutionary guerrilla insurgency can triumph without a long period of political preparation before the shooting begins. While Mao's protracted war doctrine has guided most revolutionary guerrilla insurgencies in Asia and Africa, Castro's theory of the revolutionary *foco* has been much more influential in Latin America.

A *foco* is a small group of armed men who move into a remote rural area and immediately begin killing agents of the government. Che Guevara, Castro's right-hand man, held that a single *foco* could start a revolution:

A nucleus of 30 to 50 men ... is sufficient to initiate an armed fight in any country of the Americas with their conditions of favorable territory for operations, hunger for land, repeated attacks upon justice, etc.[4]

The *foco* can be even smaller than that; Castro himself began operations in Cuba's Sierrra Maestre mountains with 11 followers.

The *foco* begins to win support and build an organization after it arrives in its area of operations. It spreads oral and written propaganda among the peasants it encounters but it also relies very heavily on "propaganda of the deed." The latter consists of armed attacks on suitable targets designed to show the peasantry that their oppressors can and should be overthrown. Killing or driving away landlords, policemen, and corrupt officials shows not only that they are not invulnerable, but also that life is much better for the peasantry in their absence. Propaganda of the deed wins popular support, attracts recruits into the guerrilla force, and elicits the food, shelter, and information about government activities that the guerrillas need. It soon becomes possible to establish more *focos* in other areas. That is how a Cuban-style revolutionary guerrilla war starts; its middle and final stages are not greatly different from those of a Maoist protracted war.

Revolutionary guerrilla war has turned out to be much more difficult than Castro and his disciples anticipated. They triumphed in Cuba, more because of the internal weakness of the Batista regime than by their own efforts, but scarcely anywhere else. Che Guevara died in 1967 leading a hopeless attempt to start a guerrilla war in Bolivia; by then at least 20 *focos* had appeared in various Latin American countries and all been quickly suppressed. The *focos* were unable to win significant peasant support before government forces could run them down and destroy them.

The *focos* failed because the revolutionaries in them, almost all members of the urban, educated middle class, simply did not speak the same language as the peasantry, figuratively and sometimes even literally. (Many *focos* were made up of revolutionaries who spoke only Spanish and were trying to mobilize peasants who spoke only Quechua or some other Indian language.) One of the most important obstacles to the successful conduct of a revolutionary guerrilla war is the social, intellectual, and psychological gap between the urbanized, educated leaders of a revolutionary party and the peasants they must lead. The protracted war strategy allows the revolutionaires to close that gap before they do anything else. That is the primary purpose and achievement of the long period of political organization and indoctrination before the shooting starts.

The Urban Guerrilla and Terrorism

Guerrilla wars have traditionally been fought in the countryside. Guerrillas begin their operations in the most remote rural areas, gradually extend them to the villages, then to the provincial towns, and leave the cities for last. However, the Third World is urbanizing rapidly. Already most of the population of Latin America lives in cities, as does a rapidly growing proportion of the population of Asia and Africa. In Uruguay and Venezuela 85 percent of the population lives in cities, in Argentina and Chile over 80 percent, and in Brazil and Mexico about 70 percent. Winning control of the countryside in such heavily urbanized countries simply is not decisive. The guerrillas must conduct their war in the cities if they are to mobilize more than a small minority of the population. That is the conclusion many revolutionaries have reached during the past 20 years, particularly in Latin America.

Urban guerrilla warfare has been most common in Latin America. Latin American revolutionaries usually have had to work in urbanized countries, and they have had to face the repeated failures of rural *focos* on their continent. During the 1960s and early 1970s there were urban guerrilla insurgencies in Brazil, Venezuela, Argentina, and Uruguay. By far the strongest of these was led by the Tupamaros in Uruguay; in 1971–2 it seemed very possible that the Tupamaros would overthrow the Uruguayan government.

Urban guerrilla warfare in Latin America is basically Cuban-style revolutionary guerrilla warfare carried on in the cities rather than in the countryside. It begins with a *foco*: a small band of armed men who begin the shooting without waiting to propagandize and organize the people of the city. These urban guerrillas attack policemen, military posts, and foreign-owned businesses; kidnap businessmen, diplomats, and political leaders; obtain and disseminate information discreditable to the government; and rob banks. The purpose of most of these actions is to win popular support – they are propaganda of the deed. The guerrillas seek to draw attention to their grievances, to demonstrate that the government and whatever it seeks to protect can be attacked, and to demoralize soldiers, policemen, and civil servants. Attacks on diplomats, other foreigners, and foreign-owned businesses dramatize the guerrillas' claim that the government is a servant of foreign interests. Many urban guerrilla operations are undertaken largely to generate publicity and permit the guerrillas to harangue a captive audience. The guerrillas also try to provoke the government into overreacting and taking harsh repressive measures such as curfews, mass arrests, and the imposition of martial law. These, the guerrillas hope, will antagonize the people and inspire sympathy and support for their cause. It is particularly important to induce a democratic government to take repressive measures so as to discredit what one urban guerrilla leader termed "this election farce."[5] With luck the democracy will be replaced by an unpopular military dictatorship which will be easier to overthrow. The urban guerrillas' ultimate objective is a mass uprising in the cities.

Urban guerrilla tactics have one great strength: it is very easy for a small band of guerrillas to conceal themselves among the teeming masses of a large city. In the words of Carlos Marighella, leader of the Brazilian urban guerrillas:

Familiar with the avenues, streets, alleys, ins and outs, and corners of the urban center, its paths and shortcuts, its empty lots, its underground passages, its pipes and sewer system, the urban guerrilla safely crosses through the irregular and difficult terrain unfamiliar to the police, where they can be surprised in a fatal ambush or trapped at any moment.[6]

The guerrillas can easily buy or steal the weapons and other supplies they need. They can find apartments and other dwellings where they will be safe from the police. It is easy for them to observe their targets' activities and very difficult for the police to observe them. Small, wary *focos*, constantly moving from one hideout to another, can be very difficult to locate and destroy.

The great weakness of any urban guerrilla insurgency is that it cannot get very large. Even the Tupamaros, at the height of their power, had no more than 4,000 active members at most. Urban guerrillas cannot establish base areas. They may – conceivably – take over part of a city, but they cannot hold it for more than a few days; the army can easily concentrate its forces against that area and retake it. Without a base area they cannot establish a strong administrative apparatus, or raise, train, and equip military units of any size. They cannot have an army and without an army they have little hope of seizing power.

Urban guerrillas have so far been less successful at winning popular support than their rural counterparts. Most people do not like bombings, kidnappings, assassinations, and other forms of propaganda of the deed. Guerrillas of any kind, if they have any sense, employ these measures only against carefully-selected targets. In a rural area it is possible to use violence without endangering those against whom it is not directed; it is not possible to do so in a big city. Urban guerrilla warfare often drives governments into taking very harsh repressive measures, but those measures usually receive popular support. The freely-elected Uruguayan legislature voted in 1972 to suspend important civil rights in order better to counter the Tupamaros. Most people will choose to give up their freedom, at least temporarily, rather than endure anarchy and continual violence.

Because they cannot establish base areas or an administrative apparatus, urban guerrillas are essentially terrorists rather than warriors. "The term 'urban guerrilla,' in short, is a public relation's term for terrorism."[7] Terrorism is not a form of war although it is politically-motivated violence. Wars are waged by sovereign states. Even a guerrilla war is waged by an organization – the revolutionary party – which in the course of the struggle more and more comes to resemble a sovereign state. Terrorist, or urban guerrilla, bands never at any stage of their careers resemble sovereign states.

In the past decade or two, many terrorist organizations have severed the connections between themselves and any particular country. They have become free-floating gangs operating all over the world:

> Terrorism came to resemble the workings of a multinational corporation. An operation would be planned in West Germany by Palestine Arabs, executed in Israel by terrorists recruited in Japan with weapons acquired in Italy but manufactured in Russia, supplied by an Algerian diplomat, and financed with Libyan money.[8]

A kind of "Terrorist International" has emerged – dozens of terrorist groups helping each other out. These small groups have proven to be extremely difficult to eradicate because of their international character. There is not much, for example, that the Israeli government can do about the PLO as long as the PLO's headquarters, its sources of funding and weapons, and its members are all outside Israeli territory. On the other hand, a political organization cannot win over the people of a country and overthrow the government if it is not even located in that country.

Notes

1 *Mao Tse-tung on Guerrilla Warfare*, trans. Samuel B. Griffith (New York: Frederick A. Praeger, 1961), pp. 103–4.
2 W.W. Rostow, "Guerrilla Warfare in Underdeveloped Areas", in *The Guerrilla – And How To Fight Him*, ed. Lieutenant Colonel T. N. Greene (New York: Frederick A. Praeger, 1962), p. 56.
3 Jeffrey Race, *War Comes to Long An* (Berkeley, Ca: University of California Press, 1972), p.128. These are the words of a former high-ranking official in the Communist Party of Vietnam.
4 Che Guevara, *Guerrilla Warfare* (Lincoln, Ne.: University of Nebraska Press, 1985), p.158.
5 Carlos Marighella, "Minimanual of the Urban Guerrilla", Appendix to *Urban Guerrilla Warfare*, Robert Moss, Adelphi Paper no. 79 (London: International Institute for Strategic Studies, 1971), p. 40
6 Ibid., p. 26.
7 Walter Laqueur, *Terrorism* (Boston, Mass.: Little, Brown, 1977), p. 217.
8 Walter Laqueur, *Guerrilla* (Boston, Mass.: Little, Brown, 1976), p. 324.

17
Fighting the Guerrilla

"You know you never defeated us on the battlefield," said the American colonel.
The North Vietnamese colonel pondered this remark a moment. "That may be so," he replied, "but it is also irrelevant."

Conversation in Hanoi, April 1975[1]
from Harry G. Summers, Jr. *On Strategy: A Critical Analysis of the Vietnam War*

Counterinsurgency

In 1836 General Thomas-Robert Bugeaud arrived in Algeria to take command of the French army there. The French had been trying to suppress an Algerian guerrilla insurgency with the tactics they had used in the Napoleonic Wars; they were not doing very well. General Bugeaud assembled his officers and said to them, "Gentlemen, you will have a great deal to forget."[2] He adopted completely new tactics and won the war.

General Bugeaud's words are good advice to any army which has been trained for conventional land warfare but must deal with a guerrilla insurgency. Revolutionary guerrilla war is as different from conventional war as conventional war is from nuclear war. To wage it well an army must forget a great deal of its doctrine and organization for the conduct of other types of wars.

The way to win a conventional land war is to defeat the enemy's army. Then, and only then, one can occupy his territory and deprive him of the ability to keep an army in the field. (There are ways to strike at his territory without defeating his army – blockade, commerce-raiding, and aerial bombardment – but they are usually not decisive.) Officers who are trained to conduct conventional war tend to approach a guerrilla war as if it were a struggle between two armies. They do what they are trained to do, which is "to close with and destroy the enemy," as United States army doctrine prescribes. The obvious way to close with and destroy guerrillas is a "search and destroy" operation. The United States army in Vietnam put most of its effort into massive search and destroy operations, particularly in 1965–8. The French army had done the same thing in the same area 15 years earlier, and most other armies which have had to fight guerrillas have employed search and destroy tactics.

A search and destroy operation begins with the designation of an area to be cleared of guerrillas (see map 16.1). Perhaps there are many guerrillas in the area and the objective is to destroy them; perhaps the area itself is strategically important and the objective is to seize it. The next step is to throw a cordon of troops around the perimeter of the area. This cordon must be thick enough that the guerrillas cannot break or infiltrate it at any point. Also, the cordon must be formed quickly, before the guerrillas can escape from the area. Airborne and helicopter-borne units are very useful in search and destroy operations because they can move into position very quickly. Marines may be sent in to seal off the beaches if the operation is conducted along a coastline. Once established, the cordon starts moving towards the center, driving the enemy before it, rather like a net closing around fish. Eventually the guerrillas are caught, forced to fight, and destroyed. In the course of the operation the troops in the cordon make a thorough search of the area and interrogate its inhabitants; they look for concealed guerrillas, arms caches, and members of the revolutionary party's administrative apparatus. What happens next depends on the objective of the operation. If it is simply to destroy the guerrillas, the troops pull out to conduct other search and destroy operations elsewhere. Perhaps they will come back to the same place and do the same thing again several months later. If the area is to be held permanently, the government forces must garrison it so the guerrilla army cannot return, strengthen or re-establish the government's administrative apparatus, and induce the inhabitants to obey its orders.

Search and destroy operations are arduous and frequently fruitless. The forces in the cordon must be many times stronger than the guerrillas they attempt to trap; even a ten-to-one numerical superiority may not be enough. The only way to assemble a sufficient force may be to strip other areas of their garrisons, leaving them undefended. It is difficult to get the entire cordon into position quickly enough, and even a small gap may permit the guerrillas to escape. Taking the guerrillas by surprise is also difficult. Time and again the Viet Cong got advance warning of American search and destroy operations, either from Vietnamese civilians with access to United States bases or from Viet Cong agents in the South Vietnamese government. The results of a major search and destroy operation the French conducted in Indochina are sadly typical:

> The steel jaws of a modern armed force, supported by naval ships, amphibious tanks, and aircraft, had slammed shut on a force of hurriedly trained farmers led by men, who, in only a few cases, had received the training of corporals and sergeants. A trap ten times the size of the force to be trapped, had shut – and had caught nothing.[3]

Even a "successful" search and destroy peration does little permanent good. As soon as a cordon force leaves, the guerrillas return, punish anybody who actively assisted their enemies, and reassert their authority. Even if the government makes an effort to hold the area after an operation is over, it is likely to be a weak effort. Almost all of the cordon force will have been withdrawn to conduct other search and destroy operations elsewhere or to protect the areas it was drawn from. The peasant over whose head this military storm has passed is left with the guerrillas still able to infiltrate his village – and often with his home and his crops destroyed and some of his neighbors dead.

Governments defeat guerrilla insurgencies the same way guerrillas defeat governments: by depriving the enemy of his ability to keep an army in the field. A revolutionary guerrilla movement is comparable to a tree whose leaves and branches are the guerrilla army, its roots and trunk the revolutionary party, and popular support the soil from which it draws nourishment. Trim back the branches and they will grow again. Sever its roots or poison the soil against it, and the tree will die. Counter-insurgency – the art of defeating guerrillas – is mostly a matter of inducing people to fight for and support the government, discouraging them from helping the guerrillas, protecting or reestablishing the government's own administrative apparatus, and rooting out that of the revolutionary party. Fighting the guerrilla army is also important, but mostly for its impact on the psychological and administrative aspects of the struggle.

The guerrillas' power over the peasantry depends on their ability to elicit two emotions: loyalty and fear. Guerrillas earn loyalty by articulating widely-felt grievances, promising to remedy those grievances, and offering a vision of a better future. They also confer concrete and immediate benefits on those who support them. They drive away landlords and tax collectors, redistribute land, and help harvest the crops. They inspire fear by killing those who defy them, inform on them, or serve in the government. To break their power, the government must inspire more loyalty than the guerrillas do and protect its people from intimidation.

A government can attempt to win the loyalty of its people by conferring material benefits on them. It can build roads, schools, and irrigation systems, train schoolteachers and doctors, and distribute food. These "civic action" measures are commendable in themselves, but they rarely persuade anybody to support the government or not to help the guerrillas. They fail to affect people's behavior partly because, unlike the benefits the guerrillas bestow, they usually are not made conditional on how people behave. If the government builds a road to a village *everybody* can use the road whether he is loyal to the government or not. They fail also because they do not address the fundamental grievances that made the insurgency possible. Peasants do not revolt simply because they are poor. They take up arms because they feel themselves to be victims of injustice and powerless to control their lives. Assuaging their grievances requires a good deal more than simply showering material and other benefits on them. There must be a fundamental shift in the balance of wealth and power. If the allocation of land is an important grievance, the government must carry out a serious land reform program. If people feel powerless, then local government must be made stronger, more democratic, and independent of central government control. If the children of the poor can make careers for themselves only by joining the guerrilla army or the revolutionary party, the government must make it possible for them to become officers or civil servants. Building schools and roads helps to defeat an insurgency only if the people want them and if building them is clearly a response to popular initiatives. It is much more important that people feel they control their lives than that they get a new road.

It is, however, much easier for the ruling elite of a Third World country to confer material benefits on people than to give up a share of its power – particularly if the material benefits are paid for through American foreign aid. The major obstacle to winning popular support is neither knowing what has to be done nor paying for it. The problem is summoning up the will to do it. Third World governments – particularly those threatened by strong guerrilla insurgencies – tend to be weak and unrepresentative. Usually they are controlled by the groups that will have to give up some of their power and wealth to defeat the insurgency. It is difficult for these people to do what must be done or even to perceive it. A national leader who realizes what defeating the insurgency requires faces a baffling dilemma: if he does not carry out the necessary reforms, the guerrillas will eventually conquer his country; if he does carry out those reforms he will anger the powerful elites whose support he must have and therefore will be overthrown very quickly. It should surprise nobody if a leader in such a situation worries mostly about the immediate threat from his colleagues and neglects the long-term threat posed by the guerrillas.

Even if the people think well of their government they will not do much to help it if they fear the guerrillas. As Machiavelli said:

> Men are less concerned about offending someone they have cause to love than someone they have cause to fear. Love endures by a bond which men, being scoundrels, may break whenever it serves their advantage to do so; but fear is supported by the dread of pain, which is ever present.[4]

None of the reforms necessary to win popular support can be carried out unless peasants and civil servants are reasonably safe from the guerrillas. The government cannot carry out a land reform program if its representatives are afraid to venture into the villages. It will not matter how much power a village government enjoys, or how democratically it was chosen, if the mayor is compelled to collaborate with the guerrillas or is afraid to stay in his village overnight. Building roads and schoolhouses does little good if they are blown up as soon as they are built.

The government's most important military task is to protect its villages. This can only be done by stationing troops in the villages themselves, not in isolated fortresses where they protect only themselves, or in the cities. Dividing an army into thousands of tiny village garrisons, each of no more than a few dozen men, contradicts some important principles of conventional war doctrine, but it must be done to defeat a guerrilla insurgency. Each garrison must be not only in, but also a part of, the village it protects. It should be permanently assigned to that village and perhaps recruited from among the villagers. Soldiers in the Third World are much more willing to fight in defense of their homes and for people they know than they are for strangers. The soldiers in the village defense force can be militiamen, part-time soldiers called on only to fight in defense of their villages. This greatly reduces the cost per soldier and allows the government to maintain a force large enough to protect its entire population. The mayor, other civil servants, and anybody else who is a likely target of guerrilla violence, should be armed and trained in self-defense.

The government will also need a regular army, trained and equipped for conventional war. Its missions are rapidly to reinforce village garrisons under attack, to defend the cities, and occasionally to conduct search and destroy operations. These forces need to be mobile, fairly well armed, and each unit able to operate independently. During its struggle against the Communist-led Huk guerrillas (1946–57), the Philippine army was organized into battalion combat teams, each an independent infantry battalion with its own artillery, heavy weapons, and combat service support detachments. They proved well-suited for counterinsurgency campaigning.

Soldiers fighting against guerrillas must be very restrained in the use of their firepower, very careful not to inflict casualties on civilians unless it is absolutely necessary. Guerrillas often fire on aircraft or troops from within a village, bringing down a heavy retaliatory barrage on that village, and then disappear. This is an excellent way to convince the villagers that the government and its army are their enemies.

The police force plays an important role in counterinsurgency. Any government facing a guerrilla insurgency needs a strong police force, sometimes several times as strong as its army. The police bear the primary responsibility for hunting down and eliminating members of the guerrilla administrative apparatus. (Sometimes, though, special military units may be formed just for this purpose.) The police also gather information about other aspects of guerrilla activity, popular attitudes, and potential causes of trouble. Information is as important to the counterinsurgency forces as it is to the guerrillas; whoever can best induce people to tell him what his enemy is doing will probably win a guerrilla war.

The United States and Revolutionary Guerrilla Warfare

A revolutionary guerrilla war often leads to the replacement of an anti-Communist regime with a Communist one. That is what happened in China, Vietnam, Cuba, and Angola; it would have happened in Malaya, Greece, Thailand, and the Philippines had the guerrillas in those countries won; it may be happening today in El Salvador. The United States has a clear interest in preventing the triumph of Communist-led insurgencies in the Third World. However, it should not try to suppress every revolutionary guerrilla war everywhere. Some are led by revolutionary parties with which the United States can establish a tolerable relationship. Others are fighting regimes that do not deserve to be helped.

Neither the Soviet Union nor any other Communist-ruled country has ever caused a guerrilla war, even though the Soviets often aid the guerrillas and profit if they triumph. Guerrilla wars happen because a substantial number of people are angry enough at their government to take up arms against it. Any regime that is threatened by a guerrilla insurgency has mostly brought it on itself, by what it has done or failed to do. The regime may still deserve help, it may still have or may regain enough popular support to stay in power, but the very existence of the insurgency raises some doubts about its legitimacy.

Before intervening in a guerrilla war, American decision-makers should measure very carefully how much popular support the present regime really has. They should also think about how hostile to the United States the guerrillas are and how important their country is to the United States.

Even if the United States does help a Third World regime to defend itself against guerrillas, the regime itself must do most of the work: it should not rely on the Americans to keep it in power. One reason why not is suggested by something one United States officer said about the Vietnam War: "I'll be damned if I permit the United States Army, its institutions, its doctrine, and its traditions to be destroyed just to win this lousy war."[5] The officer had a point. The United States armed forces are not prepared to conduct counterinsurgency campaigns. They have a small, specialized counterinsurgency organization, the army's Special Forces, but are mostly trained, equipped, and organized for the conduct of nuclear and conventional warfare. To make the armed forces a counterinsurgency force would unfit them for their more important missions.

Counterinsurgency, like guerrilla war, cannot be carried on by foreigners. Even if a powerful ally could win the military aspect of the struggle, it could not achieve the essential administrative and psychological successes. Well-trained, empathetic, dedicated American counterinsurgency troops may persuade Asian or Latin American peasants to trust them, possibly even to trust the United States; but the peasant's behavior depends far more on what he thinks about the government of his own country than on what he thinks about the Americans. That government itself must win his loyalty. That government must also win the administrative struggle: recruit and train its own civil servants to go into the villages, stay there in the face of guerrilla terrorism, and govern those villages honestly and effectively. The United States government cannot run another government's country for it.

The United States therefore cannot defeat a guerrilla insurgency all by itself, but it can help an ally to do so. American economic aid can ease the social and political changes that defeating guerrillas requires. Aid money can be used to compensate landlords for giving up their land and to finance economic development projects that respond to popular initiatives. Redistributing wealth is easier if it is not a zero-sum game, if what the poor obtain has not been taken from the rich or the middle class. Foreign aid thus makes it easier for a Third World government to redistribute the wealth of its country. (Redistributing power is another question – and far more difficult.) American aid can also finance an ally's counterinsurgency effort without the ally having to impose heavy, unpopular taxes. What foreign aid cannot do is save the ally from having to make reforms. These reforms may be painful – which is why they were not made before the fighting started – and the country's ruling elite may well think of American aid as a way to escape, rather than facilitate, reform. The United States must discourage this illusion.

The United States armed forces can help an ally protect his people against guerrillas – to some degree. During the Vietnam War the Marines organized a network of Combined Action Platoons (CAPs) to protect the South Vietnamese villages in their area of responsibility. Each platoon, consisting of

15 United States Marines and 35 South Vietnamese village militia, was assigned to defend a village and was permanently stationed in that village. The CAPs proved to be quite successful in defending the villages, denying the enemy food and recruits, and collecting information. The army and the CIA had similar programs. The army Special Forces trained Montagnard tribesmen, inhabitants of the mountains of South Vietnam, to fight the Viet Cong. The CIA organized the Meo in Laos into a dedicated and, for a time, successful anti-Communist guerrilla army. In Vietnam, however, these well-conceived efforts were not enough to avert defeat, perhaps because they were not supported as they should have been. The CAPs, the Special Forces, and the CIA combined comprised only a few thousand out of more than half a million Americans in Southeast Asia. Perhaps dividing the entire United States army in Vietnam into CAPs and putting one in every village would have been the most effective strategy. It might even have won the war. (It might also have made the army unfit to fight other wars.)

Protecting the villages is the most important military task in a counterinsurgency effort, but there are other roles United States ground forces and tactical air power can play without changing the way they fight. They can intervene decisively in the final stage of a guerrilla war. When the guerrillas come out in the open, mass, and fight as a conventional army, they become vulnerable. Now, for the first time in the war, the United States armed forces can really use their tremendous firepower; in a few days they can shatter a guerrilla army that thought it was about to win the war. American intervention at this stage may not destroy the insurgency, but it certainly can avoid defeat. In 1965 and again in 1972 American intervention saved South Vietnam. It might well have done so again in 1975 had Congress permitted the United States to re-enter the war.

Conventional ground forces can also be used to seal the border of a country and prevent guerrillas moving across it. If this is done effectively the guerrillas are denied sanctuaries in neighboring countries, as well as supplies and reinforcements. This strategy was used during the 1954–62 guerrilla war in Algeria. The French army constructed a fortified barrier along the Tunisian border which seriously impeded, although it never completely stopped, infiltration. The idea might be worth trying again. How effective it would be depends on how impermeable the barrier could be made and how much the guerrillas depend on help from the outside.

Notes

1 Five days later the government of South Vietnam surrendered unconditionally.
2 "Messieurs, vous aurez beaucoup à oublier.' Walter Laqueur, *Guerrilla* (Boston, Mass.: Little Brown, 1976) p. 70.
3 Bernard B. Fall, *Street Without Joy* (New York: Schocken Books, 1967), p. 168. The entire chapter on "Operation Camargue' is an often-cited, classic picture of guerrilla war.
4 Niccolo Machiavelli, *The Prince* (New York: Bantam Books, 1966), p.60.
5 Guenther Lewy, *American in Vietnam* (Oxford, Oxford University Press, 1978), p. 138.

18
Territorial Defense

*The mobilization of the common people throughout the country will create a
vast sea in which to drown the enemy.*

Mao Tse-tung, "On Protracted War"

Guerrilla war can be a means of defense. Most guerrilla wars before the
twentieth century were fought by the people of an invaded country against
their invader. The term "guerrilla" was first applied to the savage struggle
the people of Spain waged from 1808 to 1814 to expel Napoleon's armies.
There also were defensive guerrilla struggles in Russia and the Austrian Tyrol
during the Napoleonic Wars, in France during the Franco-Prussian War
(1870–1), in southern Africa during the Boer War (1899–1902), and all over
Europe and Asia during World War II. From 1979 to 1988 the people of
Afghanistan fought a defensive guerrilla war against an invading Soviet army.
The Soviets began to pull out in the summer of 1988 but the war continues
against the Communist government they left behind.

Often the victims of an invasion rise up against the invader while their
country or its allies still have an army in the field. This is what the Spanish
did during the Napoleonic Wars and what the Chinese and the Russians did
in World War II. The invader then has to fight two wars at once, a guerrilla
war and a conventional war. The conventional forces compel him to concentrate
his army at the front and leave most of the territory he has seized undefended.
The guerrillas, then, can attack his supply lines, harass his reserves, and
prevent him using the economic resources of the occupied areas. Even if he
does smash the defender's conventional forces and overrun the entire country,
he has not yet won the war; there are still the guerrillas to deal with. The
combination of a conventional defense and a defensive guerrilla war is much
more formidable than either would be alone.

None of the peoples who have fought defensive guerrilla wars in the past
had prepared for guerrilla war beforehand; they raised and organized their
guerrilla armies after they were invaded. Since World War II, though, several
countries have prepared in peacetime to wage a defensive guerrilla war in the
event of an invasion. This is the "territorial defense" strategy; the Chinese
call it "People's War."

Yugoslavia, Romania, and Sweden rely heavily on territorial defense. China
now relies much less on this strategy than it did ten years ago, but still devotes
some effort to preparing to wage "People's War" against an invader. These
four countries all have the same difficult military problem to solve. All have

good cause to fear a Soviet invasion. China is probably the Soviet Union's most bitter enemy; Yugoslavia rebelled against Soviet control even before China did; Romania has asserted its independence; and Sweden is a liberal democracy. China and Romania have common borders with the Soviet Union, while Yugoslavia and Sweden are within easy reach of Soviet-controlled territory. At the same time, none is a member of NATO or of any other alliance that would help them against a Soviet invasion; they might have to stand alone against the Soviet Union. They need to be able to defeat the Soviet armed forces, or at least to make a Soviet victory so costly as to deter aggression. Perhaps – just conceivably – they can do this if they mobilize enough of their people.

The foundation-stone of territorial defense is universal military service: "Everyone a soldier," as the Chinese slogan proclaims. Everyone is expected to help defend his or her country. In Yugoslavia every citizen between the ages of 16 and 60 – men and women alike – serves in the regular army, the army reserves, the territorial defense force, or the civil defense organization. All Swedish males serve in the regular armed forces and then in the reserves until the age of 47. In China and Romania service in the militia is "voluntary," but a large percentage of those qualified to serve join up. Yugoslavia, Romania and Sweden can mobilize in wartime five or ten times as many troops as they have in their peacetime regular armies. China at one time had as many as 20 million soldiers in its Armed Militia – trained, organized, and equipped to fight as light infantry – and another 100 million receiving regular military training in the Ordinary Militia. It still has some 4.3 million Armed Militia and another 6 million Ordinary Militia.

Switzerland and Israel can also mobilize very large reserve forces if they go to war, but they do not have territorial defense strategies. In those countries the reserves, once mobilized, join the regular army and fight as conventional forces. Neither Switzerland nor Israel plans to wage guerrilla war. In contrast, China, Yugoslavia, and Romania have two kinds of military force, each with its own mission. They have regular armies, manned by active-duty troops and intended for conventional warfare, and they also have territorial defense forces, manned by reservists and intended for guerrilla war.

In the event of a Soviet invasion the regular army defends the border as long as possible. The Chinese People's Army could hold out for a fairly long time, the Romanian or Yugoslav armies for a few days. This gives the territorial defense force time to mobilize. The Yugoslavs expect to mobilize half their territorial defense troops in six hours, and all of them within a day or two. As long as the regular army is still unbroken, the territorial defense force moves supplies to it and guards the rear areas. If the army is forced to retreat, territorial defense troops stay behind and wage guerrilla war. Even if the army is destroyed and the country's entire territory occupied by the invader, territorial defense guerrillas are to continue the struggle. Remnants of the regular army may join them.

The Chinese have added a new wrinkle to territorial defense doctrine. The Chinese People's Militia plans to conduct guerrilla warfare in the cities. There are very extensive tunnel systems underneath Chinese cities, intended partly as bomb shelters but also as bases in which Chinese guerrillas could shelter and from which they could contest an invader's control of the cities. The Chinese hope to make each of their cities a "poisoned shrimp," impossible for the invader to digest.

Sweden's organization for territorial defense is different from that of China, Yugoslavia, and Romania. Sweden does not have a territorial defense force separate from the regular army. Swedish reservists join regular army units when mobilized. The entire Swedish army is prepared to employ guerrilla tactics, as well as conventional war tactics, as appropriate.

The Chinese People's Army and the People's Militia together probably could defeat a Soviet invasion if the Soviets were to try to occupy the entire country. The People's Army does not have anywhere near as many tanks, aircraft and other heavy weapons as the Soviet armed forces, but it is a very large army. The Chinese have a tremendous guerrilla war tradition and have devoted very considerable resources to territorial defense. The Chinese Communist Party would be (as it once was) a superb framework around which to organize a guerrilla resistance. Also, there are a billion Chinese. The Soviets' difficulties in Afghanistan are nothing compared with what they would face trying to hold down a country with sixty times Afghanistan's population. The Soviet Union probably could seize and pacify some of the outlying areas of China such as Sinkiang and Inner Mongolia. There are not enough people in those areas to sustain a guerrilla war, and many of them are not loyal to the Chinese government.

Yugoslavia, Sweden and Romania on the other hand are small countries. If the Soviet Union wants them, it can take them. However, by employing a territorial defense strategy they can force the Soviets to fight a long, unpleasant war. As long as the Soviet Union has powerful enemies in Western Europe, this prospect should be enough to deter Soviet aggression. A prolonged war in Sweden or Yugoslavia would very likely escalate into a war with NATO, possibly even a nuclear war. Nothing the Soviets could get by invading Sweden or Yugoslavia is worth a nuclear war, or even a serious threat of one. (The Soviets could invade Romania at much less risk because Romania is behind the Iron Curtain and NATO has little ability or inclination to help it.) Territorial defense thus protects Sweden, Yugoslavia and – to some extent – Romania by deterring Soviet aggression, but it only works because there is a balance of power between NATO and the Warsaw Pact. If the NATO countries were ever conquered, every non-NATO country in Europe would soon be subjugated as well.

PART VII
The Future of War

From this time onward the danger of mankind perishing by human action will always be with us – it will never vanish again ... and it is under this pressure that man can rise to his highest potentialities. The moment he relaxes in the illusion of final success, the extreme menace will once more be real.

Man either grows in freedom, and maintains the tension of this growth, or he forfeits his right to live. If he is not worthy of his life, he will destroy himself.

Karl Jaspers, *The Future of Mankind*

19
Nuclear War in the Twenty-first Century

The Future of Mutual Assured Destruction

Mutual Assured Destruction is a *faute de mieux* strategy. Nobody has found a better alternative, but few are really comfortable with it. None of the defects MAD was born with have been corrected. MAD continues to rely on the brutal threat to destroy an entire people for the actions of their government. It inevitably depends on the perceptions and actions of those who rule the Soviet Union and the other nuclear powers. Because no American can really know what the Soviet leaders think, or what they may think in the future, no American can be really sure that the American deterrent is adequate. Finally, as long as nuclear weapons exist, there could be a nuclear war. For the past 40 years, year by year we have managed to prevent a nuclear war, but one could still happen next year or the year after:

> It is a bit as if members of a basketball team insisted they could keep the ball in the air forever, if they were only serious enough about the effort, kept in training, and regularly rotated players. Common sense says someone would eventually drop the ball.[1]

MAD, like any other military strategy, rests on certain assertions about what weapons can and cannot do. It makes sense only if those assertions are correct. MAD assumes that:

1 Strategic nuclear weapons can be built to survive even the heaviest possible counterforce attack.
2 Cities cannot survive a strong countervalue attack no matter what measures are taken to protect them.

These were indisputable axioms in the mid-1960s, they have become debatable in the mid-1980s, and they may become exploded fallacies in the course of the twenty-first century.

Strategic nuclear weapons are somewhat less survivable today than they were a decade or two ago and may become even more vulnerable in the future. There are already serious doubts about the ability of ICBMs to survive a strong counterforce attack. Modern ballistic missiles are already so accurate that either superpower could disable 80 or 90 percent of the other's ICBMs. They can be made even more accurate. The warhead designed for the American Pershing II IRBM has a guidance system similar to TERCOM and can maneuver as it approaches the target. Its CEP may be as low as 65 feet. If the superpowers fit this kind of warhead to their ICBMs, each might well be able to destroy virtually all of its opponent's missiles in a preemptive strike.

The great threat to bomber survivability is SLBMs with very short flight times. These missiles can get to an airfield before the bombers on it can take off. They flight time of a ballistic missile can be greatly shortened by launching it on a "depressed trajectory." A missile on a depressed trajectory flies at a much lower altitude than it would on a normal "maximum range trajectory" and reaches its target sooner. Firing it in this manner reduces the missile's range but the most modern Soviet SLBMs have sufficient range even on depressed trajectories to reach a United States bomber base in Kansas from a Soviet missile submarine in the Atlantic Ocean. Their flight times on depressed trajectories may be short enough to destroy even bombers on ground alert at many United States bases. So far, however, the Soviets are not known actually to have tested any of their SLBMs on depressed trajectories.

Missile submarines will probably still be very survivable when the twenty-first century arrives. There are several types of non-acoustic ASW sensors which might someday make a submarine underwater as easy to locate as an ICBM silo. It is possible to build sensors able to locate and track a submerged submarine by the heat it gives off or by the minute swelling on the surface of the ocean caused by its passage underwater. Non-acoustic sensors can be mounted on reconnaissance satellites; unlike sonar they need not be in the water to function. A network of satellites equipped with non-acoustic sensors could search every ocean in the world in a few minutes. On the other hand, missile submarines can cruise under the Arctic ice cap which conceals them almost completely from non-acoustic sensors.

Despite the growing vulnerability of strategic nuclear weapons, a first strike on a superpower would still be a very inadvisable move. It should remain so for the rest of this century and well into the next. However, the balance of terror is becoming less stable, which serves to strengthen the case against MAD. That case would be even stronger if the superpowers really could defend their cities against nuclear weapons rather than just threaten to destroy each other's cities. Perhaps the superpowers can.

Ten to fifteen years ago the vast majority of American nuclear strategists had given up on anti-ballistic missiles (ABMs). It seemed that no "ballistic missile defense" (BMD) could stop a determined Soviet attack on American cities. Even Damage Limitation advocates relied primarily on counterforce strikes. Since them, though, great progress has been made in BMD weapons technology and in the development of space satellites.

The best-known and most spectacular new BMD weapon is the laser. A laser projects a beam of light whose very unusual physical properties enable it to destroy solid objects. (Among other peculiarities, a laser beam can be focused so narrowly that the entire beam falls on a spot several inches in diameter several thousand miles away.) Weapons scientists hope to develop a laser powerful enough to disable an ICBM, by burning a hold through its skin, and then lock on to another one in less than a second.

New and greatly improved ABM rockets have also been developed. The first generation ABMs, built during the 1960s, were so inaccurate they had to be nuclear-armed; the guidance systems of those early ABMs could not put them close enough to an incoming enemy warhead for anything less than a nuclear explosion to destroy the warhead. Consequently, they were large, expensive weapons, some nearly as large and expensive as ICBMs. Once large numbers of ICBMs had been fitted with MIRVs, neither superpower could afford enough ABMs to shoot down all the warheads the other could send at it. It is now possible to build effective ABMs that do not have explosive warheads of any kind, nuclear or conventional. These "kinetic-energy" ABMs have guidance systems so refined they can home in on an enemy warhead and collide with it. The sheer force of the collision, occurring at several thousand miles an hour, destroys the enemy warhead. ABMs of this type can be made much smaller and cheaper than the old nuclear-armed ABMs.

Aerospace technology has been so far perfected that defense technologists can now contemplate mounting many parts of a BMD system on space satellites. Battle satellites, armed with lasers and orbiting over enemy missile fields, are the most dramatic and best-known example of the new BMD devices. The laser-armed battle satellite may not be feasible even in the early twenty-first century: it may be that lasers powerful enough to destroy ballistic missiles will still be too heavy to be put into orbit. If so, another option is to equip battle satellites with large numbers of small, short range, kinetic-energy ABM rockets. A third option is to place lasers on the ground and put mirrors into orbit to guide their beams. A ground-based laser could project its beam to a relay mirror in a geosynchronous orbit, about 22,300 miles above the surface of the earth. That mirror would reflect the beam back down to a lower-altitude battle mirror passing over an enemy missile field, which in turn would direct the beam at an enemy missile rising from its silo.

Mounting BMD weapons on satellites permits attacking a ballistic missile at any point in its flight, from the moment it is launched until it is about to hit the target. The defender no longer has to destroy every warhead the attacker has put up in the last minute before the warheads explode. He has about 30 minutes to destroy all the warheads (still not much time). A battle satellite can attack ballistic missiles at the beginning of their flight, while their rocket motors are still firing. This is the best time to attack ballistic missiles. Finally, the defender can now deploy a multi-layered defense: several sets of BMD weapons and sensors, each independent of the others and each designed to attack a ballistic missile in a different phase of its flight path. No one layer of defense could conceivably destroy all the weapons sent at it, but three or four layers together, each highly effective, may be able to approach the elusive goal of a leak-proof defense.

Defending an entire country against ballistic missiles is, to say the very least, extremely difficult. Not long ago it seemed impossible. Many respected strategists and scientists think it still is, but the new BMD technologies offer some hope that a "comprehensive" BMD system (one able to protect an entire country) can be built early in the twenty-first century. Because of this possibility, and because of all the doubts about the stability of the balance of terror, the United States is now engaged in a long-term effort to develop an effective defense against ballistic missiles. This is the Reagan administration's Strategic Defense Initiative – which its critics call "Star Wars."

The Strategic Defense Initiative

On March 23, 1983 President Ronald Reagan repudiated four decades of United States strategic nuclear war doctrine:

> I've become more and more deeply convinced that the human spirit must be capable of rising above dealing with other nations and human beings by threatening their existence. ... What if free people could live secure in the knowledge that their security did not rest upon the threat of instant U.S. retaliation to deter a Soviet attack, that we could intercept and destroy strategic ballistic missiles before they reached our own soil or that of our allies?
> ... I am directing a comprehensive and intensive effort to define a long-term research and development program to begin to achieve our ultimate goal of eliminating the threat posed by strategic nuclear missiles.[2]

A year later the Reagan administration announced, and Congress voted the initial funding for, the "Strategic Defense Initiative' (SDI).

The SDI is a very long range plan, still tentative in most aspects. The 1980s are being devoted to finding out whether a BMD system able to protect the entire United States can be built. If the initial feasibility studies show that it can be, the decision to deploy a full-scale system could be made soon after 1990. The entire, country-wide BMD system might be complete and in operation by the year 2010 or 2020.

To protect the United State's population and economy a BMD system must stop a very high percentage of the ballistic missile warheads the Soviet Union might send at it: both ICBM warheads, and those from the SLBMs on Soviet missile submarines. They system does not have to stop every single warhead. Even several dozen warheads will not destroy the United States, although they could inflict great damage. Most of the warheads in a Soviet attack would probably be aimed at American missile silos or other military targets. The attacker's first objective would be to disable as many American strategic nuclear weapons as possible before they were launched in retaliation. (Even if the Soviets were to have their own BMD system, they could not rely on it stopping every American warhead.) The American BMD system does, though, have to stop over 90 percent of the warheads coming at it, probably 99 percent or more. Even 1 percent of all Soviet ICBM and SLBM warheads is more than 90 warheads.

Ballistic Missile Defense Technologies, a report prepared for Congress by the Office of Technology Assessment, describes a country-wide BMD system that the United States might build. Many combinations of BMD devices are under consideration, but the BMD system described in *Ballistic Missile Defense Technologies* is as viable as any other.

This system consists of four layers of defense, each with its own weapons, sensors, and command and control system. The first layer is designed to destroy enemy ballistic missiles during the boost phase of their flight (see figure 19.1). This is immediately after they have been launched, while their rocket motors are firing. The boost phase of an ICBM's flight last 3–5 minutes, that of an SLBM 2 or 3 minutes. It is comparatively easy to locate and track a ballistic missile during the boost phase because of the great heat given off by its rocket motors. This is also when it is easiest to destroy the missile; simply burning a small hole in its large, lightly-constructed booster does the job. The defender must destroy the vast majority of the enemy missiles during this phase, while they are most vulnerable, if he is to conduct a successful defense.

The first layer of defense is armed with ground-based lasers, relay mirrors in geosynchronous orbits, and lower-altitude battle mirrors. The system needs only a few lasers and relay mirrors. If, as the weapons scientists hope, each laser can destroy a missile a second, then ten lasers can in 4 minutes destroy every ICBM and SLBM the Soviet Union has. However, the lasers and relay mirrors must be extremely large, sophisticated, expensive devices. Each laser must be powerful enough to project its beam nearly 50,000 miles, out to its relay mirror and almost all the way back to earth again. Each relay mirror must be about 100 feet in diameter. The battle mirrors are smaller, 16 feet in diameter, but the system needs a fairly large number of them.

Enemy missiles which penetrate the boost phase defenses then enter a post-boost phase during which the bus on a MIRVed missile launches its MIRVs and decoys. Then, during the mid-course phase, the warheads and decoys glide through outer space almost all the way to their targets. The post-boost phase last 2–5 minutes and the mid-course phase 15–20 minutes for an ICBM and 7–10 minutes for an SLBM. The second layer of defense attempts to destroy the warheads during the post-boost and mid-course phases. It is not easy. Warheads and decoys gliding through space are very difficult to locate and track because they are small and give off little heat. Also, there are a lot of them, even if the first layer of defense has done its work well. Every missile that survives the boost phase defenses may release ten or more warheads and more than a hundred decoys. On the other hand, the second layer of defense has more time in which to do its work than all the others put together.

The second layer of defense is battle satellites armed with kinetic-energy ABM rockets and equipped with sensors to locate and track enemy warheads and guide the ABM rockets to them. These satellites use rockets because missile warheads are too strongly-constructed for a laser to destroy quickly. The sensors on the battle satellites must be extremely capable, able not only to track warheads and decoys but also to distinguish one from the other. As warheads and decoys look and behave almost exactly the same during the mid-course phase, designing sensors able to do this is one of the most difficult tasks weapons scientists working on the SDI must solve.

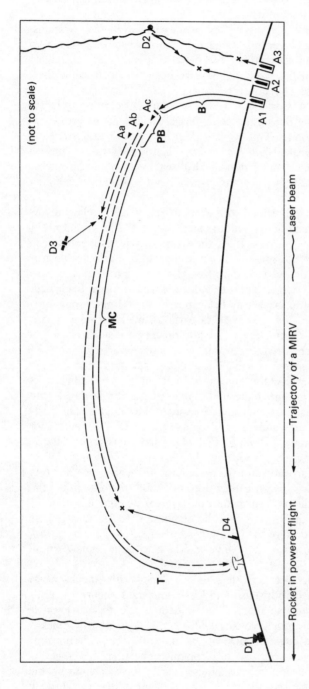

The attacker launches three ICBMs (A1, A2, and A3). The defender destroys A2 and A3 during the boost phase (B) of their flight with a laser beam. The beam is generated by a ground-based laser (D1) and projected to a relay mirror (not shown) in geosynchronous orbit, which directs it to a low-orbit battle mirror (D2), which in turn aims it at the missiles. A1 survives attack and during the post-boost (PB) phase of its flight releases three MIRVs: (Aa, Ab, and Ac). Aa is destroyed during the mid-course phase of its flight (MC) by a kinetic-energy rocket fired from a battle satellite (D3). Ac is destroyed during the terminal phase of its flight (T) by a ground-based ABM rocket (D4). Ab survives attack and explodes over the defender's territory. (A full-scale attack would employ about 3,000 – not three – missiles.)

Figure 19.1 *An ICBM attack countered by a comprehensive ballistic missile defence (BMD) system*

The third layer of defense attacks warheads towards the end of the mid-course phase and as they reenter the atmosphere. This third layer is armed with several hundred very long range, land-based ABM rockets, and has its sensors mounted on a hundred or more satellites. It relies heavily on information relayed to it by the second layer's command and control system as to which objects are decoys and which warheads.

The fourth layer of defense attacks warheads during the terminal phase of their flight, in the last minute as they plunge through the atmosphere towards their targets. It has several thousand short range ABM rockets and ground-based tracking radars. This last layer of defense resembles the 1960s-type ground-based ABM systems discussed in chapter 7. It could be constructed before any of the others because the technology it requires is already well developed. A purely ground-based system would not, by itself, give cities much protection but it could defend ICBM silos, command posts, and other hard targets.

There are about 9,800 warheads on Soviet ICBMs and SLBMs. What would happen if the Soviets were to launch this entire force against the BMD system described above and each defensive layer were to prove 80 percent effective? After the boost phase defenses had done their work there would be 1,960 warheads left. Of these, 392 would get past the second layer of defense. Of these 78 would survive the third layer, and 16 would penetrate the fourth layer and hit the United States. This would not be a perfect defense – never in history has there been a perfect defense – but it should enable the United States to survive a nuclear total war.

A country-wide, multi-layer BMD system would be extremely expensive, perhaps costing as much as all the strategic nuclear weapons the United States has hitherto built put together. Tentative estimates of its cost range between 100 billion and one trillion dollars. Is the SDI worth that much? Many knowledgeable people hold that it is not. They argue that the objective is unattainable, no country can protect its people against a full-scale nuclear strike, and that attempting to do so destabilizes the balance of terror, stimulates the arms race, and makes the world more dangerous.

It may be prohibitively difficult to build an effective defense. The BMD system described above requires:

1 Lasers several orders of magnitude more powerful than any of the same type available today.
2 Sensors greatly more discriminating than current models.
3 Computer software to direct and coordinate the system three to five times as complex as the largest military software system ever before written.
4 Satellite-mounted instruments and other devices able to survive for years in outer space without maintenance and ready to work instantly when called upon.
5 Other equipment requiring fundamental technological breakthroughs before it can even be designed.

The experts working on these problems have been unable to promise that all, or any, of them will be solved. The best they can say is that there are no *known* insurmountable obstacles to solving them.

Even today, decades before a comprehensive BMD system could be in operation, numerous counters to it have been proposed. One is to fit ballistic missiles with fast-burn boosters – rockets that do their work and burn out in less than a minute rather than the 3–5 minutes that contemporary ICBM boosters require. Fast-burn boosters make the task of the boost phase defenses vastly more difficult by reducing the time they have available.

An even more effective counter might be to attack the satellite-mounted components of the BMD system. Battle satellites and battle mirrors could be hit from the ground as their orbits carry them over enemy territory. They could be destroyed by space mines: explosive charges put into orbit, maneuvered near a satellite, and then detonated. They might also be attacked by enemy satellites armed with lasers or rockets. Both the United States and the Soviet Union have already tested anti-satellite (ASAT) weapons. If they develop effective BMD weapons they will also almost certainly be able to destroy satellites; ASAT and BMD technologies are so closely related that a good boost-phase or mid-course phase BMD is almost by definition a good ASAT weapon. There are ways to protect satellites against the attentions of ASAT weapons, but none that cannot be penetrated. An aggressor could begin by attacking the defender's satellite-mounted BMD devices and then, having torn gaping holes in the boost-phase and mid-course defenses, fire his missiles through the gaps.

The best counter to BMD may be one that has not even been thought of yet. If it will take 30 years to develop and deploy a comprehensive BMD system, weapons scientists will have 30 years to find ways to defeat it. They can also continue their efforts after the system is deployed. Discussions of BMD should avoid the "fallacy of the last move" – the illusion that technological progress and weapons development will stop the moment a comprehensive BMD system is in place.

Even if BMD does everything its most enthusiastic supporters expect, it will not completely protect the United States from nuclear attack. The attacker could still use strategic bombers and cruise missiles. Some elements of the BMD system might be effective against these particular weapons, but the United States probably would have to build a separate anti-bomber defense. If the anti-bomber defense cannot be made very nearly leak-proof, American cities will be vulnerable no matter how strong the BMD shield is.

When President Reagan left office the Strategic Defense Initiative had only been begun. His successors and Congress must decide whether to cancel it, deploy a weak BMD system designed just to protect strategic nuclear weapons and command posts, or build a comprehensive system designed to protect the entire United States.

American weapons scientists may not overcome the technological barriers to an effective BMD system; the Soviets may develop cheap and effective countermeasures; or, the American economy may not be able to pay for the SDI. If so, the program should be cancelled.

The United States may end up with a BMD system that protects its missile silos, bomber bases, and command posts but leaves its cities essentially unprotected. The United States government may choose this option deliberately and deploy just one layer of BMD, a terminal-phase system. The government may also stumble into this option, deploy a comprehensive BMD system that does not work very well. (A BMD system that does not work well enough to protect cities almost certainly can protect strategic weapons and command posts.) A weak American BMD shield would, on balance, work to stabilize the balance of terror especially if the Soviet Union deploys a similar system. It would lessen the danger that either superpower would launch a counterforce first strike. The SDI might in this manner justify at least part of its cost, even if it fails to meet its most ambitious objective.

Perhaps – just conceivably – President Reagan's dream will come true. Perhaps the United States – and the Soviet Union – will achieve effective, comprehensive BMD systems. It would liberate the people of the United States and the Soviet Union from the fear of nuclear total war. There would no longer be a balance of terror, nor would there need to be. The superpowers would no longer have to rely for their security on the threat of mutual suicide. Nuclear disarmament would become a real possibility.

The building of comprehensive BMD systems would force major readjustments on the countries of Western Europe and on China. The countries of Western Europe, and perhaps China as well, should be able to build their own BMD shields. However, the French, British, and Chinese nuclear deterrents would become virtually useless. A Soviet BMD shield strong enough to defeat the United States's strategic nuclear force would be more than strong enough to defeat any other country's strategic nuclear force. NATO could no longer employ the European Nuclear Deterrent strategy, nor would the American Massive Retaliation pledge offer much protection. NATO would have to adopt the Flexible Response strategy to the exclusion of any other and do what that strategy requires: deploy land forces and tactical air forces strong enough to defeat a Warsaw Pact invasion.

The Strategic Defense Initiative is a gamble at long odds for a very high stake. It probably will not lead to effective comprehensive BMD, but if it did it would be worth whatever it cost. Opponents of the SDI must find answers to two questions:

1   Are you prepared to live with the balance of terror indefinitely?
2   Can you suggest a better way to escape from the balance of terror?

Notes

1 Thomas Powers, *Thinking About the Next War* (New York: Alfred A. Knopf, 1982), pp. 149–50.
2 President Reagan, "Peace and National Security", *Department of State Bulletin*, 83 (April 1983), pp. 13–14.

20
To Put an End to War?

Last night I had the strangest dream
I'd ever dreamed before,
I dreamed the world had all agreed
To put an end to war.
Ed McCurdy, "Last Night I Had the Strangest Dream"*

Nuclear Disarmament

We will not put an end to war as long as the sovereign state exists. Sovereign states must have armies, if for no other reason than because many regimes need armies to control their subjects. In many parts of the world armies fight civil wars and suppress or conduct *coups* more often than they wage war against foreign countries. Even if every army in the world were disbanded there would still be police forces; in the absence of armies, police forces take their place. There are about half a million police officers in the United States – armed, trained, disciplined men and women under government control. They could not fight an army, but they probably could defeat, say, the Mexican police and conquer Mexico if Mexico did not have an army.

It may be possible to put an end to strategic nuclear war. (One hopes that no regime will ever feel the need to employ nuclear weapons against its subjects.) Two ways to do this are comprehensive BMD and complete nuclear disarmament. BMD deprives nuclear weapons of their sting and disarmament does away with them altogether.

Nuclear disarmament is certainly the more complete solution and perhaps the more satisfactory. Complete nuclear disarmament, in principle, makes any use of nuclear weapons impossible: there are none to use. Even a very strong BMD shield lets through some missile warheads and offers little protection against battlefield nuclear weapons. Also, a BMD system primarily protects the country that has it while nuclear disarmament protects the entire human community.

On the other hand, nuclear disarmament is even harder to achieve than comprehensive and effective BMD. The superpowers can each build BMD

* Words and music by Ed McCurdy, TRO © Copyright USA 1950 (renewed 1978), 1951 (renewed 1979), and 1955 (renewed 1983), Almanac Music, Inc., New York, NY. ©Copyright UK 1950, 1955 Almanac Music Inc. Assigned to Kensington Music Ltd. Reprinted with permission.

shields for themselves by their own unilateral efforts, but all the countries of the world would have to agree and act together to achieve nuclear disarmament. Forty years of protracted and generally unsuccessful disarmament negotiations have shown how difficult that is to do.

Complete nuclear disarmament is unattainable without an international inspection service able to locate nuclear weapons wherever they might be hidden. No government will give up its own nuclear weapons unless it is certain that every other nuclear power will do likewise. The inspection service must have access to the territories of all the countries of the world, and must also be able to search the deepest reaches of the oceans, outer space, and perhaps even the moon. The inspection service must also have a large, well-qualified staff, its members able to rise above any national loyalties that might compromise the performance of their duties. BMD does not require any such tremendous work of institution-building.

There are also technological prerequisites for complete nuclear disarmament, and these are even more demanding than those for comprehensive and effective BMD. The international inspection service must have instruments that can locate nuclear weapons wherever they might be hidden. At present nobody has any idea, even in principle, how to design such technology. Nuclear warheads are not very large; the United States W78 warhead, which has a yield of 335 kilotons, is about 6 feet long. Nuclear warheads buried underground might not be found by even the most careful yard-by-yard inspection of a country's territory. The Soviet Union has 8,649,490 square miles in which to hide things, the United States has 3,623,420 square miles.

What would happen if the international inspection service did find that the nuclear disarmament treaty had been violated? Disarmament advocates assert that, if a country were to violate the agreement, other countries would feel free to declare the agreement abrogated and begin rearming. The fear that they might do so would keep the agreement unviolated. The most likely responses, though, to minor violations would be to ignore them, to claim they had not occurred, or to negotiate about them at length, hoping they would go away. Governments would be very reluctant openly to abrogate an agreement which had been achieved with so much difficulty and in which so much hope had been invested. The likely responses to a major violation would be just as feeble, but for a different reason: fear of the violator. A violator who has nuclear weapons in hand can threaten very serious damage to whoever calls him to account.

Complete nuclear disarmament requires that governments and peoples learn to trust one another. Unfortunately, many governments do not deserve to be trusted. The Cold War is not the fruit of mutual American and Soviet misconceptions. The ruling elites of the United States and the Soviet Union understand each other rather well. Each knows that the other cherishes objectives that in the long run are incompatible with its own. Negotiations, personal meetings, and other "conflict-resolution" measures can help both sides to keep their conflict under control, but they cannot talk the conflict out of existence. Until there is a revolution in at least one of the superpowers, and far-reaching changes in other countries, the major governments will not be able to trust one another.

Preventing countries from rearming with nuclear weapons may be more difficult than disarming them in the first place. No disarmament treaty can extinguish the knowledge of how to build nuclear weapons. A country that had once had nuclear weapons could soon have them again if it were to break the treaty. Nuclear disarmament would not prevent – and might even invite – conventional wars, including wars between countries that had once been nuclear powers. The first casualty of a major conventional war would be the international inspection service and the second casualty might thus be nuclear disarmament itself, with both side racing to acquire nuclear weapons. One of them would win the race – and the war. So not only is a nuclear disarmament treaty likely to break down, but it is likely to break down in such a way that one side in a major war would have nuclear weapons and the other side not. Once before, when that was the case, the country that had nuclear weapons used them – on Hiroshima and Nagasaki. Complete nuclear disarmament might well cause the nuclear war that the balance of terror has so far prevented.

The way to put an end to strategic nuclear war might be to combine strategic nuclear disarmament with comprehensive BMD. Prohibit possession of strategic (and perhaps also theater) nuclear weapons and at the same time allow, and even encourage, comprehensive BMD systems.

The combination of strategic nuclear disarmament and BMD is stronger and more stable than either measure can be alone. Each remedies the other's defects. While locating every nuclear warhead in the world is probably beyond the capabilities of any inspection service, locating every strategic nuclear weapon probably is not. ICBMs, strategic bombers, and missile submarines are large objects. An inspection service should be able to find almost all of them, and thus greatly weaken the threat BMD systems must counter. It prevents the deployment of any new offensive weapon able to penetrate BMD shields after they are in place. At the same time, BMD reduces to military insignificance the violations of a nuclear disarmament treaty that the international inspection service cannot prevent. A secret cache of a dozen ICBMs does not enable its possessor to intimidate or destroy a country protected by a strong BMD shield. The treaty violator must have thousands of strategic nuclear weapons even to make a dent in the defenses, more than he could ever hope to conceal. If he openly repudiates the treaty and starts rearming it will be years before he has a sufficiently capable nuclear strike force. The international community will have ample time to bring him to heel. Even if there is a major war, and the belligerents start building nuclear weapons, the war will probably be over before either has a nuclear strike force able to penetrate the other's defenses.

No treaty, or weapons system, can by itself put an end to nuclear war. The way to do it is a sequence of arms control measures, defensive weapons deployments, and disarmament. The first step should be an arms control treaty strictly forbidding the development, manufacture, or deployment of any new strategic or theater nuclear weapons. It must go beyond the SALT II treaty. The new treaty must prohibit *all* building of new ballistic missiles, missile submarines, long range bombers and cruise missiles. Also, unlike SALT II, it must be ratified by all the nuclear powers and potential nuclear powers, it must continue in force indefinitely, and it must be observed. The purpose of the treaty would be, not just to stabilize the balance of terror, but eventually to make possible the abolition of strategic nuclear war. The United States, the Soviet Union, and other countries can then start building comprehensive BMD systems without fearing that a new leap forwards in offensive weapons technology might nullify their efforts. (This would of course require abrogating or renegotiating the 1972 ABM treaty which prohibits the deployment of comprehensive BMD systems.) The Soviet Union has for years kept up a significant BMD research effort and the United States has begun the SDI. Two decades more of well-funded effort might see both countries and their major allies safe from strategic nuclear weapons. It might take many years of living this way for people to realize that the great nuclear arsenals had been rendered impotent. Gradually the effort to keep those arsenals in readiness would seem more and more pointless, then become more and more perfunctory, and finally cease altogether. The missiles would be left to rust in their silos. Some day the governments of the world would sign a complete nuclear disarmament treaty and the missiles be carted away as scrap metal. As long as these weapons are the key to national security and the great symbol of military power, no government will give them up. When keeping them becomes pointless, it should be possible to get rid of them.

Battlefield nuclear weapons are harder to eliminate than are strategic nuclear weapons. The delivery vehicles employed in the former, mainly artillery and fighter-bombers, are harder for an inspection service to find than are ICBMs and strategic bombers. Also, artillery and fighter-bombers are very widely used in conventional warfare, and the countries of the world are very unlikely to give them up until war itself has been abolished. Finally, there is no way to give a treaty banning battlefield nuclear weapons the crucial hedge against evasion that comprehensive BMD systems would give a strategic nuclear disarmament treaty. Nobody has even envisaged a missile able to shoot down a nuclear artillery shell in flight.

A World without War

The way to put an end to war is to put an end to its constant companion, the sovereign state. The peoples of the world must institute a world government which will supplant the sovereign states that rule them now or at least keep those states from waging war. This is most certainly easier said than done but down the ages it has appealed to scholars, statesmen, and visionaries as the ultimate solution to the problem of war. Nearly seven centuries ago the poet Dante wrote *De Monarchia*, an argument that only world government can ensure peace and the unity of mankind.

In our own day the best-known of the various proposals for a world government is the one offered by Grenville Clark and Louis B. Sohn in *World Peace through World Law*. Clark and Sohn propose that a reformed and greatly strengthened United Nations be the guarantor of peace. Every country in the world is to destroy its weapons, armaments factories, and weapons development centers, and disband its armed forces. Each may keep only a lightly-armed police force just strong enough to maintain law and order within its boundaries. At the same time the United Nations is to raise and have at its command an international army, the United Nations Peace Force. The Peace Force must keep the various countries disarmed, prevent any of them using their police forces against other countries, and enforce the United Nations' decrees. There are also to be an International Court of Justice and a World Equity Tribunal to resolve diplomatic disputes peacefully. The 165 sovereign states that exist today are to remain but not as fully sovereign bodies; they will not wage or prepare for war. The United Nations is to be a very limited government, possessing only those powers necessary to prevent war and peacefully to resolve diplomatic disputes, but it is to be a world government, which the present United Nations is not.

Clark and Sohn make two promises on behalf of their proposed world government, one about what it would do, and the other, what it would not do. They promise that world government will put an end to war. A sovereign state today compels its citizens to settle their disputes in a court of law and not take the law into their own hands; the world government of the future will compel its subjects, the no-longer-sovereign states, to settle their disputes in the International Court of Justice and not go to war. Clark and Sohn promise also that a world government will *not* threaten the vital interests of any country or the fundamental rights of any person. Power in their remodelled United Nations is to be carefully balanced and limited so that no one country or bloc can dominate the institution. Also, the United Nations Charter is to include a Bill of Rights, a statement of certain fundamental rights which the United Nations may not take away from anyone.

There are many other world government proposals and all of them stand or fall on the credibility of those same two promises: that a world government will put an end to war and that a world government will not be a global tyranny. If world government does not make good these promises it will not be worth the tremendous effort required to build it. If its advocates cannot demonstrate, with a very high degree of certainty, that it will make good these promises, there will not even be a serious attempt to build a world government. Perilous as "the condition of war" may be, governments and peoples are not yet desperate enough to try a really risky alternative. "Better the devil you know than the devil you don't know."

How credible are these promises? They are not easy for governments at any level to make good. Many fail to do so even within the narrow bounds of an individual country. Civil wars, military *coups*, and large-scale banditry are common occurrences. Some countries have suffered four, five, even six civil wars since World War II. During that period of time there have been about 80 civil wars in various countries and several times as many military *coups*. The sovereign state's human rights record is even more depressing. Four-fifths of the governments in the world today are dictatorships, many of them brutal oppressors of political or ethnic minorities and individuals under their control. Governments do at least try to keep the peace within their borders, but many do not even pretend to respect any rights their citizens might have.

The mere existence of a government clearly does not guarantee either peace or justice within the area it rules. There are, though, some governments which do deal justly with their citizens and maintain good order within their boundaries. Their existence inspires hope that a world government could put an end to war and yet not become a global tyranny; it shows peace and justice are possible. What enables them to be what they are?

Just and order-keeping governments are not based on coercion. They occasionally employ it, but the vast majority of their citizens almost all the time accept their decisions willingly. No government can please everybody, but the decisions of these governments are at least acceptable to the vast majority, because they reflect a nationwide community of values. Some of these values are procedural: for example, a legitimate decision is one that is made by majority rule and with careful attention to the views and rights of the minority. Other values are substantive and a legitimate decision is one that reflects the community's principles of social justice, national purpose or ethics. There are both procedural and substantive elements in a real community of values, although one element may be more important than the other.

Unless its citizens share and its actions reflect such a community of values, a government can be strong enough to keep the peace only by being tyrannical. This is as true of any future world government as it is of the national governments that exist today:

> It would be very dangerous to create a machinery of central force before one created a machinery of central justice. For a machinery of central justice to work satisfactorily, its judgments would have to be based upon a world-wide community of values.[1]

Before they institute a world government, peoples and ruling elites must do more than just agree to abstain from making war. They must also reach a world-wide consensus on all the issues which now impel them to fight, and that consensus must be so far-reaching that they can accept the world government's arbitration of those issues. Putting an end to war requires creating a substitute for war. The world government must itself decide the issues which countries have for so long decided by waging war – and those countries must accept its decisions as just, or at least as final. There may always be outlaw countries which must be coerced, but no world government that is not imposing global tyranny will live for very long if it must go to war very often. And, a world government frequently at war with its subjects would be just as dangerous and destructive as the condition of war that now prevails.

There is no world-wide community of values, neither today, nor on the far horizon. Marxist-Leninists and believers in liberal democracy each reject the other's values and fear their power. Muslim fundamentalists reject and fear both liberal democracy and Marxism-Leninism. Adherents of each of those systems of belief control certain countries. Sovereign states would have conflicting interests even if they all shared the same system of belief, but their quarrels are greatly exacerbated by the differing values and perceptions of those who rule them. That is not only why the sovereign state is so dangerous but also why it is so difficult for a state to renounce its sovereignty. Each people or ruling elite clings to the state it controls because that state embodies and protects its values. Each must fear that a world government would be dominated by others who would use the institution to impose different values. World government is a threat to believers in liberal democracy because most of the world is ruled by dictatorships, to a Marxist-Leninist because most of the world is non-Communist, and to a Muslim fundamentalist because most of the world is non-Muslim.

The peoples of the world are not now even capable of forming a community of values. Governments prevent the search for areas of agreement, often with the approval of those over whom they rule. People everywhere must learn to reason together, to unite in what the philospher Karl Jaspers terms "the community of reason." It is not enough for a few good men and women in some countries to reason together. It is not even enough for the ruling elites reach a consensus; elites can be overthrown and their places taken by others that have not been initiated into the community of reason. "Reason must pervade the nations." [2] All the peoples of the world must join together in a search for common values, and agreement on the means by which those values may be realized. There must be no Iron Curtains, no barriers to the free flow of ideas throughout the world. Governments must not prevent the formation of a global community of reason by walling their peoples off from alien influences. The peoples must rule and all governments reflect their will. Only then will the common values formed by their community of reason lead to a common effort to realize those values. Jaspers prescribes strong medicine to cure the condition of war, but so grave an illness surely requires strong medicine.

As soon as the community of reason exists, the search for a world-wide community of values may begin; when the community of values exists, the construction of a just and order-keeping world government may begin. Unfortunately, nobody knows how to take even the first step on this long road. Many powerful governments fiercely resist any attempt to draw their peoples into any global community of reason. To the leaders of a Communist-ruled or a fundamentalist Muslim state, the free flow of ideas across their borders is subversion. It is a fundamental threat to all they have and are, an act of aggression when sponsored by foreign governments and scarcely less threatening when it occurs spontaneously. They cannot be forced to accept the free flow of ideas by any force short of war. The war required to tear down the Iron Curtain might put an end to the human race as well. One can at least chart a path that leads to nuclear disarmament, difficult and improbable as many of the steps along that path are. There may be no path to a just and order-keeping world government.

World government is out of the question at this stage in history. Even if, by a miracle, a world government were instituted within our lifetimes, it would be a failure. That world government would not put an end to war, but would gravely threaten the rights and interests of entire peoples, and would quickly fall apart. Someday the world may be ready for a just and order-keeping world government, but it is impossible to predict when that may be or how it may come about.

Living with War

George Orwell once wrote:

> To survive you often have to fight, and to fight you have to dirty yourself. War is evil, and it is often the lesser evil. Those who take the sword perish by the sword, and those who don't take the sword perish by smelly diseases.[3]

This is bleak wisdom, but not even the invention of nuclear weapons has refuted it. Disarmament and world government are our hopes for the future – but we live today. In the world as it is today sovereign states protect their vital interests by waging war, by threatening to wage war, and by preparing for war. Those that do not may cease to be sovereign states.

That does not mean that everything can go on as it always has. Nuclear weapons have radically changed the meaning of war and the ways governments may use war. Nuclear powers dare not wage war against other nuclear powers. They prepare for war and threaten war, but no nuclear power has ever yet gone to war with another nuclear power. A nuclear power may still fight a country that does not have nuclear weapons, but when that happens other nuclear powers stay out of the conflict even if they must abandon an ally. The Soviet Union invades Hungary, Czechoslovakia, Afghanistan – and the United States stands aside. The United States intervenes in Korea and Vietnam, attacks Libya – and the Soviet Union stands aside. The superpowers might be partners in a dance, each adroitly stepping back as the other steps forwards. It is a cautious, careful dance, each participant aware of what might happen if one day it were to step forward and the other not step back. Four decades of this deadly dance have left the superpowers bitter enemies in most respects but also collaborators in a common effort to avert nclear war. They are, as Raymond Aron has termed them, "the enemy partners."

The Cold War dance allows many geopolitical oddities to persist. West Berlin, part of West Germany, sits 70 miles deep inside East Germany. Germany and Korea remain severed nations. Taiwan is independent of China. Cuba safely defies the United States, 90 miles from Florida and 5,000 miles from its protector, the Soviet Union. Behind each of these geopolitical oddities is a bitter conflict which has not been resolved by war or in any other way. Why not?

Two factors have kept the conflict over West Berlin alive and unresolved: nuclear weapons and the sovereign state. Because there are about 10,000 nuclear weapons in Europe, the Soviets dare not touch West Berlin even though the Warsaw Pact forces surrounding the city could overrun it in a day. On the other hand, as long as the parties to the conflict are sovereign states, West Berlin's status cannot be changed in any other way. The Warsaw Pact countries will not accept the reunification of Germany, NATO will not accept the absorption of West Berlin into East Germany, and nobody can force a settlement on either side. All that can be done is to legalize and manage West Berlin's strange status, as the 1971 Quadripartite Agreement does.

The other conflicts are kept alive and unresolved by the same two factors: nuclear weapons and the sovereign state. In the past they probably would have been resolved by war, in the future they may be adjudicated by an international court of justice, but in the era of the nuclear-armed sovereign state they often do not get settled by any means. The same is true of the Cold War, the global conflict into which these lesser conflicts have been absorbed. War no longer serves to settle conflicts between nuclear powers, but nothing has taken its place.

Statesmen must navigate these dangerous waters as best they can. There is no chart to tell them exactly which course to follow or any guarantee that even the most carefully chosen course will not lead to disaster. They have, though, some seamarks to guide them, rules of conduct which wise men have given to the world over the centuries. The most important of these rules can be summed up in a word: prudence.

Raymond Aron describes prudence as "the statesman's supreme virtue."

> To be prudent is to act in accordance with the particular situation and the concrete data, and not in accordance with some system or out of passive obedience to a norm or pseudo-norm; it is to prefer the limitation of violence to the punishment of the presumably guilty party or to a so-called absolute justice; it is to establish concrete accessible objectives conforming to the secular law of international relations and not to limitless and perhaps meaningless objectives, such as "a world safe for democracy" or "a world from which power politics will have disappeared."[4]

The first rule of prudent conduct is that there are no unbreakable rules. Every challenge that might provoke the use of military force must be judged on its merits and as a unique event. United States policy-makers in particular should resist their perennial tendency to deal with each crisis the way they should have dealt with the last one. The United States's successful defense of South Korea did not show that South Vietnam could have been defended against a very different kind of threat. Failure in South Vietnam does not show that the American counterinsurgency effort in El Salvador will fail, nor does the successful invasion of Grenada show that an invasion of Nicaragua would succeed. General principles are no substitute for accurate knowledge and careful judgement of particular cases.

Prudence also means setting realistic objectives. Not even a superpower can achieve everything its leaders want and think they deserve. The first objective of United States national security policy must be to keep the United States safe from nuclear war and free of communist domination. Americans should pursue other objectives only after assuring the survival and freedom of their own country. The United States cannot protect other countries, negotiate nuclear disarmament, or help put an end to war if it no longer exists.

Finally, prudence means pushing a defeated opponent no further than to achieve one's objectives requires. Do not back him into a corner, especially if he or his ally has nuclear weapons. Leave him room to retreat. As the proverb has it: "If your enemy turns to flee, give him a silver bridge." An opponent trapped is an opponent who may start a nuclear war.

It is easy enough to recognize and expound these rules of prudent conduct. The difficulty is in living by them. Any statesman works in a "fog of policy" not so very different from the fog of war which besets the soldier. One of the elements beclouding his view is – inevitably – ignorance. His response to a challenge often is dictated by instinct and general principles because he has no accurate knowledge of the unique circumstances from which it arose. We live in a very complex and rapidly-changing world. Even if United States policy-makers were much better informed than they are, there would still be many things they do not know.

Human emotions also contribute to the "fog of policy":

> In the field of international politics one is dealing with the very fundamentals of life and death: with the beliefs, the habits, the structures which shape moral communities and for which it is considered appropriate to die – and, worse, to kill.[5]

Issues of such magnitude inevitably rouse strong passions in the minds of statesmen and among the peoples they lead. Strong emotions impede clear thinking and prudent behavior in any sphere of life. It is customary in war to hate the enemy, often for very good reason. It is difficult to pursue limited objectives with limited means in combat against a hated enemy. Statesmen and peoples engaged in a savage war are always tempted to say "Let justice be done though the world perish." Nevertheless, they must remember that in the nuclear age "though the world perish" is no longer just a figure of speech.

Notes

1 William T. R. Fox, "International Control of Atomic Weapons", in *The Absolute Weapon*, edited by Bernard Brodie, (New York: Harcourt, Brace and Company, 1946) p. 174.

2 Karl Jaspers, *The Future of Mankind* (Chicago: University of Chicago Press, 1961), p. 291.

3 "Looking Back on the Spanish War", in *The Collected Essays, Journalism and Letters of George Orwell*, ed. Sonia Orwell and Ian Angus (New York: Harcourt Brace Jovanovich, 1968), vol. 2, pp. 251–2.

4 Raymond Aron, *Peace and War: A Theory of International Relations* (New York: Frederick A. Praeger, 1967), p. 585.

5 Michael Howard, "Ethics and Power in International Policy", in *The Causes of War* (Cambridge, Mass.: Harvard University Press, 1983), p. 49.

Glossary

ABM (antiballistic missile) A missile designed to destroy enemy ballistic missiles or their warheads in flight.

ABM system A weapons system that employs ABMs for defense against ballistic missiles.

Active defense Forces or weapons employed to defend a country against aerial attack by destroying enemy missiles and bombers in flight. They include ABMs, anti-aircraft weapons, and fighter planes. *See also* Civil defense.

Active sonar A device which locates an object in the water by sending out pulses of sound which bounce off the object and back to the sonar operator. *See also* Passive sonar; Sonar.

Administrative apparatus The bureaucracy which the government of a sovereign state uses to govern its citizens and obtain resources from them. Also, the very similar organization which a guerrilla insurgency uses to govern people and obtain resources in the area under its control.

Air burst A nuclear explosion in the air, high enough that its fireball does not touch the earth's surface.

Air superiority The ability to fly one's aircraft in a particular area without prohibitive risk and denial to the enemy of the ability to do so.

Airborne forces Light infantry trained and equipped to be carried to the battlefield in transport airplanes and parachute into battle, and the airplanes used to carry them. *See also* Light infantry.

Airborne alert The practice of keeping a number of bombers in the air and with nuclear weapons on board around the clock.

Aircraft Flying machines generally: fixed-wing airplanes, helicopters, gliders, dirigibles, etc.

Airlift Transport airplanes and helicopters used to move troops, equipment, and supplies by air.

Amphibious forces Ground forces trained and equipped to land on a hostile shore and the amphibious warfare vessels used to transport, land, and support them.

Amphibious landing A landing carried out by amphibious forces.

Amphibious warfare vessel A ship or landing craft designed to transport, land, and support ground forces making an amphibious landing.

APC (armored personnel carrier) An armored vehicle used to transport infantry on the battlefield.

Arms control Diplomatic measures or agreements to limit military forces or weapons, particularly nuclear weapons.

Assault helicopter A helicopter designed to move troops on or to the battlefield and resupply them.

ASUW (anti-surface warfare) Ships, weapons, forces, or tactics dedicated to destroying enemy surface ships.

ASW (anti-submarine warfare) Ships, weapons, forces, or tactics dedicated to destroying enemy submarines.

ATGM (anti-tank guided missile) A type of ground-launched missile designed primarily for use against tanks and other vehicles.

Attack submarine A submarine designed to destroy enemy submarines and surface ships and armed primarily with torpedoes.

Attrition warfare A way to wage war that relies primarily on concentrating maximum firepower on the enemy and physically destroying him.

Auftragstaktik A way to command a military force that assigns maximum freedom of action and responsibility to lower-level officers and men.

Balance of terror That condition which deters both of the superpowers from starting a strategic nuclear war because each is equally and deeply fearful of its consequences.

Ballistic missile A missile that is powered for only part of its flight time and can fly outside the earth's atmosphere. *See also* ICBM; IRBM; MRBM; SLBM.

Base area An area totally dominated by a guerrilla army, to the extent that the guerrillas are the effective government of that area.

Battalion A ground forces military unit of approximately 500–1,000 troops.

Battlefield nuclear weapon A nuclear weapon designed for use against enemy front-line troops and the support forces immediately behind them. Battlefield nuclear weapons have ranges of less than 100 miles. *See also* Nuclear weapon; Theater nuclear weapons.

Blast shelter A shelter strongly built to shield its occupants against the explosive blast from a nuclear detonation. *See also* Fallout shelter.

Blitzkrieg A type of land warfare attack which employs armored and mechanized forces to penetrate deeply behind the enemy's front. It exemplifies maneuver warfare to a high degree.

BMD (ballistic missile defense) All measures to destroy enemy ballistic missiles or warheads in flight, including lasers, battle satellites, and ABMs. *See also* ABM system; Comprehensive BMD.

Carrier battle group In the United States navy, a small fleet consisting of an aircraft carrier and the other half dozen or dozen warships required to protect and provision it.

CEP (circular error probable) The radius of a circle around a target within which half of a given missile's warheads will land if the missile's guidance system has been programmed with perfect accuracy.

Civil defense Measures employed to protect the people and economic resources of a country against the effects of nuclear or conventional weapons exploding on its territory. They include blast shelters, fallout shelters, and the evacuation of people from cities. *See also* Active defense; Blast shelter; Fallout shelter.

Clausewitz, Carl von (1780–1831) Prussian officer and military theorist, author of *On War*.

Close support airplane An airplane designed to attack enemy forces on the battlefield.

Coast defense Forces or tactics dedicated to defending the waters immediately adjacent to the defender's shoreline.

Combat service support Military units and equipment dedicated to sustaining the combat forces: supply, transportation, maintenance, and medical services.

Command and control Personnel, equipment, and facilities used by commanders to obtain information, make decisions, and issue orders.

Command of the sea The ability to sail one's ships upon the various seas and oceans of the world without prohibitive risk and denial to the enemy of the ability to do so.

Command post An installation at which the commander of a military unit or the leader of a country carries out his functions while protected from attack.

Commerce-raiding Air, submarine, and surface ship attacks on enemy merchant ships. This is a sea-denial strategy, employed by a country whose navy does not have command of the sea. *See also* Sea-denial.

Company A ground forces military unit of approximately 100–200 troops.

Comprehensive BMD A BMD system strong enough effectively to protect the entire territory of a country against the kind of ballistic missile attack a superpower could launch. *See also* BMD.

Condition of war, the The circumstance that war is a normal and common part of human life because of the existence of sovereign states and their ability to wage war.

Conventional war A war fought with conventional weapons alone. *See also* Strategic nuclear war; Tactical nuclear war.

Conventional weapon Any device employed in combat to destroy, injure or defeat an enemy that is not a nuclear, chemical, or biological warfare weapon.

Counterforce An attack, strategy, or weapon designed to destroy enemy military forces – in particular, to destroy enemy strategic nuclear weapons, before they are launched. *See also* Countervalue.

Counterinsurgency Forces, tactics, and government policies dedicated to repressing a guerrilla insurgency.

Countervalue An attack, strategy, or weapon designed to kill and injure enemy civilians and destroy economic resources. *See also* Counterforce.

Credibility The power or nature of a threat that convinces those against whom it is made that it is likely to be carried out.

Cruise missile A pilotless airplane fitted with a warhead. It flies only within the earth's atmosphere.

Cruise missile submarine A submarine armed primarily with cruise missiles.

Damage Limitation Any of several United States strategies intended to limit the damage the United States would suffer if there were to be a strategic nuclear war. They propose to employ counterforce attack, active defense, and civil defense measures.

Defense Any strategy of countering the threat of aggression by limiting the damage suffered by the defender in the event of an attack. *See also* Deterrence.

Defense professional A professional student of military affairs. This term is usually applied to civilians rather than to professional military officers.

Deterrence Any strategy of preventing aggression by threatening to inflict on the aggressor damage greater than whatever he hopes to gain by his aggression. *See also* Defense.

Direct radiation Nuclear radiation emitted from a nuclear explosion at the instant the explosion occurs.

Division A ground forces military unit of approximately 10,000–20,000 troops.

Early-warning satellite A type of reconnaisance satellite which detects the launching of enemy ballistic missiles by the heat from their rocket motors. *See also* Reconnaissance satellite.

ECM (electronic countermeasures) Measures to impede or shut down enemy radio transmitters, radars, and other devices which depend on use of the airwaves.

EMP (electromagnetic pulse) An intense burst of radio waves, given off by a nuclear explosion, capable of damaging or destroying many types of electrical equipment.

ER (enhanced radiation) warhead A nuclear warhead which releases much more of its energy in the form of direct radiation and much less in its explosive blast and thermal flash than do other types of nuclear warhead.

Escalation The phenomenon of a war becoming larger, more serious, and more destructive, or the policy of deliberately making it so.

Escalation control Preventing a war from escalating.

Escalation dominance The ability to employ more military force than the enemy at any level of violence to which he may escalate a conflict.

European Nuclear Deterrent A strategy which proposes to deter a Soviet invasion of Western Europe by the threat to use against the Soviet Union nuclear weapons in the hands of some or all of the countries of Western Europe themselves.

Explosive blast One of the effects of a nuclear explosion. It consists of a blast wave, followed by very high winds, travelling through the air and (from a ground burst) a shock wave travelling through the ground.

Fallout shelter A shelter built to shield its occupants against the radioactive fallout from nuclear explosions. It is less strongly built than a blast shelter. *See also* Blast shelter.

Ferret satellite A type of reconnaissance satellite which intercepts radio transmissions and the radio waves emitted by radar stations. *See also* Reconnaissance satellite.

Fighter-bomber An airplane which can be used either to shoot down enemy aircraft or to attack targets on land with bombs and missiles.

Fireball The extremely bright, extremely hot sphere of gases formed by a nuclear explosion.

First strike An attack, using strategic nuclear weapons, which a country inflicts upon another which has not yet launched weapons against that country. A preemptive attack is a first strike.

First-strike capability The amount of damage a strategic nuclear force could inflict if it were employed to make a first strike.

Flexible Response A strategy which proposes to defend Western Europe against Soviet aggression with the same weapons the Soviet Union uses, and to escalate (e.g., from conventional to tactical nuclear weapons) only if the Soviets do.

Foco A small group of armed revolutionaires who seek to instigate a revolutionary guerrilla war by violence.

Fog of war, the The confusion, illusions, and lack of information inherent in the conduct of war.

Friction The inefficiency, lack of coordination, and failure to perform assigned tasks inherent in all human enterprises, but particularly in war.

Geosynchronous orbit An orbit approximately 22,300 miles above the surface of the earth. A satellite in a geosynchronous orbit remains continually directly over the same point on the earth.

Ground alert The practice of keeping bombers continually armed, fueled, and with their crews nearby, so that they could take off and launch an attack with only a few minutes' warning.

Ground Zero The point on the surface of the earth on which, or directly over or under which, a nuclear explosion occurs.

Guerrilla insurgency Any revolt, whatever its objectives, which takes the form of a guerrilla war.

Gunship helicopter A helicopter armed with guns and missiles and designed to attack targets on the ground.

Hamburg grab Any of several possible attacks with limited objectives (such as Hamburg, West Germany) which Warsaw Pact forces might conceivably launch in Europe.

Helicopter-borne forces Light infantry carried into battle in assault helicopters. *See also* Light infantry.

High seas All of the world's oceans and seas not immediately adjacent to a shoreline.

Hydrophone An underwater microphone used, as part of a passive sonar system, to detect sounds passing through the water. *See also* Passive sonar.

ICBM (intercontinental ballistic missile) A land-based ballistic missile with a range of 3,400 miles or more. *See also* ballistic missile.

IRBM (intermediate range ballistic missile) A land-based ballistic missile with a range of 1,690–3,400 miles. *See also* Ballistic missile.

Kiloton A measure of the power of a nuclear warhead: the equivalent in yield of 1,000 tons of TNT.

Landing craft Boats designed to carry troops and infantry right up on to the beach in an amphibious landing. They are small enough to be carried on, and launched from, large amphibious warfare ships.

Launch-on-warning The practice of planning to launch ballistic missiles as soon as the launching of enemy ballistic missiles has been detected, while the enemy missiles are still in flight.

Light infantry Ground forces with few or no vehicles, guns, or other heavy weapons, and those they have frequently lighter and more easy to transport than the standard models. Marine, airborne, helicopter-borne, air-portable, and guerrilla units are all light infantry.

MAD *See* Mutual Assured Destruction.

Main battle tank The principal type of tank in modern armies typically weighing 35–60 tons.

Maneuver warfare A way to wage war that relies primarily on outmaneuvering the enemy, confusing and demoralizing him by striking where he is weakest.

Marine forces Light infantry trained and equipped to make amphibious landings. *See also* Light infantry.

Maritime strategy, the A United States military strategy that places heavy reliance on naval power projection attacks on the Soviet mainland to deter or defeat Soviet aggression in Eurasia.

Maritime trade Exports and imports transported by ship across the sea.

Massive Retaliation A strategy which proposes to deter a Soviet invasion of Western Europe by the threat that the United States will launch a massive nuclear attack against the Soviet Union.

Massive Retaliation pledge The United States's promise to launch a massive nuclear attack in retaliation for a Soviet invasion of Western Europe.

Medium-rank nuclear power A country which has a strategic nuclear force much weaker than that of a superpower but capable of inflicting great damage on a superpower, at least in a first strike. These countries are, at present, France, China, and the United Kingdom.

Megaton A measure of the power of a nuclear warhead; the equivalent in yield of 1,000,000 tons of TNT.

Military objective What a country's armed forces must achieve – in territory conquered or defended, casualties inflicted, etc. – for it to attain the political goal for which it is fighting a war.

MIRV (multiple independently targetable reentry vehicle) Each of the individual warheads of a ballistic missile which carries two or more warheads and can aim each of them at a separate target.

Missile submarine A submarine armed primarily with ballistic missiles.

Moment of truth, the When a war breaks out or a decision must be made that may lead to a war.

Mosquito fleet A fleet of numerous small, short range warships suited primarily for coastal defense.

MRBM (medium range ballistic missile) A land-based ballistic missile with a range of 695–1,690 miles. *See also* Ballistic missile.

Mutual Assured Destruction A United States strategy of preventing a strategic nuclear war by maintaining the balance of terror and ensuring that both superpowers would be destroyed in a strategic nuclear war between them. *See also* Balance of terror.

National security policy All the measures a government takes to provide for itself the armed forces it needs and to use those forces to support its policies.

NATO (North Atlantic Treaty Organization) A military alliance among the United States, Belgium, Canada, Denmark, France, the Federal Republic of Germany, Greece, Iceland, Italy, Luxembourg, the Netherlands, Norway, Portugal, Spain, Turkey and the United Kingdom. (France and Spain do not fully participate in NATO.) It is led by the United States and directed against the Warsaw Pact. *See also* Warsaw Pact.

Nuclear predicament, the The danger that the human race faces because of the existence of nuclear weapons and their possession by mutually-hostile sovereign states.

Nuclear total war A war between the United States and the Soviet Union in which they use all the nuclear weapons they have and hit all the targets they can reach.

Nuclear weapon A nuclear warhead or bomb together with the delivery vehicle (rocket, airplane, gun, or torpedo) required to get it to its target. *See also* Battlefield nuclear weapon; Strategic nuclear weapon; Tactical nuclear weapon; Theater nuclear weapon.

OMG (operational maneuver group) In the Soviet ground forces, a self-contained, all-arms raiding party able to operate behind enemy lines for several days. It usually has the strength of a division.

Order of battle A tabulation of the military units, and their strength, command structure, deployment, and equipment, in a military force.

Passive sonar A device which locates an object in the water by using hydrophones to listen for the sounds it makes. *See also* Active sonar; Hydrophone; Sonar.

Penetrativity The ability of a bomber or a missile to penetrate active defenses, to survive attempts to destroy it in flight.

Photoreconnaissance satellite A type of reconnaissance satellite which collects information by photographing objects and territory it passes over. *See also* Reconnaissance satellite.

Political goal What a government goes to war to achieve.

Power projection The use of naval and amphibious forces to attack targets on land and seize territory.

Preemptive attack An attack upon an enemy who himself is just about to attack or is very likely to do so. *See also* First strike; Preventive war.

Preventive war A war of aggression justified by the assertion that the enemy against whom it is waged might attack the aggressor some day. *See also* First strike; Preemptive attack.

Propaganda of the deed An act of violence committed by guerrillas or terrorists in order to attract attention to their cause and win recruits.

Protracted war Mao Tse-tung's strategy for the conduct of a revolutionary guerrilla war. It emphasizes extensive propaganda and organization-building before as well as after beginning the military struggle.

Psi overpressure A measure of the force of the explosive blast generated by an explosion. "Psi" is "pounds per square inch" and "overpressure" is pressure in excess of the normal atmospheric pressure. *See also* Explosive blast.

Radioactive fallout Particles of dust and debris made radioactive by a nuclear explosion; also, the process of these particles falling to earth.

Reconnaissance satellite A man-made earth satellite used to collect information on the objects and territory it passes over. *See also* Early-warning satellite; Ferret satellite; Photoreconnaissance satellite.

Revolutionary guerrilla war A guerrilla war which the guerrillas wage in order to seize power from the established government of the country in which the war is being waged.

Revolutionary party A political organization which seeks to seize power from the government of its country and which employs revolutionary guerrilla war or other violent means to do so.

SAM (surface-to-air missile) A ground-launched missile used to destroy aircraft in the air.

SDI *See* Strategic Defense Initiative.

Sea-denial Any naval strategy, such as commerce-raiding, employed by a country whose navy does not have command of the sea to deny use of the sea-lanes to the enemy. *See also* Commerce-raiding.

Sea-lanes The routes across the seas and oceans of the world commonly used by merchant shipping.

Sealift Transport ships and amphibious warfare vessels used to move troops, equipment, and supplies by sea.

Search and destroy operation A counterinsurgency tactic that entails throwing a cordon around an area in which enemy guerrillas are believed to be, drawing the cordon in towards the center, and eliminating any guerrillas encountered.

Second strike An attack, using strategic nuclear weapons, launched by a country against another which has just attacked it with nuclear weapons.

Second-strike capability The amount of damage a strategic nuclear force could inflict after it had suffered a heavy counterforce attack.

Self-propelled gun A gun mounted on a tracked, lightly armored chassis resembling that of a tank.

Silo A thick-walled concrete tube set vertically into the ground with its upper edge flush with the surface and provided with a heavy, movable cover on top. It shelters a ballistic missile.

SLBM (submarine-launched ballistic missile) Any ballistic missile, whatever its range, launched from a submarine.

Snorkel A tube which a submerged submarine or a tank at the bottom of a river sticks up above the surface to suck in air for its engines and crew.

Sonar Any of several means of locating an object in the water by the use of sound waves. *See also* Active sonar; Passive sonar.

SOSUS (Sound Surveillance System) A United States reconnaissance system which employs hydrophones to search portions of the North Atlantic, North Pacific, and Arctic Oceans for other countries' submarines.

Sovereign state A political unit whose government exercises exclusive authority over a given population and territory and is not subject to any other government.

Strategic Defense Initiative A United States weapons development program, first proposed by President Reagan in 1983, intended to provide the United States with a comprehensive BMD system early in the twenty-first century.

Strategic mobility The ability of a military unit or a weapon to move or be transported to a distant area of the world.

Strategic nuclear force A military force which a nuclear power maintains to deliver a nuclear attack on the territory of another nuclear power or to deter such an attack. The superpowers have strategic nuclear weapons for this purpose but the medium-rank nuclear powers rely primarily on theater nuclear weapons.

Strategic nuclear war A war in which strategic nuclear weapons are employed. *See also* Conventional war; Tactical nuclear war.

Strategic nuclear weapon A nuclear weapon with a long enough range that it can be based in Eurasia and hit a target in the continental United States, or be based in the continental United States and hit a target in Eurasia. *See also* Nuclear weapon; Tactical Nuclear weapon.

Strategy The art of applying military force, or the threat of force, to achieve political goals.

Superpowers, the The United States and the Soviet Union.

Surface burst A nuclear explosion on the surface of the earth or at so low an altitude that its fireball touches the surface.

Survivability The ability of a bomber or a missile to survive a counterforce attack, i.e., to survive attempts to destroy it before it has been launched.

Tactical mobility The ability of a military unit or a weapon to move on the battlefield.

Tactical nuclear war A war in which tactical nuclear weapons, but not strategic nuclear weapons, are employed. *See also* Conventional war; Strategic nuclear war.

Tactical nuclear weapon Any nuclear weapon that is not a strategic nuclear weapon. It may be a battlefield nuclear weapon or a theater nuclear weapons.

Tactics The art of employing military units on the battlefield.

TERCOM (terrain contour matching) An extremely accurate guidance system used in United States cruise missiles.

Territorial defense A strategy for deterring or defeating an invasion by preparing to wage, or waging, a guerrilla war on the territory occupied by an invader.

Theater nuclear weapon A nuclear weapon with a range of more than 100 miles but not enough range to be considered a strategic nuclear weapon. *See also* Battlefield nuclear weapon; Nuclear weapon; Strategic nuclear weapon.

Thermal flash (referred to in technical writings as "thermal radiation") The heat and light emitted from a nuclear explosion.

Third World The underdeveloped, non-Communist countries of Africa, Asia, and Latin America.

Trailing tactics A technique for destroying enemy submarines by stationing attack submarines outside the ports they use and having one or several attack submarines follow each enemy submarine that emerges wherever it goes, prepared to destroy it when the word is given.

Trip-wire force A military force placed on the territory of an ally primarily to make credible a promise to come to that ally's defense or to retaliate for aggression against the ally.

Urban guerrilla war Guerrilla war waged primarily in large cities rather than in rural areas; the tactics employed in such a war, or one who wages it.

VSTOL (vertical or short take-off and landing) **airplane** A fixed-wing airplane able to take off straight into the air or after a very short roll and to land straight down on to the ground.

War-fighting strategy Any military strategy that places primary reliance on being able to wage a war successfully and does not depend on deterring the war.

Warsaw Pact, the A military alliance among the Soviet Union, Bulgaria, Czechoslovakia, East Germany, Hungary, Poland, and Rumania. It is led by the Soviet Union and directed against NATO. *See also* NATO.

Yield The total energy, or explosive power, released in a nuclear explosion, customarily expressed in kilotons or megatons.

A Reader's Guide to the Conduct of War

The first 20 sections of this guide correspond to the chapters of the book and each lists readings relevant to the subject matter of that chapter. The last four sections list readings on topics that extend outside the confines of any one chapter.

1 A Glance into the Abyss

Daugherty, William, Barbara Levi, and Frank von Hippel, "The Consequences of 'Limited' Nuclear Attacks on the United States", *International Security*, 10 (Spring 1986).

Glasstone, Samuel and Philip J. Dolan (eds), *The Effects of Nuclear Weapons*, 3rd edn. Washington, DC: Government Printing Office, 1983. The standard text on the physical effects of nuclear detonations, much of it comprehensible by non-scientists.

Katz, Arthur M., *Life After Nuclear War*. Cambridge, Mass.: Ballinger, 1982. Covers both limited and full-scale nuclear attacks on the United States.

Leaning, Jennifer and Langley Keyes (eds), *The Counterfeit Ark*. Cambridge, Mass.: Ballinger, 1984. Why city evacuation won't work.

Peterson, Jeannie (ed.), *The Aftermath: The Human and Ecological Consequences of Nuclear War*. New York: Pantheon Books, 1983. What a nuclear total war would do to the entire world.

Sagan, Carl, "Nuclear War and Climatic Catastrophe: Some Policy Implications", *Foreign Affairs*, 62 (Winter 1983/84). The "nuclear winter" case stated.

Thompson, Starley L. and Stephen H. Schneider, "Nuclear Winter Reappraised", *Foreign Affairs*, 64 (Summer 1986). The "nuclear winter" case refuted.

US Congress, Office of Technology Assessment, *The Effects of Nuclear War*. Washington, DC: Government Printing Office, 1979. A balanced, politically neutral assessment.

2 The Nuclear Predicament

Johnson, James Turner, *Can Modern War Be Just?* New Haven, Conn.: Yale University Press, 1984. Derives from traditional just-war doctrine answers to the most significant issues raised by the existence of nuclear weapons.

National Conference of Catholic Bishops, *The Challenge of Peace: God's Promise and Our Response.* Washington, DC: US Catholic Conference, 1983. An authoritative statement of the American Catholic Church's position on the morality (or lack thereof) of nuclear strategy.

Nye, Joseph S. Jr, *Nuclear Ethics.* New York: Free Press, 1986. A valuable essay on how to think about the moral aspects of nuclear strategy.

Powers, Thomas, *Thinking About the Next War.* New York: Alfred A. Knopf, 1982. Serious, pessimistic essays on the nuclear predicament.

Russett, Bruce M., "Ethical Dilemmas of Nuclear Deterrence", *International Security*, 8 (Spring 1984). An exposition and defense of the American Catholic bishops' 1983 statement on nuclear war.

Wieseltier, Leon, *Nuclear War, Nuclear Peace.* New York: Holt, Rinehart and Winston, 1983. A strongly-felt, well-argued argument for the middle way between nuclear disarmament and Damage Limitation.

3 The Study of War

Allen, Thomas B., *War Games: The Secret World of the Creators, Players and Policy Makers Rehearsing World War III Today.* New York. McGraw-Hill, 1987. A balanced survey of wargaming and game theory in the United States.

Cohen, Eliot, "Guessing Game: A Reappraisal of Systems Analysis", in *The Strategic Imperative*, ed. Samuel P. Huntington, Cambridge, Mass.: Ballinger, 1982. Strongly and convincingly critical of systems analysis.

Ellis, John, *The Sharp End: The Fighting Man in World War II.* New York: Charles Scribner's Sons, 1980. A grim, moving account of what combat is like.

Gray, Colin S., *Nuclear Strategy and National Style.* Lanham, Md: Hamilton Press, 1986. Gray argues that United States and Soviet strategists think about nuclear war in very different ways.

——, *Strategic Studies: A Critical Assessment.* Westport, Conn.: Greenwood Press, 1982. Primarily a vindication of the work of United States defense professionals.

Herken, Greg, *Counsels of War.* New York: Oxford University Press, 1987. A somewhat negative survey of the activities and ideas of United States defense professionals from the end of World War II up to the mid-1980s.

Kaplan, Fred, *The Wizards of Armageddon.* New York: Simon and Schuster, 1984. Similar to the Herken book.

Marshall, S. L. A., *Men Against Fire.* Magnolia, Mass.: Peter Smith Publishers, n.d. (first published 1947). A classic study of the behavior of United States soldiers in combat during World War II.

Rosen, Stephen, "Systems Analysis and the Quest for Rational Defense", *The Public Interest*, 76 (Summer 1984). Balanced and sensible.

4   The Condition of War

Primitive Peoples and War

Bramson, Leon, and George W. Goethals (eds), *War: Studies from Psychology, Sociology, Anthropology*. New York: Basic Books, 1969. The anthropologists (particularly Malinowski) contribute the most to understanding the causes of war.

Turney-High, Harry Holbert, *Primitive War: Its Practice and Concepts*. Columbia, SC: University of South Carolina Press, 1971 (first published 1949). Probably the best work available on how war is waged before the invention of the sovereign state.

The Sovereign State and War

Aron, Raymond, *Peace and War: A Theory of International Relations*. Melbourne, Fla: Robert E. Krieger Publishing, 1981 (first published 1966). Aron's *magnum opus*, the fruit of a lifetime's study of war and politics.

Brodie, Bernard, *War and Politics*. New York: Macmillan, 1974. Brodie's last book, perhaps his best, certainly his most ambitious.

Brown, Seyom, *The Causes and Prevention of War*. New York: St Martin's Press, 1987. A good introduction to the subject.

Howard, Michael, *The Causes of Wars*. Cambridge, Mass.: Harvard University Press, 1983. Humane, perceptive essays on a variety of topics.

5   Military Power as an Instrument of Politics

Kahn, Herman, *On Escalation: Metaphors and Scenarios*. Westport, Conn.: Greenwood Press, 1975 (first published 1965). Written around Kahn's well-known 44-rung escalation ladder: from "Ostensible crisis" to "Spasm or Insensate War."

Schelling, Thomas C., *Arms and Influence*. New Haven, Conn.: Yale University Press, 1966. A readable, influential exposition of how military power can be used to achieve political goals.

Smoke, Richard, *War: Controlling Escalation*. Cambridge, Mass.: Harvard University Press, 1978. Uses five case studies to bring out the full complexity of escalation and the factors which cause or limit it.

Snyder, Glenn H., "Deterrence and Defense: A Theoretical Introduction", chaper 1 of *Deterrence and Defense*, Westport, Conn.: Greenwood Press, 1975 (first published 1961). A complete theoretical analysis of the differences between deterrence and defense.

6 United States National Security Policy

The Soviet Union's national objectives
Brzezinski, Zbigniew, "Tragic Dilemmas of Soviet World Power", *Encounter*, December 1983.

Kaplan, Stephen S., *Diplomacy of Power*. Washington, DC: Brookings Institution, 1981. Case studies in the use of Soviet military power to advance Soviet political interests, 1944–79.

Lenczowski, John, *Soviet Perceptions of US Foreign Policy*. Ithaca, NY: Cornell University Press, 1982. An ominous, well-substantiated picture of the Kremlin mind.

United States national security policy
Blechman Barry M., and Stephen S. Kaplan, *Force Without War*. Washington, DC: Brookings Institution, 1978. A typically exhaustive (and slightly exhausting) Brookings Institution study: the use of United States military power to advance American political interests without going to war, 1946–75.

Cohen, Eliot A., "Constraints on America's Conduct of Small Wars", *International Security*, 9 (Fall 1984).

Gray, Colin S., *Maritime Strategy, Geopolitics, and the Defense of the West*. New York: Ramapo Press, 1986. A short, meaty exposition of the impact of geography on United States national security policy and the role of sea power in achieving American objectives.

Osgood, Robert E., *Limited War Revisited*. Boulder, Col.: Westview Press, 1979. A short overview of the use of United States military power to protect Western Europe and advance American interests in the Third World.

Smoke, Richard, *National Security and the Nuclear Dilemma*, 2nd edn. New York: Random House, 1987. The role of nuclear weapons in the evolution of United States national security policy since World War II.

Snow, Donald M., *National Security: Enduring Problems of US Defense Policy*. New York: St Martin's Press, 1987. Short but comprehensive: covers political goals, military strategy, and technology.

7 Strategic Nuclear Weapons

ICBMs, Bombers, and Missile Submarines
Garwin, Richard L., "Will Strategic Submarines Be Vulnerable?", *International Security*, 8 (Fall 1983). Garwin does not think they will.

Polmar, Norman, *Strategic Weapons: An Introduction*, rev. edn. New York: Crane, Russak and Company, 1982. A short, entirely non-technical account of the development of ICBMs, bombers and missile submarines since World War II.

Tsipis, Kosta, *Arsenal: Understanding Weapons in the Nuclear Age*. New York: Simon and Schuster, 1984. A well-known physicist explains how strategic nuclear weapons work and what they can do.

Strategic Command and Control Systems

Blair, Bruce G., *Strategic Command and Control*. Washington, DC: Brookings Institution, 1985. The United States's strategic command and control system.

Bracken, Paul, *The Command and Control of Nuclear Forces*. New Haven, Conn.: Yale University Press, 1985. An interesting use of management theory to evaluate command and control systems.

Burrows, William E., *Deep Black*. New York: Random House, 1987. A very informative account of the United States's reconnaissance satellites and other information-gathering technology – whose capabilities Burrows may overestimate.

Carter, Ashton B., John D. Steinbruner, and Charles A. Zraket (eds), *Managing Nuclear Operations*. Washington, DC: Brookings Institution, 1987. Excellent essays on all aspects of strategic command and control systems.

8   Soviet Nuclear Strategy

See also section 21, "Soviet National Security Policy".

Carter, Ashton B., "Soviet Nuclear Operations", in *Managing Nuclear Operations*, ed. Ashton B. Carter, John D. Steinbruner, and Charles A. Zraket, Washington, DC: Brookings Institution, 1987.

Douglass, Joseph D. Jr and Amoretta M. Hoeber, *Soviet Strategy for Nuclear War*. Stanford, Ca: Hoover Institution Press, 1979. Somewhat overstated but based on extensive reading in the Soviet military press.

Lambeth, Benjamin S., "Has Soviet Nuclear Strategy Changed?' in *The Logic of Nuclear Terror*, ed. Roman Kolkowicz, Boston, Mass.: Allen and Unwin, 1987. Lambeth says it has not changed.

Lee, William T. and Richard F. Staar, *Soviet Military Policy since World War II*. Stanford, Ca: Hoover Institution Press, 1986. A fairly hostile survey of Soviet nuclear strategy.

Meyer, Stephen M., "Soviet Perspectives on the Paths to Nuclear War", in *Hawks, Doves, and Owls*, ed. Graham T. Allison, Albert Carnesale, and Joseph S. Nye, Jr, New York: W.W. Norton, 1985. Stephen Meyer's essays offer the soundest, most sophisticated analysis of Soviet nuclear strategy yet published.

9   United States Nuclear Strategy

Brodie, Bernard, *Strategy in the Missile Age*. Princeton, NJ: Princeton University Press, 1959. Probably the best book of its time on nuclear strategy: still illuminating.

Harvard Nuclear Study Group [Albert Carnesale, Paul Doty, Stanley Hoffmann,

Samuel P. Huntington, Joseph S. Nye, Jr, and Scott D. Sagan], *Living with Nuclear Weapons*. Cambridge, Mass.: Harvard University Press, 1983. A balanced, well-written exposition on various aspects of the nuclear predicament.

Mandelbaum, Michael, *The Nuclear Question*. Cambridge: Cambridge University Press, 1979. The evolution of United States nuclear strategy from 1946 to 1976.

Pranger, Robert J. and Roger P. Labrie (eds), *Nuclear Strategy and National Security: Points of View*. Washington, DC: American Enterprise Institute, 1977. Articles, speeches, and documents reflecting various positions on United States nuclear strategy.

Snow, Donald M., *Nuclear Strategy in a Dynamic World: American Policy in the 1980s*. Alabama: University of Alabama Press, 1981. Discusses both weapons and strategies.

Mutual Assured Destruction

Jervis, Robert, *The Illogic of American Nuclear Strategy*. Ithaca, NY: Cornell University Press, 1985. A defense of MAD against Damage Limitation strategies.

Kolkowicz, Roman (ed.), *The Logic of Nuclear Terror*. Boston, Mass.: Allen and Unwin, 1987. A critical analysis of MAD.

Panofsky, Wolfgang K. H., "The Mutual-hostage Relationship between America and Russia", *Foreign Affairs*, 52 (October 1973). A clear, even somewhat extreme, statement of the MAD doctrine.

Wieseltier, Leon, "When Deterrence Fails", *Foreign Affairs*, 63 (Spring 1985). An advocate of MAD wrestles with the question that is hardest for advocates of MAD to answer.

Damage Limitation

Clark, Ian, *Limited Nuclear War*. Princeton, NJ: Princeton University Press, 1982. Considers, and rejects, the possibility of fighting a limited strategic nuclear war.

Iklé, Fred Charles, "Can Nuclear Deterrence Last Out the Century?", *Foreign Affairs*, 51 (January 1973). A classic early statement of the Damage Limitation position.

Kahn, Herman, *On Thermonuclear War*. Westport, Conn.: Greenwood Press, 1978 (first published 1960). Still controversial and also one of the most important books on nuclear strategy ever written.

Slocombe, Walter, "The Countervailing Strategy", *International Security*, 5 (Spring 1981). An exposition and defense of the Carter administration's version of Damage Limitation.

10 The Nuclear Deterrents of France, China, and the United Kingdom

Freedman, Lawrence, *Britain and Nuclear Weapons*. London: Macmillan, 1980. Mostly about the debates in the United Kingdom over what kind of strategic nuclear force to build.

Segal, Gerard, "Nuclear Forces", in *Chinese Defense Policy*, ed. Gerard Segal and William T. Tow. Urbana, Ill.: University of Illinois Press, 1984. Summarizes most of what is publicly known about China's strategic nuclear forces.

Yost, David S., *France's Deterrent Posture and Security in Europe*. Part I: *Capabilities and Doctrine*. Part II: *Strategic and Arms Control Implications*. Adelphi Papers, nos 194, 195. London: International Institute for Strategic Studies, 1984/85. A recent thorough discussion of French strategic and battlefield nuclear forces.

11   The Weapons and Conduct of Land War

Bellamy, Chris, *The Future of Land Warfare*. New York: St Martin's Press, 1987. An interesting survey, covering both weapons and tactics, the present and the future.

Keegan, John and Richard Holmes, *Soldiers*. New York: Viking Penguin, 1985. Land warfare through the centuries.

Weapons and Arms of the Service

Armitage, M. J. and R. A. Mason, *Air Power in the Nuclear Age*, 2nd edn. Urbana, Ill.: University of Illinois Press, 1985. The history of tactical air power since 1945.

English, John A., *A Perspective on Infantry*. New York: Praeger, 1984. Infantry organization and tactics at the small-unit level.

English, J. A., J. Addicott, and P. J. Kramers (eds), *The Mechanized Battlefield*. Washington, DC: Pergamon-Brassey's, 1985. Interesting essays on small-unit tactics.

Lee, R. G., *Introduction to Battlefield Weapons Systems and Technology*, 2nd edn. Elmsford, NJ: Pergamon Press, 1985. A careful explanation of the design and employment of land warfare weapons.

Mason, R. A., *Air Power: An Overview of Roles*. Elmsford, NJ: Pergamon Press, 1987. An authoritative exposition of each of the roles played by land-based combat aircraft.

Reed, Fred, "Tanked: Test-Driving the Army's M1", *Harper's*, February 1986. Sensitively conveys what tanks are like.

Simpkin, Richard E., *Mechanized Infantry*. Oxford: Brassey's, 1980. How the tank–infantry team works.

Strategy and Tactics

Lind, William S., *Maneuver Warfare Handbook*. Boulder, Col.: Westview Press, 1985. How the United States Marines do it.

Luttwak, Edward N., "The Operational Level of War", *International Security*, 5 (Winter 1980/81). Explains and advocates maneuver warfare.

Mearsheimer, John J., "Maneuver, Mobile Defense, and the NATO Central

Front", *International Security*, 6 (Winter 1981/82). On the defects of maneuver warfare, particularly for the defense of West Germany.

US Department of the Army, *FM 100-5, Operations*. Washington, DC: Headquarters, Department of the Army, 1982. (Available from US Army Adjutant General Publications Center, 2800 Eastern Boulevard, Baltimore, Md 21220) The United States army's basic manual for the conduct of land war.

Chemical and Tactical Nuclear Weapons

Hoeber, Amoretta M. and Joseph D. Douglass, Jr, "The Neglected Threat of Chemical Warfare", *International Security*, 3 (Summer 1978). The authors are worried about the Soviet Union's chemical warfare capabilities.

Kaplan, Fred M., "Enhanced-radiation Weapons", *Scientific American*, May 1978. How the "neutron bomb" works and why it should not be built.

12 The Soviet Ground Forces

See also section 21, "Soviet National Security Policy".

Erickson, John, Lynn Hansen, and William Schneider, *Soviet Ground Forces*. Boulder, Col.: Westview Press, 1986. On Soviet land war tactics and command style.

Isby, David C., *Weapons and Tactics of the Soviet Army*, 2nd edn. London and New York: Jane's Publishing Company, 1988. More than a guidebook, a real analysis of Soviet weapons and their performance in combat.

Meyer, Stephen M., *Soviet Theatre Nuclear Forces*. Part I: *Development of Doctrine and Objectives*. Part II: *Capabilities and Implications*. Adelphi Papers, nos 187, 188. London: International Institute for Strategic Studies, 1983/84.

Suvorov, Viktor [pseud.], *Inside the Soviet Army*. New York: Macmillan, 1982. Written by a one-time officer in a Soviet motor rifle division.

Vigor, P.H., *Soviet Blitzkrieg Theory*. New York: St Martin's Press, 1983. Points out the danger of a Soviet Hamburg grab surprise attack.

13 The Defense of Europe

Bundy, McGeorge, George F. Kennan, Robert S. McNamara, and Gerard Smith, "Nuclear Weapons and the Atlantic Alliance", *Foreign Affairs*, 60 (Spring 1982).

Kaiser, Karl, Georg Leber, Alois Mertes, and Franz-Josef Schulze, "Nuclear Weapons and the Preservation of Peace: A German Response", *Foreign Affairs*, 60 (Summer 1982).

de Rose, François, "Inflexible Response", *Foreign Affairs*, 61 (Fall 1982).

These three articles, taken together, make up a debate between American advocates of Flexible Response and European advocates of something like Massive Retaliation.

European Security Study, *Strengthening Conventional Deterrence in Europe*. New York: St Martin's Press, 1983. See especially Part IV on the potential contribution of high-technology conventional weapons to the defense of Western Europe.

Garnham, David, "Extending Deterrence with German Nuclear Weapons", *International Security*, 10 (Summer 1985).

Hamilton, Andrew, "Redressing the Conventional Balance: NATO's Reserve Military Manpower", *International Security*, 10 (Summer 1985). Advocates building up NATO's reserve forces so NATO could defeat a Soviet invasion without using nuclear weapons.

Mearsheimer, John J., *Conventional Deterrence*. Ithaca, NY: Cornell University Press, 1983. Chapters 6–8 discuss the defense of Western Europe today.

von Mellenthin, F. W. and R. H. S. Stolfi with E. Sobik. *NATO under Attack*. Durham, NC: Duke University Press, 1984. They propose to employ defensive maneuver tactics to defeat a Soviet invasion without using nuclear weapons.

Yost, David S., *France and Conventional Defense in Central Europe*. Boulder, Col.: Westview Press, 1985. What the French could do to help defend West Germany and how likely they are to do it.

14 Command of the Sea

Naval Strategy and Nuclear War at Sea

Ball, Desmond, "Nuclear War at Sea", *International Security*, 10 (Winter 1985/86). Mostly about the danger of a conventional war at sea escalating to a nuclear war.

Brodie, Bernard, *A Guide to Naval Strategy*, 4th edn. Westport, Conn.: Greenwood Press, 1977 (first published 1958). Most of it written during World War II but still very relevant to naval strategy today; a model of how an introductory work should be written.

Till, Geoffrey, *Modern Sea Power*. London: Brassey's Defence Publishers, 1987. A concise but comprehensive guide to the subject.

——, et al. *Maritime Strategy and the Nuclear Age*, 2nd edn. New York: St Martin's Press, 1984. Excellent.

Naval Weapons

Friedman, Norman, *Carrier Air Power*. New York: Rutledge Press, 1981.

——, *Modern Warship Design and Development*. New York: Mayflower Books, 1979.

——, *Submarine Design and Development*. Annapolis, Md: Naval Institute Press, 1984.

The Norman Friedman books are thorough, authoritative surveys of naval technology.

Lautenschläger, Karl, "Technology and the Evolution of Naval Warfare", *International Security*, 8 (Fall 1983). A valuable effort to consider battlefleets as integrated organisms rather than simply as collections of weapons.

Moore, John E. and Richard Compton-Hall, *Submarine Warfare*. Bethesda, Md: Adler and Adler, 1986. Submarines, submarine tactics, and the submarine fleets of the world's navies.

The Soviet Navy

See also section 21, "Soviet National Security Policy".

Gorshkov, S. G., *The Sea Power of the State*. Oxford: Pergamon Press, 1979. Admiral Gorshkov was Commander-in-Chief of the Soviet navy from 1956 to 1985.

Watson, Bruce W. and Susan M. Watson (eds), *The Soviet Navy*. Boulder, Col.: Westview Press, 1986. A collection of papers on every important aspect of the subject.

The United States Navy

Beatty, Jack, "In Harm's Way", *The Atlantic*, May 1987. Argues against the maritime strategy.

Brooks, Linton F., "Naval Power and National Security: The Case for the Maritime Strategy", *International Security*, 11 (Fall 1986). As good an argument for the maritime strategy as can be made.

George, James L. (ed.), *The U.S. Navy*. Boulder, Col.: Westview Press, 1985. Papers on United States naval weapons and theaters of operations.

Mearsheimer, John J., "A Strategic Misstep: the Maritime Strategy and Deterrence in Europe", *International Security*, 11 (Fall 1986). A careful dissection and refutation of the several versions of the maritime strategy.

15 Light Infantry for Power Projection

Bartlett, Merrill L. (ed.), *Assault from the Sea*. Annapolis, Md: Naval Institute Press, 1983. A collection of articles about amphibious warfare, mostly accounts of amphibious assaults over the centuries.

Cordier, Sherwood S., *U.S. Military Power and Rapid Deployment Requirements in the 1980s*. Boulder, Col.: Westview Press, 1983. A useful handbook that places the airborne and marine forces of the superpowers in a broad power projection context.

Weeks, John, *The Airborne Soldier*. Poole, Dorset: Blandford Press, 1982. Mostly about the specialized equipment used by airborne and helicopter-borne forces.

16 Revolutionary Guerrilla War

Chaliand, Gérard (ed.), *Guerrilla Strategies*. Berkeley and Los Angeles: University of California Press, 1982. Descriptions and analyses of recent guerrilla wars.

Guevara, Che, *Guerrilla Warfare*. Lincoln, Ne.: University of Nebraska Press, 1985. The Cuban style of revolutionary guerrilla war, as expounded by its leading practitioner.

Laqueur, Walter, *Guerrilla*. Boston: Little, Brown, 1976. The history of guerrilla war from biblical times to the recent past.

Mao Tse-tung, *Mao Tse-tung on Guerrilla Warfare*, trans. Samuel B. Griffith. New York: Frederick A. Praeger, 1961. A short, well-known exposition of his strategy by the leading guerrilla warrior of the twentieth century.

Moss, Robert, *Urban Guerrilla Warfare*. Adelphi Papers, no. 79. London: International Institute for Strategic Studies, 1971. Includes Carlos Marighella's "Minimanual of the Urban Guerrilla".

O'Neill, Bard E., William R. Heaton, and Donald J. Alberts (eds), *Insurgency in the Modern World*. Boulder, Col.: Westview Press, 1980. Mostly case studies; chaper 1 is an excellent analytic introduction to the subject.

Pike, Douglas, *PAVN: People's Army of Vietnam*. Novato, Ca: Presidio Press, 1986. The history, organization and strategy of one of the most formidable guerrilla armies in history.

Race, Jeffrey, *War Comes to Long An*. Berkeley and Los Angeles: University of California Press, 1972. A detailed, well-researched picture of both insurgency and counterinsurgency efforts in a South Vietnamese province, 1954–1965.

Wolf, Eric R., *Peasant Wars of the Twentieth Century*. New York: Harper and Row, 1969. On the political and social causes of revolutionary guerrilla wars.

17 Fighting the Guerrilla

Beckett, Ian F., and John Pimlott (eds), *Armed Forces and Modern Counter-Insurgency*. New York: St Martin's Press, 1985. Case studies of how various armies have attempted to suppress guerrilla insurgencies.

Blaufarb, Douglas S., *The Counterinsurgency Era: U.S. Doctrine and Performance*. New York: Free Press, 1977. An excellent analytic history of United States counterinsurgency efforts from World War II through the Vietnam War.

Krepinevich, Andrew F., Jr, *The Army and Vietnam*. Baltimore, Md: Johns Hopkins University Press, 1986. Why the United States army failed to carry out its counterinsurgency mission in Vietnam.

Thompson, Robert, *Defeating Communist Insurgency*. London: Chatto and Windus, 1966. Still an excellent guide to the conduct of counterinsurgency warfare.

18 Territorial Defense

Roberts, Adam, *Nations in Arms*, 2nd edn. New York: St Martin's Press, 1986. The standard work on territorial defense.

Roberts, Thomas C., *The Chinese People's Militia and the Doctrine of People's War*. Washington, DC: National Defense University Press, 1983.

19 Nuclear War in the Twenty-first Century

Nuclear Weapons and Nuclear Strategy in the Future

Dyson, Freeman, *Weapons and Hope*. New York: Harper and Row, 1985. An interesting, sometimes loosely argued, attempt to find a way out of the nuclear predicament.

Snow, Donald M., *The Nuclear Future: Toward a Strategy of Uncertainty*. University, Al: University of Alabama Press, 1983. Argues that strategic nuclear offensive weapons are becoming less survivable and strategic defenses more effective, thus making MAD a weaker strategy.

The Strategic Defense Initiative

Brzezinski, Zbigniew (ed.), *Promise or Peril*. Washington, DC: Ethics and Public Policy Center, 1986. Thirty-five articles on the SDI and related issues, most written by supporters of the SDI.

Glaser, Charles S., "Why Even Good Defenses May Be Bad", *International Security*, 9 (Fall 1984). An interesting argument against the SDI.

Payne, Keith B., *Strategic Defense: "Star Wars" in Perspective*. Lanham, Md: Hamilton Press, 1986. Probably the best defense of the SDI.

Schroeer, Dietrich, *Directed-energy Weapons and Strategic Defence: A Primer*. Adelphi Papers, no. 221. London: International Institute for Strategic Studies, 1987. An up-to-date guide to laser and particle beam weapons.

US Congress, Office of Technology Assessment, *Strategic Defenses: Ballistic Missile Defense Technologies; Anti-Satellite Weapons, Countermeasures, and Arms Control*. Princeton, NJ: Princeton University Press, 1986. Balanced and authoritative.

US Department of Defense, *Report to the Congress on the Strategic Defense Initiative*. Prepared by the Strategic Defense Initiative Organization, April 1987. The official report on what has been achieved so far and what is planned for the future.

Weapons in Space, vol. I: "Concepts and Technologies", Vol. II: "Implications for Security", Daedalus, 114 (Spring 1985; Summer 1985). Most of the articles in these two issues of *Daedalus* oppose the SDI.

20   To Put an End to War?

Brown, Seyom, *The Causes and Prevention of War*. New York: St Martin's Press, 1987. See Part III, "The Prevention and Control of War".

Clark, Grenville and Sohn, Louis B., *World Peace through World Law*, 2nd edn. Cambridge, Mass.: Harvard University Press, 1960. Still the most complete and best-known plan for a world government.

Claude, Inis L. Jr, *Power and International Relations*. New York: Random House, 1962. A classic statement of why world government will not lead to world peace.

Falk, Richard, *A Study of Future Worlds*. New York: Free Press, 1975. This book's strength is its recognition of the importance of a world-wide community of values as the foundation of a world government; its weakness is Falk's great overestimate of the extent to which a world-wide community of values already exists.

Foell, Earl W. and Richard A. Nenneman (eds), *How Peace Came to the World*. Cambridge, Mass.: MIT Press, 1986. Forty-nine people from several countries and many walks of life envisage how world peace might come about by the year 2010: charming, stimulating, occasionally useful.

Howard, Michael, "The Concept of Peace", *Encounter*, December 1983. Not everybody defines it the way European pacifists do.

Jaspers, Karl, *The Atom Bomb and the Future of Man*. Chicago: University of Chicago Press, 1984 (first published 1958). Perhaps the only author who approaches the problem with the depth and seriousness it deserves.

Schell, Jonathan, *The Abolition*. New York: Avon Books, 1984. An interesting attempt to find a way out of the nuclear predicamant.

21   Soviet National Security Policy

Cockburn, Andrew, *The Threat*. New York: Random House, 1983. Cockburn argues that the Soviet military machine is not too much of a threat to the West.

Erickson, John and E. J. Feuchtwanger (eds), *Soviet Military Power and Performance*. Hamden, Conn.: Archon Books, 1979. Primarily on the organization and personnel of the Soviet armed forces.

Holloway, David, *The Soviet Union and the Arms Race*. New Haven, Conn.: Yale University Press, 1983. Takes rather a benign view of Soviet military strategy and capabilities.

Leebaert, Derek (ed.), *Soviet Military Thinking*. London: George Allen and Unwin, 1981. Essays on Soviet nuclear strategy, land warfare, and naval strategy.

Scott, Harriet Fast and William F. Scott (eds), *The Soviet Art of War: Doctrine,*

Strategy, and Tactics. Boulder, Col.: Westview Press, 1982. Articles written by Soviet military leaders, 1917–1981.

Sokolovskiy, V. D., *Soviet Military Strategy*, 3rd edn. ed. Harriet Fast Scott. New York: Crane, Russak and Company, 1975. The Soviet armed force's standard work on the subject.

22 Military Leadership, Recruitment, and Institutions

Cohen, Eliot A., *Citizens and Soldiers*. Ithaca, NY: Cornell University Press, 1985. About the different ways to recruit soldiers, particularly in a liberal democracy.

van Creveld, Martin, *Command in War*. Cambridge, Mass.: Harvard University Press, 1985. Conveys well how confused even the most astute commanders and best-organized armies are in the heat of battle.

Hadley, Arthur T., *The Straw Giant*. New York: Avon Books, 1987. The condition of the United States military today, particularly at the small-unit level: based on four decades of personal observation.

Luttwak, Edward N., *The Pentagon and the Art of War*. New York: Simon and Schuster, 1986. An incisive critique of the organization and organizational behavior of the United States military establishment today.

23 Strategy and Strategists

Bond, Brian, *Liddell Hart*. New Brunswick, NJ: Rutgers University Press, 1977. Probably the best introduction to Liddell Hart: no one of the many books Liddell Hart himself wrote covers the entire range of his thinking as this one does.

von Clausewitz, Carl, *On War*, ed. and trans. Michael Howard and Peter Paret. Princeton, NJ: Princeton University Press, 1976. The most important book on war ever written.

Howard, Michael, *Clausewitz*. Oxford: Oxford University Press, 1983. A short, readable introduction to von Clausewitz.

Luttwak, Edward N., *Strategy*. Cambridge, Mass.: Harvard University Press, 1987. Provocative.

Paret, Peter (ed.), *Makers of Modern Strategy from Machiavelli to the Nuclear Age*. Princeton, NJ: Princeton University Press, 1986. Better than the 1943 version – which itself was one of the most important books about military strategy.

Sun Tzu, *The Art of War*, trans. Samuel B. Griffith. Oxford: Oxford University Press, 1963. Written by a Chinese general in the fourth century BC, it was an important influence on both Liddell Hart and Mao Tse-tung – which shows how much maneuver warfare and guerrilla war have in common.

24   Handbooks and General Works

Allison, Graham T. Albert Carnesale and Joseph S. Nye Jr. (eds), *Hawks, Doves, and Owls*. New York: W.W. Norton, 1985. What might cause a nuclear war and how to prevent it.

Bishop, Chris and David Donald (eds), *The Encyclopedia of World Military Power*. New York: Military Press, 1986. A lavishly illustrated guide to all kinds of weapons, from ICBMs to pistols.

Cochran, Thomas B., William M. Arkin, and Milton M. Hoenig. *U.S. Nuclear Forces and Capabilities*, vol. I of *Nuclear Weapons Databook*. Cambridge, Mass.: Ballinger, 1984. Information on every type of American nuclear weapon or warhead in service.

Dunnigan, James F., *How to Make War*. New York: William Morrow, 1983. Modern warfare from a wargamer's perspective: much interesting statistical data.

Freedman, Lawrence, *The Evolution of Nuclear Strategy*. New York: St Martin's Press, 1983. The nuclear strategies of all the world's nuclear powers.

The Military Balance 1987–1988. London: International Institute for Strategic Studies, 1987. The most complete and authoritative guide to the manpower strengths and weapons inventories of the world's armed forces. A new edition is issued every year.

Schroeer, Dietrich, *Science, Technology, and the Nuclear Arms Race*. New York: John Wiley and Sons, 1984. Discusses, with authority, nuclear warheads, nuclear weapons, nuclear strategy, arms control, and disarmament.

Index

Note: numbers in bold refer to figures and maps.

Index by Justyn Balinski